Stability and Stable Oscillations in Discrete Time Systems

Advances in Discrete Mathematics and Applications

A series edited by *Saber Elaydi, Trinity University, San Antonio, Texas, USA* and *Gerry Ladas, University of Rhode Island, Kingston, USA*

Volume 1
Analysis and Modelling of Discrete Dynamical Systems
Edited by Daniel Benest and Claude Froeschlé

Volume 2
Stability and Stable Oscillations in Discrete Time Systems
Aristide Halanay and Vladimir Răsvan

This book is part of a series. The publisher will accept continuation orders which may be cancelled at any time and which provide for automatic billing and shipping of each title in the series upon publication. Please write for details.

Stability and Stable Oscillations in Discrete Time Systems

Aristide Halanay

Bucharest University, Romania

and

Vladimir Răsvan

Craiova University, Romania

CRC Press
Taylor & Francis Group
Boca Raton London New York

CRC Press is an imprint of the
Taylor & Francis Group, an **informa** business

CRC Press
Taylor & Francis Group
6000 Broken Sound Parkway NW, Suite 300
Boca Raton, FL 33487-2742

First issued in hardback 2019

ISBN-13: 978-90-5699-671-0 (hbk)

Visit the Taylor & Francis Web site at
http://www.taylorandfrancis.com

and the CRC Press Web site at
http://www.crcpress.com

British Library Cataloguing in Publication Data

Halanay, Aristide
    Stability and stable oscillations in discrete time systems.
    - (Advances in discrete mathematics and applications ; v.
    2)
    1.Stability - Mathematical models 2.Oscillations -
    Mathematical models 3.Discrete-time systems - Mathematical
    models
    I.Title II.Rasvan, Vladimir
    515.3'52

Professor Aristide Halanay
(1924–1997)

The distinguished Romanian mathematician Professor Aristide Halanay, one of the founders of the modern Romanian school of ordinary differential equations, was born in the small Romanian town Ramnicu-Sarat. In October 1946 he was appointed as an instructor in algebra at Bucharest University where he later obtained a Diploma in Mathematics in 1947. From 1949–1952, Aristide was a graduate student at the "Lomonosov" State University in Moscow under the supervision of Professor V. V. Nemytskii, one of the pioneers of the qualitative theory of ordinary differential equations. In 1952 Halanay became a Candidate in Physico-Mathematical Sciences (the Russian equivalent of a PhD) with a thesis on 2nd order linear differential equations with almost periodic coefficients. He was an associate professor from 1953–1968 and professor from 1968–1989 on the Faculty of Mathematics at Bucharest University.

Aristide was a member of the editorial board of the *Journal of Differential Equations* from its inception in 1965. His scientific activity started in 1947 with papers in algebra, and saw immense growth exemplified by production of approximately 200 papers and 12 books and monographs. He closely followed new trends in mathematical research and, over the years, made important contributions to the foundation of the theory of differential equations with delayed argument; stability and oscillations; singular perturbations; absolute stability of control systems; qualitative theory of discrete and stochastic systems; and optimal control of delay and discrete systems. In the last quarter of the twentieth century, Halanay showed an almost exclusive interest in problems that are most directly connected with practical applications, for example, stability of synchronous machines, dynamics of hydropower plants, and so on. The relevance and importance of Halanay's results are perhaps best illustrated by the fact that his monograph *Differential Equations, Stability, Oscillations, Time Lags*, first published in Romanian in 1963, was later published in English and Japanese and is one of the most cited references in the field.

Whether teaching or doing scientific research, he showed tremendous energy in everything he did. Amazingly creative and a hard worker, Aristide worked on mathematics until his last days. To those who knew and cherished him, he will be greatly missed.

CONTENTS

SERIES EDITORS' PREFACE

This is a new volume in the book series *Advances in Discrete Mathematics and Applications*. The series will be a forum for all aspects of discrete mathematics and will act as a unifying force in the field, presenting books in areas such as numerical analysis, discrete dynamical systems, chaos theory, fractals, game theory, stability, control theory, complex dynamics, computational linear algebra, boundary value problems, oscillation theory, asymptotic theory, orthogonal polynomials, special functions, combinatorics and functional equations. Volumes on applications of difference equations in science and engineering will also be considered for publication.

Advances in Discrete Mathematics and Applications will publish textbooks for both the upper undergraduate and graduate levels. In addition, it will publish advanced works at the research level.

We hope to meet the growing needs of the mathematical community for books in discrete mathematics.

Saber Elaydi

Gerry Ladas

PREFACE

This monograph was written as a result of the establishment of the book series devoted to discrete dynamical systems by Professors Elaydi and Ladas. They kindly invited us to participate in this project.

It was natural for us to use this opportunity to review and reevaluate our experience in the research on discrete time systems. For many years we worked jointly on several problems in stability and oscillations, both in continuous time and discrete time, combining the experience of a professional mathematician and a theoretical engineer.

In this monograph we have focussed on regular behavior related to stability and stable oscillations. While writing this book, we were cognizant of the large volume of recent literature on this subject, including contributions from Professors Elaydi and Ladas. However, it is almost impossible to thoroughly cover all areas of the subject through such contributions. By basing the book essentially on our research experience, it is our aim to present a unique perspective of this subject.

We take this opportunity to again thank Professors Elaydi and Ladas for connecting us with the series. Thanks also go to Professor Anton Batatorescu for his excellent performance in processing the manuscript and the publisher for their effective cooperation.

CHAPTER 1

INTRODUCTION

Discrete-time system dynamics is a topic of broad interest; the main reason for this interest comes from the variety of the sources of discrete-time dynamical models. We may cite:

1^0 Discrete-time models determined by the nature of the described processes: this is particularly true for economics, biology, physiology and discrete-time information processing.

2^0 Discrete-time models induced by the impulses occurring in continuous-time systems.

3^0 Discrete models occurring in controlled systems when the feedback information used in control generation is composed of output samples obtained through sampling intervals of time.

4^0 Discrete systems occurring during numerical treatment of continuous time systems.

But the interest in discrete-time systems may also be explained by the simplicity of their treatment - it requires minimal computational and graphical resources to obtain the solutions of the associated difference equations and follow this behavior. Since difference equations may be viewed as recurrence relations, their treatment seems much simpler that the one of differential equations.

In many cases the models describing biological or economic processes lead to what is now called *chaotic behavior* - a rather complicated, irregular picture of the evolution trajectories. In fact one of the sources for the widespread popularity of the discrete models lies in the fact that the simplest nonlinear ones may display such chaotic behavior that some people - mainly physicists - consider them to be some type of paradigm that may enhance the various phenomena of the physical universe.

In opposition to such behavior, the other three sources we mentioned for discrete-time systems display mainly regular behavior, stability. This fact is less obvious in the case of the systems with shocks where the impulses may generate chaotic behavior. Nevertheless, in this case also of interest are mainly those cases when the impulses, controlled or not, generate stable processes. The stability problems are especially important in the case of sampled data systems and also when the discrete-time system corresponds to a computational process.

In the following we shall illustrate these aspects using some simple examples.

1.1. MODELS WITH DISCRETE STORAGE OF THE INFORMATION

In several cases discrete-time equations are obtained when the physical parameter of interest is stored through intervals of time ("from time to time"). Following Maynard-Smith (1974), May (1995), Elaydi (1996), Kocic and Ladas (1993) let us cite the following discrete-time models in biology:

- the self-limited growth model (the logistic equation):

$$x_{t+1} = ax_t \left(1 - x_t\right), \ t = 0, 1, 2, ... \tag{1.1}$$

- the discrete-generation predator-prey model (Maynard-Smith (1974)):

$$\begin{aligned} x_{t+1} &= ax_t - bx_t y_t \\ y_{t+1} &= -cy_t + dx_t y_t \end{aligned} \tag{1.2}$$

- the host-parasitoid interaction model (May (1995)):

$$\begin{aligned} H_{t+1} &= R_0 H_t F\left(P_t, H_t\right) \\ P_{t+1} &= cH_t \left[1 - F\left(P_t, H_t\right)\right] \end{aligned} \tag{1.3}$$

where H_t and P_t represent the number (or density) of hosts and parasites in generation t.
- the Leslie model of the age structure for a population (Svirežev and Logofet (1978)):

$$\begin{aligned} x_{t+1}^1 &= \sum_1^n b_i x_t^i, \ b_i \geq 0 \\ x_{t+1}^{i+1} &= s_i x_t^i, \ 0 < s_i \leq 1, \ i = \overline{1, n-1} \end{aligned} \tag{1.4}$$

- the Kermack-McKendrick model of epidemics (Kocic and Ladas (1993), p. 195):

$$S_{t+1} = e^{-\alpha I_t} S_t$$
$$I_{t+1} = \beta I_t + \left(1 - e^{-\alpha I_t}\right) S_t \qquad (1.5)$$
$$R_{t+1} = (1 - \beta) I_t + R_t$$

- discrete competitive systems (Kocic and Ladas (1993), p. 199):

$$x_{t+1} = x_t f\left(a x_t + b y_t\right)$$
$$y_{t+1} = y_t g\left(c x_t + d y_t\right), \qquad (1.6)$$

where a, b, c, d are positive constants, f, g are decreasing on $(0, \infty)$ and there exist $\bar{x} > 0$, $\bar{y} > 0$ such that $f(\bar{x}) = g(\bar{y}) = 1$.

In order to illustrate the dynamic behavior for systems described by such models *we shall consider the logistic equation in some detail*. It is quite well known (Elaydi (1996)) that while elementary in structure, this equation displays various types of behavior, from very simple to very complicated ones.

A. We shall call solution of (1.1) a sequence $\{x_t\}_t$ defined recurrently by:

$$x_{t+1} = a x_t (1 - x_t)$$

with given x_0. In fact x_t is the t-iterate of the mapping $f : \mathbb{R} \to \mathbb{R}$ defined by $f(x) = ax(1 - x)$. Remark that $f(0) = f(1) = 0$ and $\max f(x) = f(1/2) = a/4$. It follows that $f(x)$ maps the interval $[0, 1]$ into itself provided $0 < a \leq 4$. Since in population dynamics x_t is a rated population density, we shall have for physical reasons $0 \leq x_t \leq 1$. From the mathematical point of view this means that the model may be considered correct if $x_0 \in [0, 1]$ implies $x_t \in [0, 1]$ for all positive integers t; we shall say *that $[0, 1]$ is an invariant set for* (1.1). The above considerations show that this is true for $a \in (0, 4]$.

In the following we shall assume $a \in (0, 4]$ and consider equation (1.1) only on the interval $[0, 1]$. *The constant solutions of (1.1) are called equilibria*; they are given by

$$x = f(x) = ax(1 - x)$$

being the fixed points of the mapping $f(x) : \hat{x}^1 = 0$, $\hat{x}^2 = (a - 1)/a$. Remark that $0 \leq \hat{x}^2 < 1$ only if $a \geq 1$ otherwise this equilibrium is not contained in the considered interval.

Let $0 < a < 1$. On the interval $[0, 1]$ we shall have

$$x_{t+1} - x_t = x_t \left[a \left(1 - x_t \right) - 1 \right] < 0 \, ;$$

hence any sequence x_t with $0 \le x_0 \le 1$ is decreasing. Since $x_t \ge 0$ we shall have $\lim\limits_{t \to \infty} x_t \ge 0$; from the existence of this limit we deduce that letting $t \to \infty$ in (1.1) is legitimate, the limit being a nonnegative equilibrium, i.e., $\hat{x}^1 = 0$ since $\hat{x}^2 < 0$ in this case. We obtained that the equilibrium $\hat{x}^1 = 0$ is attractive for the solutions starting in $[0, 1]$.

Let now $a > 1$; in this case both equilibria belong to the interval $[0, 1] : 0 = \hat{x}^1 < (a - 1) / a = \hat{x}^2 < 1$.

Consider the deviations with respect to the nonzero equilibrium $\hat{x}^2$ namely $z_t = x_t - \hat{x}^2$; we shall have

$$z_{t+1} = (2 - a) z_t - a z_t^2 \qquad (1.7)$$

If $1 < a < 3$ then $|2 - a| < 1$ and for z_0 small enough we shall have $z_t \to 0$ exponentially by the Liapunov theorem on stability by the first approximation (see Section 2.3 of the book). Since the deviations from the equilibrium $\hat{x}^2$ tend exponentially to zero (for $t \to \infty$) we shall say that $\hat{x}^2 = (a - 1) / a$ is an *attractor*. Under the condition $1 < a < 3$ the other equilibrium, $\hat{x}^1 = 0$ is *repulsive*. Indeed let $x_0 > 0$ be a neighborhood of $\hat{x}^1 = 0$ e.g. satisfying $x_0 < (a - 1) / a$. We have also

$$x_{t+1} - x_t = x_t \left(a - 1 - a x_t \right) > 0 \, ,$$

provided $a x_t < a - 1$. It follows that any solution sequence x_t that satisfies $0 < x_0 < (a - 1) / a$ is strictly increasing, i.e., the deviations from $\hat{x}^1$ even if small initially are increasing which shows repulsiveness of this equilibrium in the considered case.

Let now $a > 3$. In this case the equilibrium $\hat{x}^2$ is also repulsive, as follows from the equation in deviations (1.7); indeed we shall have

$$
\begin{aligned}
z_{t+1}^2 - z_t^2 &= \left[(2 - a) z_t - a z_t^2 \right]^2 - z_t^2 = \\
&= z_t^2 \left(a - 1 + a z_t \right) \left(a - 3 + a z_t \right) = \\
&= (a - 3)^2 z_t^2 + z_t^2 \left[2 (a - 3) + 2a (a - 2) z_t + a z_t^2 \right]
\end{aligned}
$$

The polynomial $a \lambda^2 + 2a (a - 2) \lambda + 2 (a - 3)$ has negative roots for $a > 3$ (its coefficients are strictly positive), hence we may take some $\delta > 0$ sufficiently small in order that for $|\lambda| < \delta$ this polynomial is positive. As long as $|z_t| < \delta$ we have

$$z_{t+1}^2 - z_t^2 > (a-3)^2 z_t^2$$

which shows that $\{z_t\}$ is an increasing sequence; hence it has to leave the set $|z_t| < \delta$. This proves our assertion. Summarizing we have the following:

1) If $0 \leq a < 1$, the equilibrium at $\hat{x}^1 = 0$ is attractive and the equilibrium at $\hat{x}^2 = (a-1)/a$ is repulsive.
2) If $1 < a < 3$, the equilibrium at $\hat{x}^1 = 0$ is repulsive while the equilibrium at $\hat{x}^2 = (a-1)/a$ is attractive.
3) If $a > 3$ both equilibria are repulsive.

B. Other types of solutions are the so-called cyclic solutions satisfying $x_{t+T} \equiv x_t$ for some positive integer T. Let $T = 2$; i.e., $x_{t+2} = x_t$. In the case of (1.1) we shall have

$$x_{t+2} = ax_{t+1}(1 - x_{t+1}) = a^2 x_t (1 - x_t)[1 - ax_t(1 - x_t)] \equiv x_t$$

The initial conditions defining cycles are the solutions of

$$x = a^2 x (1-x)[1 - ax(1-x)] \tag{1.8}$$

that is the fixed points of the mapping

$$f^2(x) = (f \circ f)(x) = a^2 x (1-x)[1 - ax(1-x)]$$

the iterated mapping. Obviously the fixed points of $f(x)$ are among the fixed points of $f^2(x)$; they define equilibria not cycles so we ignore them. It is easily seen that the solutions of interest of (1.8) are the solutions of the equation

$$a^2 x^2 - a(a+1)x + a + 1 = 0,$$

namely,

$$\hat{x}^3 = \frac{a+1 - \sqrt{(a+1)(a-3)}}{2a}, \quad \hat{x}^4 = \frac{a+1 + \sqrt{(a+1)(a-3)}}{2a}.$$

These solutions define the 2-cycle in the sense that

$$x_t = \hat{x}^3, \quad x_{t+1} = \hat{x}^4, \quad x_{t+2} = \hat{x}^3.$$

One may ask whether this 2-cycle is attractive, i.e., whether x_t with x_0 in a neighborhood of $\hat{x}^3$ or $\hat{x}^4$ will approach $\hat{x}^3, \hat{x}^4$, respectively.

This is nothing else but to ask whether $\hat{x}^3$ and $\hat{x}^4$ are attractive as equilibria of

$$x_{t+1} = f\left(f\left(x_t\right)\right) = a^2 x_t \left(1 - x_t\right)\left[1 - a x_t \left(1 - x_t\right)\right].$$

As previously, we may use the Liapunov theorem on stability by the first approximation (Section 2.3) by considering the linearized system around $\hat{x}^3$ or $\hat{x}^4$

$$x_{t+1} - \hat{x}^3 = f'\left(f\left(\hat{x}^3\right)\right) f'\left(\hat{x}^3\right)\left(x_t - \hat{x}^3\right) = f'\left(\hat{x}^4\right) f'\left(\hat{x}^3\right)\left(x_t - \hat{x}^3\right)$$

(the linearized around $\hat{x}^4$ system coincides with the above one).

The attractiveness of the 2-cycle holds if $\left|f'\left(\hat{x}^3\right) f'\left(\hat{x}^4\right)\right| < 1$ which reads, after some simple computations

$$-2 < -\left(a + 1\right)\left(a - 3\right) < 0.$$

We deduce that for $3 < a < 1 + \sqrt{6} \approx 3.449$ the 2-cycle is attractive. For $a_2 = 1 + \sqrt{6}$ the cycle doubles its period; i.e., system (1.1) will have a cycle of period equal to 4 while the 2-cycle becomes repulsive. The 4-cycle is attractive but if a increases again then the 4-cycle becomes repulsive and a 8-cycle occurs (for $a_3 \approx 3.54$). The period doubling will continue; there is no rule for the generation of the values a_i corresponding to this phenomenon but $a_{i+1} - a_i$ diminishes with increase of i. Also

$$\delta_n = \frac{a_n - a_{n-1}}{a_{n+1} - a_n} \rightarrow \delta \quad (\text{for } n \rightarrow \infty)$$

where $\delta \approx 4.66920...$ Another interesting fact is that $a_i < 3.57$. If $a > 3.57$ there exist cycles of any period and trajectories that are not attracted by them. The system is very sensitive to its initial conditions; i.e., a slight change of them implies important modifications of the trajectories. Also the system has a 2^n-cycle which is repulsive for any n. This gives a rather complicated picture of the trajectories called *chaos* and means in fact "very complicated dynamics arising from very simple models" (May (1976)). Other authors call it "stochasticity generation from deterministic models" since the dynamic model has nothing stochastic in it while the trajectories are very much similar to those generated by stochastic systems (Neimark (1978), Gumowski and Mira (1980)).

1.2. DISCRETE TIME MODELS INDUCED BY THE IMPULSES OCCURRING IN CONTINUOUS-TIME SYSTEMS

A. We shall present first in some detail a model of electrical oscillations excited by impulses (Racoveanu (1964)). Consider a linear oscillator described by

$$\ddot{y} + \gamma_1 \dot{y} + \gamma_2 y = 0 \qquad (1.9)$$

where γ_1 and γ_2 are such that $\gamma_1^2 - 4\gamma_2 < 0$; consequently the characteristic equation has the complex roots $\lambda_1 = \alpha + i\beta$, $\lambda_2 = \alpha - i\beta$, $\beta > 0$. The general solution of this equation is

$$y(t) = ce^{\alpha t} \sin \beta (t - \tau)$$

and if $\gamma_1 > 0, \gamma_2 > 0$ it will tend asymptotically to zero. The signal $y(t)$ may somehow be maintained at nonzero by applying to the oscillating device some impulses as follows: if t_k are such that $y(t_k) = 0$, where $y(t)$ is the continuous solution of (1.9) then

$$\dot{y}(t_k + 0) = \dot{y}(t_k - 0) + (\dot{y}(t_k - 0))^{-1} . \qquad (1.10)$$

Obviously between t_{k-1} and t_k the solution is given by the general solution of (1.9) which may be obtained analytically or numerically. At $t = t_k$, $\dot{y}(t_k)$ is modified by a jump $(\dot{y}(t_k))^{-1}$ and integration of (1.9) starts with this new value of $\dot{y}(t_k)$ - denoted by $\dot{y}(t_k + 0)$ - as the initial condition. It is quite clear that for a correct and complete mathematical description of the phenomena we have to describe the behavior at the jump moments.

In order to obtain this description we denote $h = 2\pi l/\beta$ with l a positive integer, $t_k = t_0 + kh$, $k = 1, 2, 3, \ldots$ and t_0 is some real number. For any $z_0 \neq 0$ we have to construct the solution of (1.9), denoted $y(t; t_0, z_0)$, which satisfies the conditions

$$y(t_0; t_0, z_0) = 0, \quad \dot{y}(t_0 + 0; t_0, z_0) = z_0$$

and also the jump conditions at $t = t_k$. On each interval (t_k, t_{k+1}) the solution verifies (1.9). Therefore

$$y(t; t_0, z_0) = c_k e^{\alpha t} \sin \beta (t - \tau_k), \quad t_k < t < t_{k+1} .$$

By assumption $y(t_k; t_0, z_0) = 0$ hence $t_k - \tau_k = m\pi/\beta$. Therefore

$$\tau_k = t_k - m\pi/\beta = t_0 + \frac{2k\pi}{\beta}l - \frac{m\pi}{\beta} = t_0 + \frac{n\pi}{\beta} .$$

It follows that

$$\sin \beta (t - \tau_k) = \sin \beta \left(t - t_0 - \frac{n\pi}{\beta} \right) = (-1)^n \sin \beta (t - t_0)$$

and, therefore,

$$y(t; t_0, z_0) = c'_k e^{\alpha t} \sin \beta (t - t_0) , \quad t_k \le t < t_{k+1}$$

This expression may be re-written as follows

$$y(t; t_0, z_0) = \frac{1}{\beta} a_k e^{-\alpha k h} e^{\alpha(t - t_0)} \sin \beta (t - t_0) , \quad t_k < t < t_{k+1}$$

provided we put $a_k = \beta e^{\alpha(t_0 + kh)} c'_k$. Since $t_0 + kh \le t < t_0 + (k+1)h$ it follows that $k \le (t - t_0)/h < k + 1$ hence $k = [(t - t_0)/h]$ and the solution becomes

$$\begin{aligned} y(t; t_0, z_0) &= \frac{1}{\beta} a_k e^{\alpha(t - t_0 - [(t-t_0)/h])} \sin \beta (t - t_0) , \\ &\qquad t_0 + kh \le t < t_0 + (k+1)h \end{aligned} \tag{1.11}$$

and it is obvious that the solution is completely determined by the sequence $\{a_k\}_k$. This sequence is defined as follows

$$\dot{y}(t) = \frac{1}{\beta} a_k e^{\alpha(t - t_0 - [(t-t_0)/h])} [\alpha \sin \beta (t - t_0) + \beta \cos \beta (t - t_0)] ,$$

$$\lim_{t \to t_k} \dot{y}(t) = \dot{y}(t_k + 0; t_0, z_0) = a_k ,$$

$$\lim_{t \to t_{k+1}} \dot{y}(t) = \dot{y}(t_{k+1} - 0; t_0, z_0) = a_k e^{\alpha h} \ne 0 \quad \text{if } a_k \ne 0 .$$

According to (1.10) we shall have

$$a_{k+1} = e^{\alpha h} a_k + \left(e^{\alpha h} a_k \right)^{-1} , \quad k = 0, 1, \dots \tag{1.12}$$

with $a_0 = z_0$ since $\dot{y}(t_k + 0; t_0, z_0) = z_0$.

It is now obvious that the solution of (1.9) is mainly determined by the solution of the recurrence (in fact a first order difference equation) that defines a_k and the study of the behavior of the solution of (1.9) reduces to the study of the difference equation (1.12).

B. We shall focus now on some properties of (1.11). Remark first that since the minimum value of $x + 1/x$ is 2, we shall have $a_k \ge 2$

provided $a_0 > 0$. Further, it follows from (1.11) that

$$a_{n+1} = \left(e^{\alpha h}\right)^n a_1 + \sum_1^n \left(e^{\alpha h}\right)^{n-k} e^{-\alpha h} \left(a_k\right)^{-1}$$

and since $a_k \geq 2$ it follows that

$$a_{n+1} \leq \left(e^{\alpha h}\right)^n a_1 + \frac{1}{2}\left(e^{\alpha h}\right)^{n-1} \sum_1^n \left(e^{\alpha h}\right)^{-k} =$$

$$= \left(e^{\alpha h}\right)^n a_1 + \frac{1 - \left(e^{\alpha h}\right)^n}{2e^{\alpha h}\left(1 - e^{\alpha h}\right)} \quad (\alpha \neq 0)$$

and

$$a_{n+1} \leq a_1 + \frac{n}{2} \quad (\alpha = 0) \ .$$

If $\alpha > 0$ then it follows from (1.12) that $a_{n+1} > a_n$ and $\lim_{n \to \infty} a_n = \infty$. The sequence $\{b_n\}_n$ defined by $b_n = a_n \left(e^{-\alpha h}\right)^n$ verifies the recurrence

$$b_{n+1} = b_n + \left(e^{-\alpha h}\right)^{2(n+1)} b_n^{-1}$$

is increasing and, taking into account the inequality for a_{n+1},

$$b_{n+1} \leq b_n \frac{1 - \left(e^{\alpha h}\right)^n}{2\left(e^{\alpha h}\right)^{n+2}\left(1 - e^{\alpha h}\right)} \ .$$

Since $\alpha > 0$, b_n has an upper bound and consequently $\lim_{n \to \infty} b_n = b < +\infty$.

If $\alpha = 0$ we have again $a_{n+1} > a_n$ and $\lim_{n \to \infty} a_n = \infty$. Also it is easily shown by induction, using the inequality for a_{n+1} and (1.12) for $\alpha = 0$, that

$$a_1 + 2\sum_1^n \frac{1}{2a_1 + j - 1} \leq a_{n+1} \leq a_1 + \frac{n}{2} \ .$$

Assume now that $\alpha < 0$ (the usual case of a damped linear oscillator). The fixed points of the mapping $f(x) = e^{\alpha h}x + \left(e^{\alpha h}x\right)^{-1}$, i.e., the solutions of the equation

$$x = f(x) = e^{\alpha h}x + \left(e^{\alpha h}x\right)^{-1}$$

are the constant solutions of the difference equation (1.12) - its equilibria. The equation has two roots of opposite sign; consider the positive one: $a = \left(\sqrt{e^{\alpha h}\left(1 - e^{\alpha h}\right)}\right)^{-1} \geq 2$. As in the case of the logistic equation one may ask whether this equilibrium is attractive. To show that the equilibrium is attractive we denote $\mu = e^{\alpha h}$ and compute

$$a_{n+1} - a = \frac{\mu}{a_n}\left(a_n - a\right)\left(a_n - \frac{1}{\mu^2 a}\right)$$

$$a_{n+1} - a_n = \frac{1}{\mu a^2 a_n}\left(a + a_n\right)\left(a - a_n\right)$$

(1.13)

$$a_{n+2} - a_n = \frac{\mu^2\left(\mu + 1\right)\left(\mu - 1\right)}{a_n\left(\mu^2 a_n^2 + 1\right)}\left(a_n^2 + \frac{1}{\mu\left(\mu + 1\right)}\right)\left(a + a_n\right)\left(a - a_n\right).$$

(1.14)

From these equalities we deduce

1. If $a_n \geq a$ and $a_n \geq \frac{1}{\mu^2 a}$ then $a \leq a_{n+1} \leq a_n$.

2. If $a \leq a_n \leq \frac{1}{\mu^2 a}$ then $a_{n+1} \leq a$; if $\frac{1}{\mu^2 a} \leq a_n \leq a$ then $a_n \leq a_{n+1} \leq a$.

3. If $a_n \leq a$ and $a_n \leq \frac{1}{\mu^2 a}$ then $a_{n+1} \geq a$.

4. If $a_n \geq a$ then $a_{n+2} \leq a_n$ and conversely. From the previous estimate for a_{n+1} in the case $\alpha \neq 0$ we deduce boundedness of a_{n+1} in the case $\alpha < 0$.

Let $\mu \geq 1/2$ hence $\mu^2 \geq 1/4$, $\mu^2 a \geq 1/2$, $1/\left(\mu^2 a\right) \leq 2$. It follows that $a_n \geq 1/\left(\mu^2 a\right)$ since $a_n \geq 2$. From the above considerations we deduce that the sequence is increasing if $a_1 \leq a$ and decreasing if $a_1 \geq a$. In any case it is convergent and from (1.13) it follows that the limit is a.

Let now $\mu < 1/2$; if $a_0 < a$ then $a_0 < 1/\left(\mu^2 a\right)$ since $\mu^2 a^2 = \frac{\mu}{1 - \mu}$, $\frac{1}{\mu^2 a^2} = \frac{1}{\mu} - 1 > 1$ hence $\frac{1}{\mu^2 a} > a$. From $a_0 < a$ and $a_0 < \frac{1}{\mu^2 a}$ we have $a_1 \geq a$. If $a_n \geq \frac{1}{\mu^2 a}$ for all n then the sequence is decreasing and convergent to a limit larger than $\frac{1}{\mu^2 a} > a$. This limit would be an equilibrium larger than a, a contradiction. This contradiction shows that if $a_1 > \frac{1}{\mu^2 a}$ there exists some N such that

$a_N \leq \dfrac{1}{\mu^2 a}$ and $a_{N-1} > \dfrac{1}{\mu^2 a}$. It follows that $\{a_{N+2n}\}_n$ is decreasing and $\{a_{N+2n+1}\}_n$ is increasing hence the two sequences are convergent. Denoting $a_{N+2n} = \alpha_n$ we obtain using (1.14)

$$\alpha_{n+1} - \alpha_n = a_{N+2n+2} - a_{N+2n} =$$

$$= \frac{\mu^2 (\mu + 1)(\mu - 1)}{\alpha_n (\mu^2 \alpha_n^2 + 1)} \left(\alpha_n^2 + \frac{\mu}{\mu(\mu + 1)} \right) (\alpha_n + a)(\alpha_n - a)$$

and since α_n is convergent it follows that $\lim\limits_{n \to \infty} \alpha_n = a$. In a similar way it is obtained that $\lim\limits_{n \to \infty} a_{N+2n+1} = a$ hence $\lim\limits_{n \to \infty} a_n = a$. We obtained in this way attractiveness (for any $a_0 > 0$) of the equilibrium a. By the same approach we can obtain attractiveness (for any $a_0 < 0$) of the equilibrium $-a$.

Additionally we may even estimate the convergence speed. Denoting $a_n - a = \mu^{\lambda n} e_n$ we find the following equation for e_n

$$e_{n+1} = \frac{1}{\mu^{\lambda - 1}} \left(1 - \frac{1}{\mu^2 a a_n} \right) e_n$$

If $\mu = 1/2$ then $\lim\limits_{n \to \infty} e_n = 0$ for any $\lambda \geq 0$ and if $\mu \neq 1/2$ then $\lim\limits_{n \to \infty} e_n = 0$ for $\lambda < \dfrac{\ln |1 - 2\mu|}{\ln \mu}$ and $\lim\limits_{n \to \infty} |e_n| = \infty$ for $\lambda > \dfrac{\ln |1 - 2\mu|}{\ln \mu}$ if $e_n \neq 0$. To prove these assertions it suffices to remark that if there exist $\varepsilon > 0$ and a positive integer N such that

$$\frac{1}{\mu^{\lambda - 1}} \left| 1 - \frac{1}{\mu^2 a a_n} \right| \leq 1 - \varepsilon, \quad n > N$$

then $\lim\limits_{n \to \infty} e_n = 0$ and if $\varepsilon > 0$ and N are such that

$$\frac{1}{\mu^{\lambda - 1}} \left| 1 - \frac{1}{\mu^2 a a_n} \right| \geq 1 + \varepsilon, \quad n > N$$

then if $e_n \neq 0$, $\lim\limits_{n \to \infty} |e_n| = \infty$. Since $\mu^\lambda > |1 - 2\mu|$ implies $\dfrac{1}{\mu^{\lambda - 1}} \left| 1 - \dfrac{1}{\mu^2 a^2} \right| < 1$ and since $\lim\limits_{n \to \infty} a_n = a$ it follows that if $\lambda < \dfrac{\ln |1 - 2\mu|}{\ln \mu}$ then $\lim\limits_{n \to \infty} \dfrac{1}{\mu^{\lambda - 1}} \left| 1 - \dfrac{1}{\mu^2 a a_n} \right| < 1$.

C. We return now to the solutions of the continuous-time oscillator (1.9). The constant solutions $a_n \equiv \pm a$ of the difference equation

(1.12) define (if $\alpha < 0$; i.e., if $\gamma_1 > 0$, $\gamma_2 > 0$) two periodic solutions

$$y^{\pm}(t) = \pm \frac{1}{\beta} a e^{\alpha \left(t - t_0 - \left[\frac{t-t_0}{h} \right] \right)} \sin \beta (t - t_0), \quad t_n \leq t < t_{n+1}$$

with the period $h = t_{n+1} - t_n$. Both solutions are attractive in the sense that $y(t; t_0, z_0) \to y^+(t)$ if $0 < z_0 < \infty$ and $y(t; t_0, z_0) \to y^-(t)$ if $-\infty < z_0 < 0$. If the convergence speed estimates are used then it is easy to see that

$$\lim_{n \to \infty} e^{-\alpha \lambda (t - t_0)} \left[y^{\pm}(t) - y(t; t_0, z_0) \right] = 0$$

for any $\lambda > 0$ if $\alpha = \dfrac{\beta}{2\pi k} \ln \dfrac{1}{2}$ and for $\lambda < \dfrac{\beta \ln \left| 1 - 2e^{\frac{2\pi k \alpha}{\beta}} \right|}{2\pi k \alpha}$ if $\alpha \neq \dfrac{\beta}{2\pi k} \ln \dfrac{1}{2}$.

It follows that the solutions of (1.9) in the stable case ($\gamma_1 > 0$, $\gamma_2 > 0$) tend exponentially to one of the two periodic solutions induced by the impulses obeying the difference equation (1.12).

D. We would like to point out that the above impulse controlled oscillator belongs to the wider class of the systems with shocks (Milman and Myshkis (1960)) which themselves belong to the systems with "interface conditions" considered by Stallard (1955, 1962), Sansone (1957), Olech (1957) and more recently by Samoilenko, Schwabik and many others; a good reference in the field is the monograph of Schwabik (1985).

In the linear case the study of systems with impulses at fixed times leads to the following:
Consider the linear system

$$\dot{x} = A(t)x, \quad x(t_0) = x_0 \tag{1.15}$$

and the sequence t_k, $k = 0, 1, 2, ...$, of increasing real numbers. On each interval (t_k, t_{k+1}) the vector solution $x(t; t_0, x_0)$ satisfies the given system (1.15) and, additionally, the interface conditions due to impulses

$$x(t_k + 0) - x(t_k - 0) = s_{k-1} \tag{1.16}$$

Now, if $X(t, s)$ is the state transition matrix of (1.15), i.e., the fundamental matrix of solutions satisfying

$$\frac{\partial}{\partial t} X(t, s) = A(t) X(t, s), \quad X(s, s) = \mathbf{I}$$

the solutions of (1.15) will have the form

$$x\left(t; t_0, x_0\right) = X\left(t, t_k\right) c_k , \quad t_k \leq t < t_{k+1}$$

where the vector sequence $\{c_k\}_k$ defined by (1.16) obeys the discrete forced system

$$c_{k+1} = X\left(t_{k+1}, t_k\right) c_k + s_k , \quad c_0 = x_0 , \ k = 0, 1, 2, \ldots \quad (1.17)$$

1.3. DISCRETE SYSTEMS OCCURRING FROM SAMPLED DATA CONTROL SYSTEMS

According to classical references (Ragazzini and Franklin (1958), Tsyp-kin (1958)) a sampled data system is a dynamical system where the signals, i.e., the time functions describing the evolution of the physical parameters are defined at discrete instants if time or their complete description is represented by the information stored at discrete instants of time. The model of a sampled-data system contains at least one "impulse element" that realizes conversion of a continuous-time signal into a sampled data one. Such impulse elements are of various types according to the method of signal processing (sometimes called pulse modulation); the sampled data signals generated by them may be described in terms that are familiar to electrical engineers (sampling period, pulse width, waveform) but we shall not insist on this matter. Instead we shall consider a simple example borrowed from radar technique - the feedback system of automatic distance tracking (Tsypkin (1958), pp. 614-629). We shall give in the following a mathematical model of this system containing two amplifiers, two integrators and an impulse element included in a feedback structure. The continuous-time part is described by

$$\begin{aligned} \dot{x} &= u\left(t\right) \\ \dot{r} &= k_1 x + k_2 u\left(t\right) , \end{aligned} \quad (1.18)$$

where the state variable x is the output of the so called correction integrator. The state variable r which is in fact the output of an actuator represents the output signal of the distance tracking. Its coincidence with r^0 - the measured distance of the tracked target - ensures the correct tracking. This correct tracking is ensured by defining the control signal $u\left(t\right)$ as a function of the tracking error $\varepsilon\left(t\right) = r^0\left(t\right) - r\left(t\right)$. The dependence of the control signal on this tracking error is realized by using an impulse element with amplitude modulation, constant sampling period and pulse width equal to the sampling period. This gives the following description of the sampling element

$$u(t) = \varepsilon(kT), \quad kT \leq t < (k+1)T$$
$$\varepsilon(kT) = r^0(kT) - r(kT)$$
(1.19)

Remark that $u(t)$ is completely defined by the samples of the reference - $r^0(t)$ - and feedback - $r(t)$ - signals at $t = kT$. This was called in some classical books on sampled data systems introduction of fictitious samplers at the input and the output of the system. It will be shown that the entire dynamics of the feedback system is determined by the values of the signals at the sampling instants. We shall have

$$x(t) = x(kT) + \int_{kT}^{t} u(\tau)\,d\tau = x(kT) + (t - kT)\varepsilon(kT) \quad (1.20)$$

$$r(t) = r(kT) + k_1 \int_{kT}^{t} x(\tau)\,d\tau + k_2 \int_{kT}^{t} x_1(\tau)\,d\tau =$$
$$= r(kT) + k_1(t - kT)x(kT) + \frac{k_1}{2}(t - kT)^2\varepsilon(kT) +$$
$$+ k_2(t - kT)\varepsilon(kT)$$
(1.21)

If we denote

$$x(kT) = x_k, \quad r(kT) = r_k, \quad r^0(kT) = r_k^0$$

and write down (1.20) and (1.21) at $t = (k+1)T$ we shall have the following discrete-time system

$$x_{k+1} = x_k - Tr_k + Tr_k^0$$
$$r_{k+1} = k_1 T x_k + \left(1 - k_2 T - k_1\frac{T^2}{2}\right)r_k +$$
$$+ \left(k_2 T + k_1\frac{T^2}{2}\right)r_k^0$$
(1.22)

In order to illustrate the problems of such systems let us consider the tracking of a target moving at constant speed. Therefore,

$$r^0(t) = a_0 + a_1 t, \quad r_k^0 = a_0 + (a_1 T)k \quad (1.23)$$

We introduce the new variable $\varepsilon_k = r_k^0 - r_k$ and (1.22) takes the form

$$
\begin{aligned}
x_{k+1} &= x_k + T\varepsilon_k \\
\varepsilon_{k+1} &= -k_1 T x_k + \left(1 - k_2 T - k_1 \frac{T^2}{2}\right)\varepsilon_k + a_1 T
\end{aligned}
\tag{1.24}
$$

The constant solution of this system is $\hat{\varepsilon} = 0$, $\hat{x} = a_1/k_1$ and it corresponds to correct tracking (the tracking error is zero). Inherent disturbances (wind, initial position of the radar station etc.) induce evolutions of (1.24) that are different of the constant solution. If the deviations from it tend to zero, i.e., if the constant solution is attractive then tracking is finally accomplished. We are thus lead to the system in deviations

$$
\begin{aligned}
\xi_{k+1} &= \xi_k + T\varepsilon_k \\
\varepsilon_{k+1} &= -k_1 T \xi_k + \left(1 - k_2 T - k_1 \frac{T^2}{2}\right)\varepsilon_k
\end{aligned}
\tag{1.25}
$$

whose solutions must tend to zero. This property, called exponential stability (see Section 2.1) is true if and only if the eigenvalues of the matrix

$$
A = \begin{pmatrix} 1 & T \\ -k_1 T & 1 - k_2 T - k_1 \dfrac{T^2}{2} \end{pmatrix}
$$

are located inside the unit disk (see Section 2.1). These eigenvalues are the roots of the characteristic equation

$$
\lambda^2 - \left(2 - k_2 T - k_1 \frac{T^2}{2}\right)\lambda + 1 - k_2 T + k_1 \frac{T^2}{2} = 0
$$

The conditions for the roots of this equation to be located inside the unit disk are the Schur-Cohn conditions (see Section 2.1) which in our case have the form

$$
k_2 T - k_1 \frac{T^2}{2} > 0 ,
$$

$$
-\left(2 - k_2 T + k_1 \frac{T^2}{2}\right) < -\left(2 - k_2 T - k_1 \frac{T^2}{2}\right) < 2 - k_2 T + k_1 \frac{T^2}{2}
$$

The first condition gives $k_1 T < k_2$ while the other ones give $k_2 T < 4$. These simple stability conditions are important in the system's design since they provide gain limitations when the sampling period T is imposed by the electronic part of the system or sampling period restrictions when the gains k_1 and k_2 are imposed by the physical realization of the system.

1.4. NUMERICAL TREATMENT OF CONTINUOUS TIME SYSTEMS

It is quite well known that specific requirements of a computational process are its convergence and absence of error accumulation. Indeed the fundamental characteristic of any approximating computational method is its precision. Increasing precision requires increasing the number of the executed elementary operations and this last fact may lead to non realizability of the computational process. In fact the error of a computing method consists of the error of the method itself (as an approximation procedure) and of the errors induced by the fact that any elementary operation is performed with limited precision. With respect to this let us point out that error accumulation during the computational process might be more important than error of the numerical procedure. To illustrate this we shall consider a simple example of integrating a first order differential equation when *we actually use an associated discrete-time system whether we realize it or not* (Elaydi (1996), p. 14). Consider the equation

$$\dot{y} + ay = f(t) , \quad y(0) = y_0 , \ a > 0 \qquad (1.26)$$

with f continuous. It is required to find the solution on $[0, T]$ satisfying (1.26) with the initial condition $y(0) = y_0$. Take $\tau = T/N$ and consider first the approximation of the derivative by the simplest difference formula of Euler

$$\dot{y}(n\tau) = \frac{y((n+1)\tau) - y(n\tau)}{\tau}$$

Denoting $y_n = y(n\tau)$, $f_n = f(n\tau)$ we shall have the discrete system

$$y_{n+1} = (1 - a\tau)y_n + \tau f_n , \quad n = \overline{0, N-1} \qquad (1.27)$$

Let $\{y_k\}_k$ be the solution sequence of (1.27) with $f_k = f(k\tau)$ as forcing term and with $y_0 = y(0)$. Consider also $\{\tilde{y}_k\}_k$ the solution sequence corresponding to input data $\tilde{f}_k, \tilde{y}_0$ affected by errors. Denoting $\varphi_k = f_k - \tilde{f}_k$ and $z_k = y_k - \tilde{y}_k$ we shall obtain the following discrete system describing error accumulation

$$z_{k+1} = (1 - a\tau) z_k + \tau\varphi_k , \quad k = \overline{0, N - 1} \qquad (1.28)$$

This system coincides with (1.27) due to linearity; only the input data are different; the input data of (1.28) are the errors. Treating (1.28) as a recurrence we shall have

$$z_k = (1 - a\tau)^k z_0 + \sum_0^{k-1} (1 - a\tau)^{k-1-i} \tau\varphi_i , \quad k = \overline{0, N - 1} \qquad (1.29)$$

Since $0 < 1 - a\tau < 1$ for $\tau > 0$ sufficiently small, the first component of the error (1.33) tends exponentially to 0 and for finite N obeys the estimate

$$\left| (1 - a\tau)^k z_0 \right| \leq (1 - a\tau)^k |z_0| \leq e^{-\frac{T}{\tau}} |z_0|$$

what shows that this error diminishes with increase of T. On the other side

$$\left| \sum_0^{k-1} (1 - a\tau)^{k-1-i} \tau\varphi_i \right| \leq \tau \sup |\varphi_k| \cdot \frac{1 - (1 - a\tau)^k}{a\tau} =$$

$$= \frac{1}{a} \left[1 - (1 - a\tau)^k \right] \sup |\varphi_k|$$

and even if k is quite large this error remains small provided the error for the forcing term remains small. This good behavior of the process of error accumulation is due to the exponential stability of the zero solution of the discrete evolution $z_{k+1} = (1 - a\tau) z_k$ provided $0 < 1 - a\tau < 1$.

Assume now that we want a better approximation for the derivative. If we know that $y(t)$ is C^3 (in our case this holds if $f(t)$ is C^3), we may write

$$y((n + 1)\tau) = y(n\tau) + \tau\dot{y}(n\tau) + \frac{\tau^2}{2}\ddot{y}(n\tau) + o(\tau^2)$$
$$y((n + 2)\tau) = y(n\tau) + 2\tau\dot{y}(n\tau) + 2\tau^2\ddot{y}(n\tau) + o(\tau^2)$$

$$\overline{-y((n + 2)\tau) + 4y((n + 1)\tau) - 3y(n\tau) = 2\tau\dot{y}(n\tau) + o(\tau^2)}$$

what gives

$$\frac{1}{2\tau}\left[-y((n + 2)\tau) + 4y((n + 1)\tau) - 3y(n\tau) \right] = \dot{y}(n\tau) + \mathcal{O}(\tau^2)$$

a better approximation for the derivative since in the previous case the error was $\mathcal{O}(\tau)$. We take

$$\dot{y}(n\tau) = \frac{1}{2\tau}\left[-y((n+2)\tau) + 4y((n+1)\tau) - 3y(n\tau)\right] \qquad (1.30)$$

and this second order difference will give a second order computing process described by

$$y_{n+2} - 4y_{n+1} + (3 - 2a\tau)y_n = -2\tau f_n \qquad (1.31)$$

The error accumulation is described by

$$z_{n+2} - 4z_{n+1} + (3 - 2a\tau)z_n = -2\tau\varphi_n \qquad (1.32)$$

Assume for simplicity that $\varphi_k \equiv 0$. Then z_n will be the general solution of the linear equation

$$z_{n+2} - 4z_{n+1} + (3 - 2a\tau)z_n = 0 \qquad (1.33)$$

For this equation we look for solutions of the form $z_k = \rho^k$. Substituting in the equation we find that ρ must be a root of the equation

$$\rho^2 - 4\rho + 3 - 2a\tau = 0$$

We have $\rho_1 = 2 - \sqrt{1 + 2a\tau}$, $\rho_2 = 2 + \sqrt{1 + 2a\tau}$ and

$$z_n = c_1\rho_1^n + c_2\rho_2^n , \quad n = \overline{0, N-1}.$$

Suppose z_0 and z_1 the initial errors are known. We shall have

$$
\begin{aligned}
z_n &= \frac{1}{2\sqrt{1 + 2a\tau}}\left\{\left[(2 - \sqrt{1 + 2a\tau})z_0 - z_1\right](2 - \sqrt{1 + 2a\tau})^n + \right.\\
&\left. + \left[(2 + \sqrt{1 + 2a\tau})z_0 - z_1\right](2 + \sqrt{1 + 2a\tau})^n\right\}
\end{aligned}
$$

It is easily seen that $|z_n| \sim (|z_0| + |z_1|)(2 + \sqrt{1 + 2a\tau})^n$, i.e., the accumulation of error is exponential and while more precise from the point of view of derivative approximation, the second method is not applicable. The explanation is given by the fact that discrete system (1.33) is unstable since its characteristic equation has at least one root outside the unit disk (see Section 2.1).

This simple example exhibits the importance of studying stability properties of discrete-time systems.

1.5. ABOUT THE BOOK

The last few decades have witnessed increased interest in discrete systems (many books and papers were published and journals and book series founded). As pointed out by some authors, we deal with "a completely new field which in most cases is very different from that of the well-known ordinary differential equations" (Sharkovskii et al. (1986)). On the other hand, it is pointed out that "with the use of a computer one can easily experiment with difference equations and one can easily discover that such equations possess fascinating properties with a great deal of structure and regularity" (Kocic and Ladas (1993)). We have seen indeed in the previous analysis that chaotic behavior is quite different from that of regular behavior not only from the philosophical point of view but from the point of view of the mathematical techniques also. Of the two cases the choice was made for the regular and stable behavior. One reason for this choice (and not the last one) has been the fact that the popularity (including the mediatic one) and increased interest for chaotic behavior somehow left aside the regular case (corresponding to stability) which, as we pointed out throughout this chapter, turns out to be the most important of the many applications of discrete-time systems. Last but not least, the authors' own research experience provided another considerable argument for this choice.

The present monograph is mainly dedicated to the idea of stability, especially the stability of the equilibrium in the sense of Liapunov (with respect to the initial conditions). Such classical problems as stability of linear systems with constant and time-varying coefficients, stability by the first approximation, stability by Liapunov functions, and the La Salle invariance principle are considered in some detail. It is worth mentioning that, while included in a pioneering paper of one of the authors (Halanay (1963)), some of these topics are now included in textbooks (e.g. Elaydi (1996)). Here they are detailed for the sake of completeness since they represent an instrument for further development. They are also illustrated by examples, among which we would like to cite the application to chemical engineering in the case of discrete storage of information (Section 2.7).

A distinct place is reserved in the book for detailed presentation of the absolute stability (the discrete time case) and of the associated techniques. After the boom of the sixties and the decreased interest, even the disregard that followed, this problem with its various connections has now received renewed attention. The book presents not only classical, by now frequency domain criteria of absolute stability (the one of Tsypkin, the modified Tsypkin criterion, the Brockett-Willems criterion with many free parameters) but also the connection between the frequency domain techniques and one of the Liapunov functions of special form - the Kalman-Szegö-Popov-Yakubovich lemma - and new criteria of absolute stability expressed in frequency domain terms and proved using Liapunov functions.

The final chapter of the book deals with stable oscillations. Several applications lead to discrete processes with oscillatory behavior. But this oscillatory behavior can be perceived (e.g. measured) only if it has some stability properties. If the oscillation is "attractive" then the evolution on some trajectory generated by a perturbed initial condition will approach asymptotically the oscillatory motion. The considered oscillations may be periodic or almost periodic. The discrete almost periodic oscillations are seldom present in the mathematical literature despite the fact that they are far from being just a mathematical curiosity. In fact, the simplest harmonic oscillation $f(t) = \sin t$ being observed through discrete intervals (samples), even of equal width, generates as a rule an almost periodic sequence; for instance, the sequence $f_k = \sin k$ is not periodic for integer k since 2π is irrational but it is almost periodic according to the general theory that states almost periodicity of the sequence $f_k = f(k)$ for (almost) periodic $f(t)$.

The part of the book dedicated to oscillations contains classical problems as existence and stability of forced periodic and almost periodic solutions for linear systems and for quasilinear ones.

Let us remember that the sources of the theory and motivation for the considered problems may be found in the mathematical models of some phenomena which belong to technology and other non-engineering fields. The interest for discrete-time models being perceived, the mathematical theory grows also independently stimulated by the analogy to the facts previously known, especially from the field of ordinary differential equations.

CHAPTER 2

STABILITY THEORY

2.1. LINEAR DISCRETE TIME SYSTEMS WITH CONSTANT COEFFICIENTS

A. It has been shown in the Introduction that various real world problems are modelled by difference equations describing the evolution of sequences of numbers that represent physical quantities (signals). In the linear case and when the only motor of the dynamics is given by internal phenomena incorporated in the initial state, the model is the autonomous linear difference system with constant coefficients

$$x_{t+1} = Ax_t \tag{2.1}$$

where x is a n-dimensional state vector, A a $n \times n$ constant matrix and t - the integer independent variable - conventionally called "*time*", although it might sometimes represent the numbering of sequence elements.

If x_0 is given we can compute the sequence of states

$$x_t = A^t x_0 , \quad t \in \mathbb{N} \tag{2.2}$$

which is called solution of (2.1). The simple fact that (2.2) verifies (2.1) can be shown by direct check. For additional information about the solutions, we need information about positive integer powers of a matrix. A convenient way of obtaining this information (for the sake of theory, not for computation) is to use Jordan form. It is well known (the reader is sent to any book on linear algebra or matrix theory) that *there exists a nonsingular matrix T such that TAT^{-1} has Jordan form.* Consequently we perform a change of coordinates

$$\hat{x} = Tx$$

and give (2.2) the form

$$\hat{x}_{t+1} = J\hat{x}_t \tag{2.3}$$

where $J = TAT^{-1}$ is composed of several Jordan cells

$$J = \begin{pmatrix} J_1 & 0 & \cdots & 0 \\ 0 & J_2 & \cdots & 0 \\ \cdots & \cdots & \cdots & \cdots \\ 0 & 0 & \cdots & J_s \end{pmatrix} \quad \text{hence} \quad J^t = \begin{pmatrix} J_1^t & 0 & \cdots & 0 \\ 0 & J_2^t & \cdots & 0 \\ \cdots & \cdots & \cdots & \cdots \\ 0 & 0 & \cdots & J_s^t \end{pmatrix}$$

where

$$J_k = \lambda \mathbf{I}_{p_k} + S_{p_k}$$

p_k being the dimension of the Jordan cell, $\mathbf{I}_{p_k}$ the identity matrix of corresponding dimension and S_{p_k} the matrix defined by

$$S_{p_k} = \begin{pmatrix} 0 & 1 & 0 & \cdots & 0 \\ 0 & 0 & 1 & \cdots & 0 \\ \cdots & \cdots & \cdots & \cdots & \cdots \\ 0 & 0 & 0 & \cdots & 1 \\ 0 & 0 & 0 & \cdots & 0 \end{pmatrix}$$

Obviously $(S_{p_k})^{p_k} = 0$. A straightforward computation will give

$$J_k^t = \begin{pmatrix} \lambda_k^t & t\lambda_k^{t-1} & \cdots & C_n^{p_k-1}\lambda_k^{t-p_k+1} \\ 0 & \lambda_k^t & \cdots & C_n^{p_k-2}\lambda_k^{t-p_k+2} \\ \cdots & \cdots & \cdots & \cdots \\ 0 & 0 & \cdots & \lambda_k^t \end{pmatrix} \tag{2.4}$$

and the transformed system (2.3) will have the following linearly independent solutions:

$$\begin{aligned}
\hat{x}_t^1 &= \lambda_1^t e_1; \\
\hat{x}_t^2 &= t\lambda_1^{t-1}e_1 + \lambda_1^t e_2; \quad \cdots \\
\hat{x}_t^{p_1} &= C_t^{p_1-1}\lambda_1^{t-p_1+1}e_1 + \cdots + \lambda_1^t e_{p_1}; \quad \cdots \\
\hat{x}_t^n &= C_t^{p_s-1}\lambda_s^{t-p_s+1}e_{n-p_s+1} + \cdots + t\lambda_s^{t-1}e_{n-1} + \lambda_s^t e_n
\end{aligned}$$

Here $p_1 + p_2 + \cdots + p_s = n$ and e_i, $i = \overline{1,n}$ are the unit vectors of the canonical basis (in fact the columns of the identity matrix $\mathbf{I}_n$). It follows that the initial system has the following solutions which are linearly independent

$$x_t^1 = \lambda_1^t u_1; \quad x_t^2 = t\lambda_1^{t-1}u_1 + \lambda_1^t u_2; \quad \cdots ;$$

$$x_t^{p_1} = C_t^{p_1-1}\lambda_1^{t-p_1+1}u_1 + \cdots + t\lambda_s^{t-1}u_{p_1-1} + \lambda_1^t u_{p_1}; \cdots ;$$
$$x_t^n = C_t^{p_s-1}\lambda_s^{t-p_s+1}u_{n-p_s+1} + \cdots + t\lambda_s^{t-1}u_{n-1} + \lambda_s^t u_n$$

where $(u_1, u_2, ..., u_n)$ is a set of linearly independent vectors. This form of the solutions allows one to obtain some information about the asymptotic behavior (when $t \to \infty$) of the solutions. Namely, if $|\lambda_k| < 1$ for any k it is obvious that all solutions tend asymptotically to zero (even exponentially because $|\lambda_k|^t = \exp(-\alpha_k t)$ where $\alpha_k = \ln|\lambda_k|^{-1} > 0$). If there exists some λ_k with $|\lambda_k| > 1$, then there exist solutions that tend exponentially to infinity because $|\lambda_k|^t = \exp(\alpha_k t)$ where $\alpha_k = \ln|\lambda_k| > 0$. If for any k we have $|\lambda_k| \leq 1$ and to all eigenvalues with $|\lambda_k| = 1$ there correspond one-dimensional Jordan cells, then all solutions of (2.2) are bounded. If at least one Jordan cell corresponding to some λ_k with $|\lambda_k| = 1$ has the dimension more than 1, then there exist unbounded solutions.

In applications the asymptotic properties described above are connected to the *stability of the linear system* as stated in the following:

Definition 2.1 The linear system (2.2) is called *exponentially stable* if all its solutions tend exponentially to zero (for $t \to \infty$). The system is called *stable* if all its solutions are bounded. If there exist unbounded solutions the system is called *unstable* and if there exist unbounded solutions that tend exponentially to infinity, *the instability is exponential.*

The above considerations connected with the eigenvalues λ_k allows one to state

Theorem 2.1 *For the exponential stability of (2.2) it is necessary and sufficient that the eigenvalues of A are inside the unit disk of the complex plane. For the stability it is necessary and sufficient that the eigenvalues of A are inside the unit disk and on the circle, those on the circle being located in one dimensional Jordan cells. The system is unstable if at least one eigenvalue on the unit circle is located in a Jordan cell of dimension larger than 1. The instability is exponential if at least one eigenvalue is located outside the unit disk.*

Remark that the unit disk plays the same role in the stability of linear discrete systems as the imaginary axis in the stability of linear continuous-time systems.

B. If we now consider the model occurring in signal processing (filtering a.s.o.), for instance, and assume the input signal to be the identically zero sequence, we obtain an autonomous higher order difference equation:

$$y_{t+n} + \alpha_{n-1} y_{t+n-1} + \cdots + \alpha_1 y_{t+1} + \alpha_0 y_t = 0 \qquad (2.5)$$

This equation can be written as a system denoting

$$x_t^1 = y_t \ , \ x_t^2 = y_{t+1} \ , \ \ldots \ , x_t^n = y_{t+n-1}$$

It follows that

$$
\begin{aligned}
x_{t+1}^1 \ &= \ x_t^2 \\
x_{t+1}^2 \ &= \ x_t^3 \\
\cdots \ &\cdots \ \cdots \\
x_{t+1}^n \ &= \ -\alpha_0 x_t^1 - \alpha_1 x_t^2 - \cdots - \alpha_{n-1} x_t^n
\end{aligned}
$$

The state vector x having $x^1, x^2, ..., x^n$ as components we obtain a system of the form (2.2) where A has a canonical form

$$
A = \begin{pmatrix}
0 & 1 & 0 & \cdots & 0 \\
0 & 0 & 1 & \cdots & 0 \\
\cdots & \cdots & \cdots & \cdots & \cdots \\
0 & 0 & 0 & \cdots & 1 \\
-\alpha_0 & -\alpha_1 & -\alpha_2 & \cdots & -\alpha_{n-1}
\end{pmatrix}
$$

For matrices of this form it can be shown (e.g., Popov (1973), Appendix A, pp. 293-310) that Jordan form is such that its cells contain distinct eigenvalues. Therefore, the set of the independent solutions of (2.5) is given by

$$y_t^1 = \lambda_1^t, \ y_t^2 = t\lambda_1^{t-1}, \ y_t^{p_1} = C_t^{p_1-1}\lambda_1^{t-p_1+1}, \ \ldots, \ y_t^n = \lambda_s^t,$$

where $p_1 + p_2 + \cdots + p_s = n$ and λ_k, $k = \overline{1, s}$ are the roots of the characteristic equation

$$\lambda^n + \alpha_{n-1}\lambda^{n-1} + \cdots + \alpha_1\lambda + \alpha_0 = 0 \qquad (2.6)$$

p_k being the multiplicity of λ_k.

2.1.1. A Stability Criterion

Both cases considered above may be characterized from the stability point of view by the location inside the unit disk of the roots of some algebraic equation - the characteristic equation of system's matrix in the first case and the characteristic equation of the difference equation in the second case (nevertheless a matrix characteristic equation too).

The algebraic criterion for polynomial root location in the left half plane is the well-known *Routh-Hurwitz criterion*. Its counterpart for the location of the roots inside the unit disk - *the Schur-Cohn criterion* - is also well known but is not very easy to write down explicitly, even for low-order polynomials. For this reason we shall take a slightly different approach, also quite well known, based on conformal mappings and on the Routh-Hurwitz criterion.

Consider the *bilinear transformation*

$$s = \frac{z-1}{z+1}, \quad z = \frac{1+s}{1-s} \tag{2.7}$$

which maps the unit disk in the z-plane onto the left half of the s-plane. Suppose we are interested in finding necessary and sufficient conditions for the location of the roots of

$$p_0(z) = a_n z^n + a_{n-1} z^{n-1} + \cdots + a_1 z + a_0 \tag{2.8}$$

inside the unit disk of the z-plane. These conditions are equivalent to Routh-Hurwitz conditions for the transformed polynomial

$$\hat{p}_0(s) = (1-s)^n p_0 \left((1+s)(1-s)^{-1} \right) \tag{2.9}$$

We may in fact state

Proposition 2.2 *The necessary and sufficient condition for a polynomial $p_0(z)$ of degree n to have its roots inside the unit disk of the complex plane is that the associated polynomial*

$$\hat{p}_0(s) = (1-s)^n p_0 \left((1+s)(1-s)^{-1} \right)$$

has its roots in the left half plane i.e. satisfies the Routh-Hurwitz conditions.

In the following we shall write these conditions explicitly for $n = 1, 2, 3$.

A. For $n = 1$ we have $p_0(z) = a_1 z + a_0$; therefore

$$\hat{p}_0(s) = a_1(1+s) + a_0(1-s) = (a_1 - a_0)s + a_1 + a_0$$

and the Routh-Hurwitz conditions will give finally

$$|a_0| < |a_1| \tag{2.10}$$

the Schur-Cohn condition that might have seen directly in this case.

B. For $n = 2$ we have $p_0(z) = a_2 z^2 + a_1 z + a_0$; therefore

$$
\begin{aligned}
\hat{p}_0(s) &= a_2(1+s)^2 + a_1(1-s^2) + a_0(1-s)^2 = \\
&= (a_2 - a_1 + a_0)s^2 + 2(a_2 - a_0)s + a_2 + a_1 + a_0
\end{aligned}
$$

and the Routh-Hurwitz conditions for $n = 2$ will give the following Schur-Cohn inequalities

$$a_2 > 0, \ |a_0| < a_2, \ |a_1| < a_2 + a_0 \tag{2.11}$$

(If $a_2 < 0$ we may consider the polynomial $-p_0(z)$.)

C. For $n = 3$ we have $p_0(z) = a_3 z^3 + a_2 z^2 + a_1 z + a_0$; therefore

$$
\begin{aligned}
\hat{p}_0(s) &= a_3(1+s)^3 + a_2(1+s)^2(1-s) + \\
&\quad + a_1(1+s)(1-s)^2 + a_0(1-s)^3 \\
&= (a_3 - a_2 + a_1 - a_0)s^3 + (3a_3 - a_2 - a_1 + 3a_0)s^2 + \\
&\quad + (3a_3 + a_2 - a_1 - 3a_0)s + (a_3 + a_2 + a_1 + a_0)
\end{aligned}
$$

and after some manipulation the Schur-Cohn conditions are the following:

$$
\begin{gathered}
a_3 + a_2 + a_1 + a_0 > 0, \ a_3 - a_2 + a_1 - a_0 > 0, \\
3(a_3 + a_0) - a_2 - a_1 > 0, \ a_3(a_3 - a_1) - a_0(a_0 - a_2) > 0
\end{gathered} \tag{2.12}
$$

2.1.2. Example: Stability of a Digital Filter

We shall consider here an example of digital implementation of a second order bandpass filter designed according to a standard procedure (Williams (1981)). The bandpass filter is described by the input/output linear differential equation

$$y'' + \frac{\omega_r}{Q}y' + \omega_r^2 y = Hu'(t) \tag{2.13}$$

and its digital implementation is described by the difference equation

$$y_{k+2} + \alpha_1 y_{k+1} + \alpha_0 y_k = \beta_2 u_{k+2} + \beta_1 u_{k+1} + \beta_0 u_k \qquad (2.14)$$

where u_k, y_k are in fact the values of the input and output signals at the sampling point $k\delta$, δ being the sampling interval. The coefficients are computed according to the following design formulae, starting from the coefficients of the continuous-time device (op. cit., pp. 11.4-11.6)

$$
\begin{aligned}
\beta_0 &= -2e^{-\omega_r \delta/2Q} \cos\left(\omega_r \sqrt{1 - \frac{1}{4Q^2}}\right), \\
\beta_1 &= 2e^{-\omega_r \delta/2Q} \left[\frac{H}{2} \cos\left(\omega_r \sqrt{1 - \frac{1}{4Q^2}}\right) + \right. \\
&\quad \left. + HQ\sqrt{1 - \frac{1}{4Q^2}} \sin\left(\omega_r \sqrt{1 - \frac{1}{4Q^2}}\right)\right], \\
\beta_2 &= H, \ \alpha_1 = \beta_0, \ \alpha_0 = e^{-\omega_r \delta/Q}
\end{aligned}
$$

The design procedure does not take into account stability and this property, as in most cases of electronic filter design, has to be checked *a posteriori*. The stability property is checked by checking location of the roots of the characteristic polynomial

$$p_0(z) = z^2 + \alpha_1 z + \alpha_0$$

inside the unit disk. Using the bilinear transformation we obtain the second order polynomial

$$\hat{p}_0(s) = (1 - \alpha_1 + \alpha_0)s^2 + 2(1 - \alpha_0)s + (1 + \alpha_1 + \alpha_0)$$

It is well known that for a second order polynomial the Routh-Hurwitz conditions have a very simple form: all coefficients have the same sign. Using the above design expressions for the digital filter we find that

$$1 - \alpha_0 = 1 - e^{-\omega_r \delta/Q} > 0$$

hence all coefficients of the transformed polynomial have to be positive in this case. Further,

$$
\begin{aligned}
1 \pm \alpha_1 + \alpha_0 &= 1 \mp 2e^{-\omega_r \delta/2Q} \cos\left(\omega_r \sqrt{1 - \frac{1}{Q^2}}\right) + e^{-\omega_r \delta/Q} = \\
&= \left(1 - e^{-\omega_r \delta/Q}\right)^2 + 2e^{-\omega_r \delta/2Q}\left[1 \mp \cos\left(\omega_r \sqrt{1 - \frac{1}{Q^2}}\right)\right] > 0
\end{aligned}
$$

The stability conditions are always fulfilled; hence, *this design always gives a stable digital filter (independently of the sampling interval δ).*

2.1.3. Example: Inherent Stability of a Digitally Controlled Thermal Process

Consider the controlled thermal process described in the book of Mahmoud and Singh (1984, pp. 2-6)

$$m_1 c_1 \left[T_1 \left((k+1)\,\delta \right) - T_1 \left(k\delta \right) \right] = \delta h_{12} \left[T_2 \left(k\delta \right) - T_1 \left(k\delta \right) \right] + Q_0 \left(k\delta \right)$$
$$m_2 c_2 \left[T_2 \left((k+1)\,\delta \right) - T_2 \left(k\delta \right) \right] = \delta h_{12} \left[T_1 \left(k\delta \right) - T_2 \left(k\delta \right) \right] -$$
$$- \delta h_{2e} \left[T_2 \left(k\delta \right) - T_e \left(k\delta \right) \right] + Q_s \left(k\delta \right)$$

$$(2.15)$$

The above model can be given the form

$$x_{k+1} = A x_k + b u_k + d_k$$

where k represents the k-th sampling instant $k\delta$; also we have

$$x = \left(\begin{array}{c} T_1 \\ T_2 \end{array} \right), \quad u = Q_s, \quad d = \left(\begin{array}{c} Q_0/m_1 c_1 \\ T_e \delta h_{2e}/m_2 c_2 \end{array} \right),$$

$$A = \left(\begin{array}{cc} 1 - \delta \dfrac{h_{12}}{m_1 c_1} & \delta \dfrac{h_{12}}{m_1 c_1} \\ \delta \dfrac{h_{12}}{m_2 c_2} & 1 - \delta \dfrac{h_{12} + h_{2e}}{m_2 c_2} \end{array} \right), \quad b = \left(\begin{array}{c} 0 \\ \dfrac{1}{m_1 c_1} \end{array} \right)$$

By *inherent stability* we mean here stability of the free system, i.e., *without external signals*. In fact, we are interested in the exponential stability of a system having the form (2.1) with A given above. The characteristic equation

$$\det \left(z\mathbf{I} - A \right) = 0$$

has the form

$$z^2 - \left[2 - \left(\frac{h_{12}}{m_1 c_1} + \frac{h_{12} + h_{2e}}{m_2 c_2} \right) \delta \right] z + \left(1 - \frac{h_{12}}{m_1 c_1} \delta \right) \times$$
$$\times \left(1 - \frac{h_{12} + h_{2e}}{m_2 c_2} \delta \right) - \frac{h_{12}^2}{m_1 m_2 c_1 c_2} \delta^2 = 0$$

Note that all coefficients in (2.15) are strictly positive. Conditions

(2.11) will give

$$\left(1 - \frac{h_{12}}{m_1 c_1}\delta\right)\left(1 - \frac{h_{12} + h_{2e}}{m_2 c_2}\delta\right) - \frac{h_{12}^2}{m_1 m_2 c_1 c_2}\delta^2 < 1$$

$$2 - \left(\frac{h_{12}}{m_1 c_1} + \frac{h_{12} + h_{2e}}{m_2 c_2}\right)\delta <$$
$$< 1 + \left(1 - \frac{h_{12}}{m_1 c_1}\delta\right)\left(1 - \frac{h_{12} + h_{2e}}{m_2 c_2}\delta\right) - \frac{h_{12}^2}{m_1 m_2 c_1 c_2}$$

The second inequality is automatically satisfied because it reduces to

$$\frac{h_{12} h_{2e}}{m_1 m_2 c_1 c_2}\delta^2 > 0$$

The first inequality is reduced to the condition

$$0 < \delta < \frac{m_1 c_1}{h_{12}} + (m_1 c_1 + m_2 c_2)\frac{1}{h_{2e}} \qquad (2.16)$$

It follows that the requirement of inherent stability introduces an upper bound for the sampling period. Such limitations are quite often met in numerical analysis where stability of computing processes limitates the step in the integration of ordinary differential equations.

2.1.4. The Liapunov Matrix Equation for the Discrete Case

In the following we shall consider the problem of finding Hermitian solutions of the matrix equation

$$A^* P A - P = -Q \qquad (2.17)$$

where $Q = Q^*$. This equation represents the discrete counterpart of the well-known Liapunov equation $A^* P + P A = -Q$.

Lemma 2.3 *If matrix A has its eigenvalues inside the unit disk, the unique solution of (2.17) is given by*

$$P = \sum_{0}^{\infty} (A^*)^k Q A^k \qquad (2.18)$$

Thus, if $Q > 0$ then $P > 0$.

Proof. We may use the following estimate for A^k : $\left|A^k\right| \le \beta_0 \rho^k$ where ρ is the spectral radius of the matrix. If the eigenvalues of A are inside the unit disk then $0 < \rho < 1$. We have $\left|(A^*)^k Q A^k\right| \le \beta_0^2 |Q| \rho^{2k}$ and this shows that the matrix series in (2.18) is convergent. It remains to check the formula for the solution

$$A^* P A - P = \sum_0^\infty (A^*)^{k+1} Q A^{k+1} - \sum_0^\infty (A^*)^k Q A^k = -Q$$

This solution is the unique solution of (2.17). Indeed, suppose there were two solutions: P defined by (2.18) and another one $\tilde{P}$. We shall have

$$
\begin{aligned}
P &= \sum_0^\infty (A^*)^k Q A^k = \sum_0^\infty (A^*)^k \left(\tilde{P} - A^* \tilde{P} A\right) A^k = \\
&= \sum_0^\infty (A^*)^k \tilde{P} A^k - \sum_0^\infty (A^*)^{k+1} \tilde{P} A^{k+1} = \tilde{P}
\end{aligned}
$$

The fact that $Q > 0$ implies $P > 0$ obviously follows from (2.18).

Proposition 2.4 *If equation* (2.17) *with* $Q > 0$ *has a solution* $P > 0$ *then the eigenvalues of* A *are located inside the unit disk.*

Proof. Let λ be an eigenvalue of A and u a corresponding eigenvector. We have $Au = \lambda u$, hence

$$u^* (A^* P A - P) u = u^* \bar{\lambda} P \lambda u - u^* P u = \left(|\lambda|^2 - 1\right) u^* P u < 0$$

and since $u^* P u > 0$, it follows that $|\lambda| < 1$ which ends the proof.

Proposition 2.5 *Let* A *have the eigenvalues located on the unit circle, with simple elementary divisors. Then there exists* $P > 0$ *such that* $A^* P A - P = 0$.

Proof. The assumption concerning the eigenvalues means that the Jordan form is diagonal hence there exists a nonsingular matrix S such that

$$SAS^{-1} = D , \quad D = \operatorname{diag}(\lambda_j)$$

We have $D^* D = \operatorname{diag}\left(|\lambda_j|^2\right) = \mathbf{I}$ hence $D^* D - \mathbf{I} = 0$. On the other hand $D = SAS^{-1}$, $D^* = \left(S^{-1}\right)^* A^* S^*$. We shall have

$$D^* D - \mathbf{I} = \left(S^{-1}\right)^* A^* S^* SAS^{-1} - \mathbf{I} = 0$$

From the above inequality it follows that

$$A^* S^* SA - S^* S = 0$$

and with $P = S^* S > 0$ the proof ends.

Combining Lemma 2.3 and Proposition 2.5 we may state

Theorem 2.6 *Assume that the eigenvalues of A are located inside the closed unit disk and that the eigenvalues on the unit circle have simple elementary divisors. Then there exists $P > 0$ such that $A^* PA - P \le 0$.*

Proof. Let S be such that

$$SAS^{-1} = \begin{pmatrix} A_1 & 0 \\ 0 & A_2 \end{pmatrix}$$

where A_1 has the eigenvalues inside the unit disk and A_2 has them on the unit circle. Let $P_1 > 0$ be such that $A_1^* P_1 A_1 - P_1 = -\mathbf{I}$ (Lemma 2.3) and $P_2 > 0$ be such that $A_2^* P_2 A_2 - P_2 = 0$ (Proposition 2.5). We may write

$$\begin{pmatrix} A_1^* & 0 \\ 0 & A_2^* \end{pmatrix} \begin{pmatrix} P_1 & 0 \\ 0 & P_2 \end{pmatrix} \begin{pmatrix} A_1 & 0 \\ 0 & A_2 \end{pmatrix} -$$
$$- \begin{pmatrix} P_1 & 0 \\ 0 & P_2 \end{pmatrix} = \begin{pmatrix} -\mathbf{I} & 0 \\ 0 & 0 \end{pmatrix} \le 0$$

Therefore,

$$\left(S^{-1}\right)^* A^* S^* \begin{pmatrix} P_1 & 0 \\ 0 & P_2 \end{pmatrix} SAS^{-1} - \begin{pmatrix} P_1 & 0 \\ 0 & P_2 \end{pmatrix} \le 0$$
$$A^* S^* \begin{pmatrix} P_1 & 0 \\ 0 & P_2 \end{pmatrix} SA - S^* \begin{pmatrix} P_1 & 0 \\ 0 & P_2 \end{pmatrix} S \le 0$$

and with

$$P = S^* \begin{pmatrix} P_1 & 0 \\ 0 & P_2 \end{pmatrix} S > 0$$

the proof ends.

2.1.5. Robust Stability and Stability Radii

No mathematical model can exactly represent the dynamics of a physical plant. So, properties established on the basis of a nominal model may not hold when applied to the real system. Roughly speaking, a property is said to be *robust* if it holds for a class of perturbed models. If this class reflects the uncertain or neglected features of the system, then one can have more confidence in the behavior of the dynamical system.

In order to make the notion of robustness more precise it is necessary to define robustness measures for a particular performance criterion with respect to a given class of perturbations.

Such work has already been performed in this area and several approaches coexist. We shall take here the approach of Hinrichsen and Pritchard (1989) which is based on state space methods, i.e., on equations rather than on frequency domain techniques that are popular in the engineering world. This corresponds better to the spirit of this book.

We already know that stability of system (2.1) can be expressed in terms of eigenvalues location. Let $\mathbb{C} = D \cup C_D$, the partition of the complex plane $\mathbb{C}$ into the "good" (from the stability point of view) region - the open unit disk D - and the "bad" region, i.e., the complementary set of D. Since we are interested in the effects of both real and complex perturbations we shall consider both fields $\mathbb{K}=\mathbb{R}$ and $\mathbb{K}=\mathbb{C}$ of real and complex numbers. We assume that the *nominal system* (2.1) has its *nominal matrix* A *with the eigenvalues inside the unit disk*

$$A \in \mathbb{K}^{n,n} , \quad \sigma(A) \subset D$$

We suppose also that the perturbed system is

$$x_{t+1} = (A + B\Delta C) x_t \tag{2.19}$$

where $B \subset \mathbb{K}^{n,m}$ and $C \in \mathbb{K}^{p,n}$ are given matrices defining the structure of the perturbations and $\Delta \in \mathbb{K}^{m,p}$ is an unknown disturbance matrix. For instance, if $B = C = \mathbf{I}$ (the so-called *unstructured case*) then all the elements of A are subject to independent perturbations whereas by a suitable choice of B, C the effect of perturbations of individual elements, rows or columns of A can be studied. To illustrate this assertion consider the difference equation (2.5) with the characteristic polynomial

$$p(z, a) = z^n + \alpha_{n-1} z^{n-1} + \cdots + \alpha_1 z + \alpha_0$$

where $a = (\alpha_{n-1}\, \alpha_{n-2}\, ... \, \alpha_1\, \alpha_0)^\top$ defines the column vector of the coefficients. To (2.5) we can associate a system of the form (2.1) with A having the companion form. The coefficients of this nominal polynomial are perturbed to

$$\alpha_{k-1}(d) = \alpha_{k-1} - \sum_1^p \delta_i c_{ik} \, ,$$

where $C = (c_{ik}) \in \mathbb{K}^{p,n}$ is a given $p \times n$ matrix while the unknown disturbance vector $d = \mathrm{col}(\delta_1, ..., \delta_p)$ accounts for the deviations of the system parameters from their nominal values. It is easy to see that

$$p(z, a(d)) = \det(z\mathbf{I} - A - bd^* C) \, ,$$

where $b = (0\, 0\, \cdots\, 1)^\top$. Given a matrix norm on $\mathbb{K}^{m,p}$ various more or less conservative bounds $\delta = \delta(A, B, C)$ have been proposed that guarantee $\sigma(A + B\Delta C) \subset D$ for all $\Delta \in \mathbb{K}^{m,p}$ with norm $|\Delta| < \delta$. The smallest norm of a "destabilizing" disturbance matrix Δ is called *stability radius* with respect to the perturbation structure (B, C) and to the considered norm.

Definitions and Elementary Properties

We suppose that $|\ |_{\mathbb{K}^m}$ and $|\ |_{\mathbb{K}^p}$ are given norms on $\mathbb{K}^m$ and $\mathbb{K}^p$ respectively and estimate the size of perturbation matrices $\Delta \in \mathbb{K}^{m,p}$ by the corresponding operator norm

$$|\Delta| = \max\{|\Delta y|_{\mathbb{K}^m} \, , \, y \in \mathbb{K}^p, \, |y|_{\mathbb{K}^p} \leq 1\}$$

Definition 2.2 The *stability radius* of A with respect to perturbations of the structure (B, C) and the unit disk D is defined by

$$\begin{aligned} r_{\mathbb{K}} = r_{\mathbb{K}}(A; B; C; D) = \\ = \inf\{|\Delta| \; ; \; \Delta \in \mathbb{K}^{m,p}, \, \sigma(A + B\Delta C) \cap \mathbb{C}\backslash D \neq \emptyset\} \end{aligned} \quad (2.20)$$

If A, B, C are real we obtain two stability radii, $r_{\mathbb{R}}$ or $r_{\mathbb{C}}$ according to whether real ($\mathbb{K} = \mathbb{R}$) or complex ($\mathbb{K} = \mathbb{C}$) perturbations are considered in (2.20). They are respectively called the *real* or the *complex* stability radii of A and are obviously related by

$$r_{\mathbb{R}}(A; B, C; D) \geq r_{\mathbb{C}}(A; B, C; D) \geq 0$$

In the unstructured case where $m = p = n$ and $B = C = \mathbf{I}$, $r_{\mathbb{K}}$ is the distance in the normed space $\mathbb{K}^{n,n}$ between A and the set of matrices in $\mathbb{K}^{n,n}$ having at least one eigenvalue outside the unit disk D.

The following properties are straightforward.

1. By definition the infimum of the empty set is infinite so that $r_{\mathbb{K}} = \infty$ iff there does not exist $D \in \mathbb{K}^{m,p}$ with $A + B\Delta C$ having at least one eigenvalue outside the open unit disk. If $r_{\mathbb{K}} < \infty$ then a compactness argument shows that there exists a destabilizing Δ with $|\Delta| = r_{\mathbb{K}}$ and in this case we can replace the *infimum* by a *minimum*.

2. The stability radius is *invariant under similarity transformations:*

$$r_{\mathbb{K}}\left(A; B, C; D\right) = r_{\mathbb{K}}\left(TAT^{-1}; TB, CT^{-1}; D\right)$$

This follows immediately from the fact that

$$TAT^{-1} + TB\Delta CT^{-1} = T\left(A + B\Delta C\right)T^{-1}$$

and the spectrum of a matrix is invariant under similarity transformations.

3. If $A + B\Delta C$ has at least one eigenvalue outside the closed unit disk, then by continuity this is true for all matrices in a small neighborhood of Δ. Therefore,

$$r_{\mathbb{K}}\left(A; B, C; D\right) = r_{\mathbb{K}}\left(A; B, C; \partial D\right)$$

where ∂D is the boundary of the unit disk that is the unit circle.

The Complex Stability Radius

At first sight it may seem artificial to consider complex perturbations of a real system but the complex stability radius - besides being a lower bound for the real one - yields valuable information about the robustness of a system with respect to wider classes of perturbations (time-varying, nonlinear - see the survey by Hinrichsen and Pritchard (1989)) - and is easier to compute due to existing software (Motscha (1988), Hinrichsen, Kelb and Linnemann (1990)). The main result is the following

Proposition 2.7   *The complex stability radius satisfies*

$$r_{\mathbb{C}}\left(A; B, C; D\right) = \left[\max_{|z|=1}\left|C\left(z\mathbf{I} - A\right)^{-1}B\right|\right]^{-1} \qquad (2.21)$$

where | | *denotes the operator (matrix) norm and by definition* $0^{-1} = \infty$.

Remark. $T(\lambda) = C(\lambda\mathbf{I} - A)^{-1}B$ is the so-called *matrix transfer function* associated to (A, B, C); it is often used in control and signal processing.

Proof. For any Δ we shall have $\sigma(A + B\Delta C) \cap \partial D \neq \emptyset$ iff there exists $x \in \mathbb{C}^n$, $x \neq 0$ and z on the unit circle such that $(A + B\Delta C)x = zx$. Since the eigenvalues of A are inside the unit disk $z\mathbf{I} - A$ is invertible on the unit circle and the above equality is equivalent to

$$x = (z\mathbf{I} - A)^{-1}B\Delta Cx,$$

where $\Delta Cx \neq 0$. Denoting $u = \Delta Cx$ we have

$$u = \Delta T(z)u \neq 0$$

which gives

$$1 \leq |\Delta T(z)| \leq |\Delta||T(z)| \leq |\Delta| \max_{|z|=1}|T(z)|\ ;$$

hence $|\Delta| \geq \left[\max_{|z|=1}|T(z)|\right]^{-1}$.

On the other hand, let $|T(z_0)| = \max_{|z|=1}|T(z)|$, $|z_0| = 1$. We chose $u \in \mathbb{C}^m$, $|u| = 1$ such that $|T(z_0)u| = |T(z_0)|$. Since $T(z_0) \neq 0$ (otherwise we have $r_\mathbb{C} = \infty$) we may apply the theorem of Hahn-Banach and take a linear form y^* in the dual space of $\mathbb{C}^p$, with the norm equal to 1 such that

$$y^*T(z_0)u = |T(z_0)u| = |T(z_0)|$$

The perturbation

$$\Delta = |T(z_0)|^{-1}uy^* \tag{2.22}$$

with u, y^* and z_0 defined above, is of norm $|T(z_0)|^{-1}$. Indeed for any $y \in \mathbb{C}^p$ with $|y| = 1$ we shall have

$$|\Delta y| = |T(z_0)|^{-1}|uy^*y| \leq |T(z_0)|^{-1}$$

which gives $|\Delta| \leq |T(z_0)|^{-1}$.

On the other hand, if $\hat{y} = |T(z_0)|^{-1} T(z_0) u$, then obviously $|\hat{y}| = 1$ since z_0 and u are such that $|T(z_0) u| = |T(z_0)|$. Further

$$\Delta \hat{y} = |T(z_0)|^{-2} u y^* T(z_0) u = |T(z_0)|^{-2} u |T(z_0) u| = |T(z_0)|^{-1} u$$

hence, $|\Delta \hat{y}| = |T(z_0)|^{-1}$. This gives

$$|\Delta| = |T(z_0)|^{-1} = \left[\max_{|z|=1} |T(z)| \right]^{-1}.$$

This perturbation is destabilizing. Indeed we may take $x = (z_0 I - A)^{-1} Bu$ with z_0, u as above. Therefore,

$$\begin{aligned} \Delta C x &= |T(z_0)|^{-1} u y^* C (z_0 I - A)^{-1} Bu = \\ &= |T(z_0)|^{-1} u y^* T(z_0) u = u \neq 0 \end{aligned}$$

what shows that $x \neq 0$. Further,

$$(z_0 I - A) x = Bu = B \Delta C x$$

what shows that z_0 with $|z_0| = 1$ is an eigenvalue of $A + B \Delta C$.

We have thus constructed a destabilizing perturbation Δ of norm $\left[\max_{|z|=1} |T(z)| \right]^{-1}$ and since we proved that in general the destabilizing perturbations have the norm larger than this value the proof ends.

Remark. If the usual Euclidean (Hilbert) norms are used, then we may take $y^* = u^* T^*(z_0) |T(z_0)|^{-1}$, where $T^*(z_0)$ denotes transpose and complex conjugate.

The Real Stability Radius

Consider again the perturbation (2.22). This is a minimum norm destabilizing perturbation for the triple (A, B, C). In general, even if matrices A, B, C are real, this Δ will be complex and it will not be possible to find a real destabilizing perturbation of the same norm. Hence, in general the obvious inequality: $r_C(A; B, C; D) \leq r_\mathbb{R}(A; B, C; D)$ will be strict. If however the above Δ is real (for real z_0), then the real and the complex stability radii of (A, B, C) coincide. To estimate the real stability radius is mildly speaking, not an easy task, both for continuous time and discrete time systems. Technical features needed for the proofs are outside of the book's mainstream and we shall not insist on this subject. The reader is sent to the appropriate publications of Hinrichsen and Pritchard (1988, 1989).

2.2. GENERAL PROPERTIES OF LINEAR SYSTEMS

The aim of this section is to study linear systems with time-varying coefficients, in fact their rather general properties which may be useful throughout the book.

A. Consider first linear time-varying systems of the form

$$x_{t+1} = A_t x_t \tag{2.23}$$

It is easily seen that the solutions of (2.23), which are sequences of states as in the previous section, define a linear space. Denoting by $x_t(t_0, x)$ the solution of (2.23) that coincides with x for $t = t_0$, the mapping $X_{t,t_0} : \mathbb{R}^n \to \mathbb{R}^n$ defined by $X_{t,t_0} x = x_t(t_0, x)$ is a *linear mapping*. To this mapping it may be associated a matrix-solution of (2.23) which for $t = t_0$ equals **I** - the identity matrix. The matrix thus defined, called also *state transition matrix*, can be expressed using matrices A_t. Indeed we have

$$X_{t_0,t_0} = \mathbf{I}, \quad \mathbf{X}_{t_0+1,t_0} = A_{t_0}, \quad X_{t_0+2,t_0} = A_{t_0+1}A_{t_0}, \quad \dots, \quad X_{t,t_0} = A_{t-1}A_{t-2}\cdots A_{t_0+1}A_{t_0} \tag{2.24}$$

From (2.24) we may deduce immediately

$$X_{t,k}X_{k,m} = X_{t,m}, \quad t \geq k \geq m$$

The state-transition matrix X_{t,t_0} might have been introduced without any reference to linearity of solution space for (2.23). Indeed direct computation shows that

$$x_t(t_0, x) = A_{t-1}A_{t-2}\cdots A_{t_0}x \tag{2.25}$$

for any solution that coincides with x for $t = t_0$. Defining

$$X_{t,t_0} = A_{t-1}A_{t-2}\cdots A_{t_0}$$

that is by (2.24), it is easily seen that X_{t,t_0} thus defined verifies (2.23). Since X_{t,t_0} is defined from (2.25) for any x and $x_{t_0}(t_0, x) = x$ it is natural to assign $X_{t_0,t_0} = \mathbf{I}$. If $x = e_i$ - the vectors of the canonical basis, having all the positions zero except position i which is 1 - it follows that the columns of X_{t,t_0} are solutions of (2.23) with e_i as initial conditions. But e_i are also columns of the identity matrix what justifies once again the above assignment for X_{t_0,t_0}.

The linear system

$$y_{t-1} = y_t A_{t-1} \tag{2.26}$$

where y_t is a row vector is called the *adjoint system* of (2.23). We may define the solution matrix Y_{t,t_0} of (2.26) as follows

$$Y_{t_0,t_0} = I , \quad Y_{t_0-1,t_0} = A_{t_0-1} , \quad Y_{t_0-2,t_0} = A_{t_0-1}A_{t_0-2} , \quad \dots ,$$
$$Y_{t,t_0} = A_{t_0-1}A_{t_0-2} \cdots A_t$$

(2.27)

From (2.24) and (2.27) it is easily seen that $Y_{t,t_0} = X_{t_0,t}$. We might have obtained the same result by considering x_t and y_t, two solutions of (2.23) and (2.26), respectively, which are defined for the same set of indices. We have

$$y_t x_t = y_t A_{t-1} x_{t-1} = y_{t-1} x_{t-1}$$

hence $y_t x_t$ is constant on that set of indices where both solutions are defined. Solution matrices $X_{t,k}$ and $Y_{t,m}$ are both defined for $k \le t \le m$ and

$$Y_{t,m} X_{t,k} \equiv const \equiv X_{m,k} \equiv Y_{k,m}$$

(we took successively $t = m$ and $t = k$).

B. Our next object of study is the general forced (affine) system

$$x_{t+1} = A_t x_t + f_t \tag{2.28}$$

We look for a representation formula for its solutions - *the variation of constants formula* in the discrete case. Let $x_t (t_0, x; f)$ be the solution of (2.28) corresponding to $\{f_t\}_t$ and that coincides with x at $t = t_0$. We shall have

$$
\begin{aligned}
x_t &= Y_{t,t} x_t = Y_{t,t} \left(A_{t-1} x_{t-1} + f_{t-1}\right) = Y_{t-1,t} x_{t-1} + Y_{t,t} f_{t-1} = \\
&= Y_{t-1,t} \left(A_{t-2} x_{t-2} + f_{t-2}\right) + Y_{t,t} f_{t-1} = \\
&= Y_{t-2,t} x_{t-2} + Y_{t-1,t} f_{t-2} + Y_{t,t} f_{t-1} = Y_{t_0,t} x + \sum_{t_0}^{t-1} Y_{k+1,t} f_k
\end{aligned}
$$

Since $Y_{k,t} = X_{t,k}$ we shall have

$$x_t (t_0, x; f) = X_{t,t_0} x + \sum_{t_0}^{t-1} X_{t,k+1} f_k \tag{2.29}$$

The above proof, based on the properties of the adjoint system, parallels the one usually performed in the theory of general linear affine systems. The same result could be obtained by direct computation in (2.28). Indeed

$$
\begin{aligned}
x_{t_0+1} &= A_{t_0}x + f_{t_0} \\
x_{t_0+2} &= A_{t_0+1}x_{t_0+1} + f_{t_0+1} = A_{t_0+1}A_{t_0}x + A_{t_0+1}f_{t_0} + f_{t_0+1} \\
\cdots \quad & \cdots \quad \cdots\cdots\cdots\cdots\cdots\cdots \\
x_t &= A_{t-1}A_{t-2}\cdots A_{t_0}x + A_{t-1}\cdots A_{t_0+1}f_{t_0} + \cdots + f_{t-1} = \\
&= X_{t,t_0}x + \sum_{t_0}^{t-1}(A_{t-1}\cdots A_{k+1})f_k = X_{t,t_0}x + \sum_{t_0}^{t-1}X_{t,k+1}f_k
\end{aligned}
$$

C. Linear canonical systems.

Definition 2.3 The linear system

$$z_{t+1} = M_t z_t \tag{2.30}$$

is called *canonical* if its matrix M_t is *simplectic*; i.e., $M_t^* J M_t \equiv J$ where $J = \begin{pmatrix} 0 & -\mathbf{I} \\ \mathbf{I} & 0 \end{pmatrix}$.

A good motivation for this definition is the following: consider the continuous-time canonical system

$$\dot{x} = JH(t)x, \quad H^* = H$$

subject to the shocks s_k at the moments t_k, $k \in \mathbb{Z}$, as follows. Between t_{k-1} and t_k the system evoluates freely, i.e., according to the above canonical equation. For $t = t_k$ the shock interface conditions are satisfied

$$x(t_k + 0) - x(t_k - 0) = s_k$$

Under these circumstances the solutions of the canonical system are

$$x(t; t_k, c_k) = X(t, t_k)c_k, \quad t_k \le t < t_{k+1}$$

where $X(t, \tau)$ is the fundamental matrix of solutions for the continuous-time canonical system satisfying

$$\frac{\mathrm{d}}{\mathrm{d}t}X(t, \tau) = JH(t)X(t, \tau), \quad X(\tau, \tau) = \mathbf{I}$$

The vectors c_k are determined from the recurrence

$$c_{k+1} = X(t_{k+1}, t_k)c_k + s_{k+1}$$

which is a straightforward consequence of the form of solutions between shocks and interface conditions. The system

$$z_{k+1} = X(t_{k+1}, t_k) z_k$$

is canonical in the sense of the above definition since $X(t_{k+1}, t_k)$ is simplectic. Indeed we have

$$\frac{\mathrm{d}}{\mathrm{d}t}\left[X^*(t, \tau) JX(t, \tau)\right] = \left(\frac{\mathrm{d}}{\mathrm{d}t} X(t, \tau)\right)^* JX(t, \tau) +$$

$$+X^*(t, \tau) J\frac{\mathrm{d}}{\mathrm{d}t} X(t, \tau) = X^*(t, \tau) \left[H(t) J^*J + JJH(t)\right] X(t, \tau)$$

But $J^*J = \mathbf{I}$, $J^2 = -\mathbf{I}$ which gives

$$H(t) J^*J + J^2 H(t) \equiv 0$$

Therefore, $X^*(t, \tau) JX(t, \tau) \equiv F(\tau)$. For $t = \tau$ it follows that $X^*(\tau, \tau) JX(\tau, \tau) = J$ hence $F(\tau) = J$. We obtained that $X^*(t, \tau) JX(t, \tau) \equiv J$, i.e., $X(t, \tau)$ is simplectic which gives that $X(t_{k+1}, t_k)$ is simplectic.

Proposition 2.8 *If (2.30) is canonical then each solution of it satisfies $z_{t+1}^* J z_{t+1} = z_t^* J z_t$ and its transition matrix (fundamental matrix of solutions) is simplectic.*

Proof. Let $Z_{t,k}$ be the state transition matrix of (2.30). We know already that $Z_{t,k} = M_{t-1} M_{t-2} \cdots M_k$ and a product of simplectic matrices is also simplectic. By a direct check we find that

$$z_{t+1}^* J z_{t+1} = z_t^* M_t^* J M_t z_t = z_t^* J z_t$$

Consider now two solutions u_t, v_t of (2.30): $u_t = Z_t u_0$, $v_t = Z_t v_0$. It is easily seen that

$$v_t^* J u_t = v_0^* Z_t^* J Z_t u_0 = v_0^* J u_0 \qquad (2.31)$$

It follows that for any two solutions of (2.30) the bilinear form (v, Ju) is invariant. It is called *Poincaré invariant* (in the discrete case).

    D. We shall consider in the following an important class of canonical systems described by

$$x_{t+1} = \left(\frac{\partial \mathcal{H}_t}{\partial y}\right)^* (x_t, y_{t+1}) = A_t x_t + B_t y_{t+1}$$

$$y_t = \left(\frac{\partial \mathcal{H}_t}{\partial x}\right)^* (x_t, y_{t+1}) = C_t x_t + A_t^* y_{t+1}$$

(2.32)

where $\mathcal{H}_t (x, y) = \dfrac{1}{2} (x^* C_t x + x^* A_t^* y + y^* A_t x + y^* B_t y)$, $B_t = B_t^*$, $C_t = C_t^*$, $\det A_t \neq 0$. System (2.32) may be written as follows

$$\begin{pmatrix} \mathbf{I} & -B_t \\ 0 & A_t^* \end{pmatrix} \begin{pmatrix} x_{t+1} \\ y_{t+1} \end{pmatrix} = \begin{pmatrix} A_t & 0 \\ -C_t & \mathbf{I} \end{pmatrix} \begin{pmatrix} x_t \\ y_t \end{pmatrix}$$

Denoting $z = \begin{pmatrix} x \\ y \end{pmatrix}$ we shall have

$$z_{t+1} = \begin{pmatrix} \mathbf{I} & -B_t \\ 0 & A_t^* \end{pmatrix}^{-1} \begin{pmatrix} A_t & 0 \\ -C_t & \mathbf{I} \end{pmatrix} z_t$$

(2.33)

and this is a canonical system in the sense of Definition 2.3. Indeed we have

$$\begin{aligned} M_t &= \begin{pmatrix} \mathbf{I} & -B_t \\ 0 & A_t^* \end{pmatrix}^{-1} \begin{pmatrix} A_t & 0 \\ -C_t & \mathbf{I} \end{pmatrix} = \\ &= \begin{pmatrix} \mathbf{I} & B_t (A_t^*)^{-1} \\ 0 & (A_t^*)^{-1} \end{pmatrix} \begin{pmatrix} A_t & 0 \\ -C_t & \mathbf{I} \end{pmatrix} = \\ &= \begin{pmatrix} A_t - B_t (A_t^*)^{-1} C_t & B_t (A_t^*)^{-1} \\ -(A_t^*)^{-1} C_t & (A_t^*)^{-1} \end{pmatrix} \end{aligned}$$

and further

$$M_t^* J M_t = M_t^* \begin{pmatrix} 0 & -\mathbf{I} \\ \mathbf{I} & 0 \end{pmatrix} M_t = \begin{pmatrix} 0 & -\mathbf{I} \\ \mathbf{I} & 0 \end{pmatrix} = J \ ;$$

hence M_t is simplectic. We have also

$$\det M_t = \frac{\det \begin{pmatrix} A_t & 0 \\ -C_t & \mathbf{I} \end{pmatrix}}{\det \begin{pmatrix} \mathbf{I} & -B_t \\ 0 & A_t^* \end{pmatrix}} = \frac{\det A_t}{\det A_t^*} = 1 \ ;$$

hence, if $Z_{t,k}$ is a state transition matrix then $\det Z_{t,k} = 1$.

E. Consider now the converse problem of the above. Let

$$M_t = \begin{pmatrix} P_t & Q_t \\ R_t & S_t \end{pmatrix}, \quad \det S_t \neq 0, \ \forall t$$

be the matrix of the canonical system in the sense of Definition 2.3. We shall prove that *there exist* A_t, $\det A_t \neq 0$, $B_t = B_t^*$, $C_t = C_t^*$ such that

$$M_t = \begin{pmatrix} I & -B_t \\ 0 & A_t^* \end{pmatrix}^{-1} \begin{pmatrix} A_t & 0 \\ -C_t & I \end{pmatrix};$$

i.e., canonical system (2.30) can be given the form (2.32) with $\mathcal{H}_t(x, y)$, a quadratic form defined by A_t, B_t, C_t.

Indeed, since M_t is simplectic we have

$$M_t^* J M_t = \begin{pmatrix} P_t^* & R_t^* \\ Q_t^* & S_t^* \end{pmatrix} \begin{pmatrix} 0 & -I \\ I & 0 \end{pmatrix} \begin{pmatrix} P_t & Q_t \\ R_t & S_t \end{pmatrix} = \begin{pmatrix} 0 & -I \\ I & 0 \end{pmatrix}$$

and from here

$$R_t^* P_t = P_t^* R_t, \quad R_t^* Q_t - P_t^* S_t = -I, \quad S_t^* Q_t = Q_t^* S_t$$

Take $A_t = (S_t^*)^{-1}$, $B_t = Q_t S_t^{-1}$, $C_t = -S_t^{-1} R_t$ and observe that from $S_t^* Q_t = Q_t^* S_t$ we have $Q_t S_t^{-1} = (S_t^*)^{-1} Q_t^*$, hence, $B_t = B_t^*$. On the other hand, we have from $R_t^* Q_t - P_t^* S_t = -I$ that $R_t^* Q_t S_t^{-1} - P_t^* = -S_t^{-1}$. Therefore,

$$C_t = -S_t^{-1} R_t = R_t^* Q_t S_t^{-1} R_t - P_t^* R_t = R_t^* B_t R_t - P_t^* R_t$$

and since $B_t = B_t^*$ and $P_t^* R_t = R_t^* P_t$ it follows that $C_t = C_t^*$.

We have further

$$P_t = (S_t^*)^{-1} + (S_t^*)^{-1} Q_t^* R_t = A_t - B_t (A_t^*)^{-1} C_t,$$
$$Q_t = B_t S_t^{-1} = B_t (A_t^*)^{-1}, \quad R_t = -S_t C_t = -(A_t^*)^{-1} C_t$$

and therefore,

$$M_t = \begin{pmatrix} A_t - B_t (A_t^*)^{-1} C_t & B_t (A_t^*)^{-1} \\ -(A_t^*)^{-1} C_t & (A_t^*)^{-1} \end{pmatrix} =$$

$$= \begin{pmatrix} I & B_t (A_t^*)^{-1} \\ 0 & (A_t^*)^{-1} \end{pmatrix} \begin{pmatrix} A_t & 0 \\ -C_t & I \end{pmatrix} =$$

$$= \begin{pmatrix} I & -B_t \\ 0 & A_t^* \end{pmatrix}^{-1} \begin{pmatrix} A_t & 0 \\ -C_t & I \end{pmatrix}$$

F. Consider the second order vector equation

$$C_t y_{t+1} = A_t y_t - C_{t-1} y_{t-1} \qquad (2.34)$$

with $C_t = C_t^*$, $A_t = A_t^*$, $\det C_t \neq 0$. Such equations can be written as canonical systems. Indeed we have

$$
\begin{aligned}
x_{t+1} &= C_t y_t \\
y_{t+1} &= -C_t^{-1} x_t + C_t^{-1} A_t y_t ,
\end{aligned}
$$

hence,

$$M = \begin{pmatrix} 0 & C_t \\ -C_t^{-1} & C_t^{-1} A_t \end{pmatrix} .$$

The simplecticity conditions are

$$
\begin{aligned}
-(C_t^*)^{-1} \cdot 0 &= -0 \cdot C_t^{-1} ; \quad -(C_t^*)^{-1} C_t - 0 \cdot C_t^{-1} A_t = \mathbf{I} ; \\
A_t^* (C_t^*)^{-1} C_t &= C_t^* C_t^{-1} A_t
\end{aligned}
$$

and are fulfilled for symmetric A_t and C_t.

Remark that when C_t is constant a suitable change of coordinates leads to a vector equation where $C = \mathbf{I}$.

2.3. STABILITY BY THE FIRST APPROXIMATION

This is the natural attempt of solving stability problems for systems describing physical phenomena that are nonlinear by local linearization around equilibria. For a linear system of the form (2.23) we defined exponential stability as the property of the solutions of (2.23) to approach zero exponentially for $t \to \infty$. The state $x = 0$ is an equilibrium of the system and any solution generated by a non-zero initial state is a deviation from the equilibrium. Exponential stability means that these deviations decrease exponentially in time. In many cases the equations of the physical systems are equations of the deviations and since before vanishing these deviations have to be "small", the linearized version of the equations may provide information about their exponential vanishing. This point of view is transparent in many classical treatises on dynamics among which we cite the "Natural Philosophy" of Thomson and Tait (1879).

A rigorous basis of this heuristic approach is given by the best known result on stability by the first approximation, belonging to

Liapunov and Poincaré. We shall give below its discrete-time version.

2.3.1. A Preliminary Result

In the following we shall need a result which is interesting in itself, being the discrete analogue of Gronwall lemma.

Proposition 2.9   *If $u_t > 0$ satisfies for $t > k$ the inequality*

$$u_t \leq \alpha u_k + \beta \sum_{k}^{t-1} u_i , \quad \alpha > 0, \ \beta > 0 , \tag{2.35}$$

then

$$u_t \leq (\alpha + \beta)(1 + \beta)^{t-k-1} u_k \tag{2.36}$$

Proof. Denoting $v_t = \alpha u_k + \beta \sum_{k}^{t-1} u_i$ we shall obviously have $u_t \leq v_t$ and also: $v_{t+1} - v_t = \beta u_t \leq \beta v_t$ hence $v_{t+1} \leq (1 + \beta) v_t$. Therefore,

$$v_t \leq (1 + \beta)^{t-k-1} v_{k+1} , \quad v_{k+1} = \alpha u_k + \beta u_k = (\alpha + \beta) u_k$$

and the required inequality is obtained.

Various generalizations are possible which represent discrete analogues of differential inequalities, but we shall not discuss this subject.

2.3.2. Stability of Equilibria for Time-Invariant Systems

We shall consider the nonlinear discrete system

$$y_{t+1} = f(y_t) \tag{2.37}$$

and let $\hat{y}$ be an *equilibrium* of (2.37). The equilibrium, called sometimes a *steady state vector* is a constant solution of (2.37), i.e., it satisfies

$$\hat{y} = f(\hat{y}) \quad (\hat{y} \text{ is a fixed point of the mapping } f) .$$

We shall call this equilibrium *exponentially stable* if the deviations from it tend exponentially to zero. There exist $\delta > 0$, $\beta \geq 1$ and $\rho \in (0, 1)$ such that if $|y_0 - \hat{y}| < \delta$ then $|y_t - \hat{y}| \leq \beta \rho^t |y_0 - \hat{y}|$ for $t > 0$.

If y_t is a solution of (2.37) we shall introduce $x_t = y_t - \hat{y}$, the deviation from the equilibrium. We shall have

$$
\begin{aligned}
x_{t+1} &= y_{t+1} - \hat{y} = f(y_t) - f(\hat{y}) = \\
&= f(\hat{y} + x_t) - f(\hat{y}) = \left(\int_0^1 \frac{\partial f}{\partial y}(\hat{y} + \alpha x_t)\, d\alpha \right) x_t = \\
&= \frac{\partial f}{\partial y}(\hat{y})\, x_t + \left[\int_0^1 \left(\frac{\partial f}{\partial y}(\hat{y} + \alpha x_t) - \frac{\partial f}{\partial y}(\hat{y}) \right) d\alpha \right] x_t
\end{aligned}
$$

Denote

$$
\frac{\partial f}{\partial y}(\hat{y}) = A \quad , \quad \left[\int_0^1 \left(\frac{\partial f}{\partial y}(\hat{y} + \alpha x) - \frac{\partial f}{\partial y}(\hat{y}) \right) d\alpha \right] x = g(x) \quad (2.38)
$$

and the system in deviations becomes

$$
x_{t+1} = A x_t + g(x_t) \tag{2.39}
$$

We shall have

$$
|g(x)| \le \sup_{0 \le \alpha \le 1} \left| \frac{\partial f}{\partial y}(\hat{y} + \alpha x) - \frac{\partial f}{\partial y}(\hat{y}) \right| |x| \; .
$$

If $\partial f / \partial y$ is continuous, it is uniformly continuous on some compact set, e.g., a ball of radius $\hat{\delta}$ centered at $\hat{y}$. Therefore, there exists ω : $\mathbb{R}_+ \to \mathbb{R}_+$ increasing, $\omega(0) = 0$ such that

$$
\sup_{0 \le \alpha \le 1} \left| \frac{\partial f}{\partial y}(\hat{y} + \alpha x) - \frac{\partial f}{\partial y}(\hat{y}) \right| \le \omega(|x|)
$$

(The function $\omega(\rho)$ is the modulus of continuity.) It follows that

$$
|g(x)| \le \omega(|x|) |x| \; , \quad \lim_{\rho \to 0} \omega(\rho) = 0 \; ,
$$

hence, for any $\gamma > 0$ there exists $\delta > 0$ such that if $|x| < \delta$ then $\omega(|x|) < \gamma$ and $|g(x)| \le \gamma |x|$. We shall use this property in the following.

Let x_0 be such that $|x_0| < \delta$ where $\delta > 0$ is determined as above, starting from some $\gamma > 0$. Consider $x_t(x_0)$ the corresponding solution of (2.39) and define

$$V_t = x_t^* P x_t \tag{2.40}$$

where the positively definite matrix P is the solution of the Liapunov equation

$$A^* P A - A = -\mathbf{I} \tag{2.41}$$

where A is defined in (2.38). If we assume that the *first approximation system*

$$x_{t+1} = A x_t \tag{2.42}$$

is exponentially stable then $P > 0$ is well defined (Lemma 2.3).

We shall have

$$
\begin{aligned}
V_{t+1} - V_t &= x_{t+1}^* P x_{t+1} - x_t^* P x_t = -|x_t|^2 + g^*(x_t) P A x_t + \\
&\quad + x_t^* A^* P g(x_t) + g^*(x_t) P g(x_t)
\end{aligned}
$$

and as long as $|x_t| < \delta$ it follows that

$$V_{t+1} - V_t \le -|x_t|^2 + 2\gamma |PA| |x_t|^2 + \gamma^2 |P| |x_t|^2$$

By choosing γ such that $|P|\gamma^2 + 2|PA|\gamma - 1/2 \le 0$ we deduce that

$$V_{t+1} - V_t \le \frac{-1}{2} |x_t|^2$$

Since $P > 0$ its eigenvalues are real and strictly positive. Let λ and Λ be the smallest and the largest of these eigenvalues; therefore,

$$0 < \lambda |x_t|^2 \le x_t^* P x_t \le \Lambda |x_t|^2 \ .$$

From here and from the inequality for $V_{t+1} - V_t$ we deduce that

$$V_{t+1} \le \left(1 - \frac{1}{2\Lambda}\right) V_t$$

and denoting $\rho^2 = 1 - \dfrac{1}{2\Lambda}$ we shall have $0 < \rho < 1$ and

$$V_t \le \rho^{2t} V_0 \quad , \quad |x_t| \le \sqrt{\frac{\Lambda}{\lambda}} \rho^t |x_0| \ .$$

The last inequality expresses, in fact, exponential stability of the equilibrium $\hat{y}$ since it reads

$$|y_t - \hat{y}| \leq \sqrt{\frac{\Lambda}{\lambda}}\rho^t |y_0 - \hat{y}| = \beta e^{-\alpha t} |y_0 - \hat{y}| \qquad (2.43)$$

where $\alpha = \ln(1/\rho) > 0$.

Remark that exponential stability is only local since we need $|x_t| \leq \delta$ for $t \geq 0$. This is achieved as follows

a) we chose $\gamma > 0$ from the inequality

$$|P|\gamma^2 + 2|PA|\gamma - 1/2 \leq 0$$

and $\hat{\delta}(\gamma)$ such that $|x| < \hat{\delta}(\gamma)$ implies $|g(x)| \leq \gamma |x|$;

b) we take $\delta \leq \hat{\delta}(\gamma)/\beta$ with β as above; with $|x_0| < \delta$ we shall have $|x_t| \leq \hat{\delta}(\gamma)$. We have thus proved the following

Theorem 2.10 *Consider the nonlinear system* (2.37) *and let* $\hat{y}$ *be an equilibrium. Assume the following: i)* f *is continuously differentiable in a neighborhood of* $\hat{y}$; *ii) matrix* $A = \dfrac{\partial f}{\partial y}(\hat{y})$ *has its eigenvalues located inside the unit disk of the complex plane. Then the equilibrium is exponentially stable, i.e., there exist* $\delta > 0$, $\beta > 1$, $\rho \in (0,1)$ *such that if* $|y_0 - \hat{y}| < \delta$ *then* $|y_t(y_0) - \hat{y}| \leq \beta\rho^t |y_0 - \hat{y}|$.

2.3.3. Stability of Motion for Time Varying Systems

Instead of the equilibrium we may consider some trajectory (motion) of the system, $\hat{y}_t$, called *basic*. We shall call it *exponentially stable* if the deviations from the basic motion that start in its neighborhood will tend exponentially to zero. There exist $\delta > 0$, $\beta \geq 1$ and $\rho \in (0,1)$ such that

$$|y_t - \hat{y}_t| \leq \beta\rho^{t-s} |y_s - \hat{y}_s| , \quad t \geq s , \text{ if } |y_s - \hat{y}_s| < \delta$$

We shall consider the following time varying system

$$y_{t+1} = f_t(y_t) \qquad (2.44)$$

and let $\hat{y}_t$ be a global trajectory of it, i.e., $\hat{y}_{t+1} \equiv f_t(\hat{y}_t)$ for all $t \geq s$. We make the following assumptions: *i)* $\hat{y}_t$ is a bounded sequence; *ii)* $f_t(y)$ is such that $\partial f_t / \partial y$ is uniformly continuous, this property being also uniform with respect to t and $A_t = \dfrac{\partial f_t}{\partial y}(\hat{y}_t)$ defines an exponentially stable evolution. There exist $\beta_0 > 0$, $\rho \in (0,1)$ such that $|X_{t,s}| \le \beta_0 \rho^{t-s}$ where $X_{t,s}$ is the fundamental matrix of solutions (state transition matrix) of A_t (see Section 2.2).

Introduce now the deviations with respect to the basic motion $x_t = y_t - \hat{y}_t$. Therefore,

$$x_{t+1} = f_t(\hat{y}_t + x_t) - f_t(\hat{y}_t) = A_t x_t + g_t(x_t) \qquad (2.45)$$

where we defined

$$A_t = \frac{\partial f_t}{\partial y}(\hat{y}_t), \quad g_t(x) = \left[\int_0^1 \left(\frac{\partial f_t}{\partial y}(\hat{y}_t + \alpha x) - \frac{\partial f_t}{\partial y}(\hat{y}_t) \right) d\alpha \right] x.$$

From the assumptions we shall have that

$$\sup_{0 \le \alpha \le 1} \left| \frac{\partial f_t}{\partial y}(\hat{y}_t + \alpha x) - \frac{\partial f_t}{\partial y}(\hat{y}_t) \right| \le \omega(|x|)$$

where $\omega(\rho)$ is the modulus of equiuniform continuity of the sequence $\partial f_t / \partial y$. Therefore, as previously

$$|g_t(x)| \le \omega(|x|)|x|, \quad \lim_{\rho \to 0} \omega(\rho) = 0,$$

hence, for any $\gamma > 0$ there exists $\delta > 0$ such that if $|x| < \delta$ then $\omega(|x|) < \gamma$ and $|g(x)| \le \gamma |x|$.

Let $x_t(s, x)$ be a solution of (2.45) such that $x_s(s, x) = x$. Denoting $\tilde{g}_t = g_t(x_t(s, x))$ we can use the variation of constants formula and have

$$x_t(s, x) = X_{t,s} x + \sum_s^{t-1} X_{t,k} \tilde{g}_k = X_{t,s} x + \sum_s^{t-1} X_{t,k} g_k(x_k(s, x)).$$

Assume x is such that $|x| < \delta$ where $\delta > 0$ is determined starting from some $\gamma > 0$. As long as $|x_k(s, x)| < \delta$, $k \ge s$, we may write, using also the estimate for $X_{t,s}$

$$|x_t(s, x)| \le \beta_0 \rho^{t-s} |x| + \sum_s^{t-1} \beta_0 \rho^{t-k} \gamma |x_k(s, x)| \qquad (2.46)$$

$$\rho^{-t} |x_t (s, x)| \leq \beta_0 \rho^{-s} |x| + \sum_s^{t-1} \beta_0 \gamma \rho^{-k} |x_k (s, x)|$$

Denoting $u_k = \rho^{-k} |x_k (s, x)|$ the above inequality becomes

$$u_t \leq \beta_0 u_s + \beta_0 \gamma \sum_s^{t-1} |u_k|$$

which has the form (2.35). Applying Proposition 2.9 it follows that

$$u_t \leq \beta_0 (1 + \gamma) (1 + \beta_0 \gamma)^{t-s-1} u_s$$

leading to

$$|x_t (s, x)| \leq \frac{\beta_0 (1 + \gamma)}{(1 + \beta_0 \gamma)} [(1 + \beta_0 \gamma) \rho]^{t-s} |x| \qquad (2.47)$$

We make now the following choices:

a) $(1 + \beta_0 \gamma) \rho < 1$, for instance $1 + \beta_0 \gamma = \dfrac{1}{\sqrt{\rho}}$ that is

$\gamma = \dfrac{1}{\beta_0} \left(\dfrac{1}{\sqrt{\rho}} - 1 \right) > 0$ (since $\rho < 1$);

b) $\hat{\delta} (\gamma)$ is taken such that $|x| < \hat{\delta} (\gamma)$ implies $|g_t (x)| \leq \gamma |x|$;

c) $\delta = \dfrac{1 + \beta_0 \gamma}{\beta_0 (1 + \gamma)} \hat{\delta} (\gamma)$.

With such choices, if $|x| < \delta$ we shall have

$$|x_t (s, x)| \leq \hat{\delta} (\gamma) \rho^{(t-s)/2} \leq \hat{\delta} (\gamma) , \ t \geq s;$$

hence, (2.46) holds and (2.47) is valid for all $t \geq s$. We have thus proved:

Theorem 2.11 *Consider the nonlinear system (2.44) and let $\hat{y}_t$ be a bounded trajectory of it. Assume the following: i) $f_t (y)$ is such that the sequence $\partial f_t / \partial y$ is uniformly continuous in a neighborhood of $\hat{y}_t$; ii) matrix $A_t = \dfrac{\partial f_t}{\partial y} (\hat{y}_t)$ defines an exponentially stable evolution. Then the bounded trajectory $\hat{y}_t$ is exponentially stable i.e. there exist $\delta > 0$, $\beta > 1$, $\rho \in (0,1)$ such that if $|y_s - \hat{y}_s| \leq \delta$ then $|y_t - \hat{y}_t| \leq \beta \rho^{t-s} |y_s - \hat{y}_s| , \ t \geq s$.*

Remark. In this case we did a different proof which unlike that of the time invariant case, does not use the quadratic form defined on the trajectories of (2.45). Nevertheless, since A_t defines an exponentially stable evolution, we might have considered the sequence of positively definite matrices

$$P_t = \sum_t^{\infty} X_{k,t}^* X_{k,t}$$

satisfying $P_t \geq \mathbf{I}$ and $|P_t| \leq \beta_0^2 \left(1 - \rho^2\right)^{-1}$ and, defining $V_t = x_t^* P_t x_t$, the proof of Theorem 2.11 might have continued along the lines of the proof of Theorem 2.10.

2.4. LINEAR DISCRETE-TIME SYSTEMS WITH PERIODIC COEFFICIENTS

The systems with periodic coefficients are the first on the way leading from the systems with constant coefficients to systems with varying coefficients. The main feature of the systems with periodic coefficients is that their solutions structure is still deducible from the characteristic equation of a constant matrix (the monodromy matrix). This feature also allows one to obtain asymptotic properties of the solutions as a consequence of the properties of the roots of the above mentioned characteristic equation.

2.4.1. Multipliers and Structure of Solutions

Consider the system

$$x_{t+1} = A_t x_t \tag{2.48}$$

and assume that the *sequence of matrices* $\{A_t\}_t$ *is T -periodic*, i.e., there exists an integer $T > 0$ such that $A_{t+T} = A_t$ for any t. Denote $X_t = X_{t,0}$ where $X_{t,k}$ is the state transition matrix previously defined; X_t is the solution matrix of (2.48) with $X_0 = \mathbf{I}$.

Denote also $X_T = U$; this matrix is called *monodromy matrix* and its eigenvalues are the *multipliers of system* (2.48). We shall have

$$X_{t+T+1} = A_{t+T} X_{t+T} = A_t X_{t+T} \ ,$$

hence, X_{t+T} is a solution of (2.48) that equals U at $t = 0$. Therefore, $X_{t+T} = X_t U$, $t > 0$.

A. Let ρ be a multiplier of (2.48) and v the corresponding eigenvector. We have $Uv = \rho v$, hence,

$$X_{t+T} v = X_t U v = \rho X_t v \ .$$

It follows that the solution $x_t(v)$ of (2.48), i.e., the solution that equals v at $t = 0$ has the property $x_{t+T} = \rho x_t$.

Conversely, consider a solution of (2.48) with the above property. In particular $x_T = \rho x_0$ hence $X_T x_0 = \rho x_0$ that is $U x_0 = \rho x_0$ and ρ is thus an eigenvalue of U. These considerations justify the term *multipliers* for the eigenvalues of U.

Let ρ be some multiplier, v an eigenvector of U and let $e_1, e_2, ..., e_s$ be the elements of the Jordan canonical basis of U corresponding to ρ and starting with the eigenvector $v = e_1$ that is:

$$U e_1 = \rho e_1 \ , \ U e_2 = \rho e_2 + e_1 \ , ..., \ U e_s = \rho e_s + e_{s-1} \qquad (2.49)$$

Let as above $x_t(v)$ be the solution such that $x_0(v) = v$ that is $x_t(v) = X_t e_1$. We have

$$x_{t+T}(e_j) = X_{t+T} e_j = X_t U e_j = X_t(\rho e_j + e_{j-1}) = \rho x_t(e_j) + x_t(e_{j-1})$$

We then show that solution $x_t(e_1)$ has the structure

$$x_t(e_1) = \rho^{t/T} p_t \ , \ p_{t+T} = p_t \qquad (2.50)$$

assuming $\rho \neq 0$. Remark that if $\rho = 0$ then $x_{t+T}(e_1) = 0$, i.e., $x_t(e_1) = 0$ for $t \geq T$ (after a finite number of steps). Since $p_t = \rho^{-t/T} x_t(e_1)$ it follows that:

$$p_{t+T} = \rho^{-(t+T)/T} x_{t+T}(e_1) = \rho^{-t/T} \rho^{-1} \rho x_t(e_1) = p_t \ .$$

To obtain the structure for $x_t(e_2)$ let us denote

$$q_t = \rho^{-t/T} x_t(e_2) - \frac{t}{T\rho} p_t$$

We shall have

$$\begin{aligned} q_{t+T} &= \rho^{-(t+T)/T} x_{t+T}(e_2) - \frac{t+T}{T\rho} p_{t+T} = \\ &= \rho^{-t/T} \rho^{-1} [\rho x_t(e_2) + x_t(e_1)] - \frac{t}{T\rho} p_t - \frac{1}{\rho} p_t = \\ &= q_t + \rho^{-1} \left[\rho^{-t/T} x_t(e_1) - p_t \right] \equiv q_t \end{aligned}$$

We have obtained the structure

$$x_t(e_2) = \rho^{t/T} \left(q_t + \frac{t}{T\rho} p_t \right) \ , \ q_t = q_{t+T} \ , \ p_t = p_{t+T} \ . \qquad (2.51)$$

Again this structure has been obtained for $\rho \neq 0$. If $\rho = 0$ we have $x_{t+T}(e_2) = \rho x_t(e_2) + x_t(e_1) = x_t(e_1)$. Hence, $x_{t+2T}(e_2) = 0$ and again in this case the solution vanishes after a finite number of steps.

These structures will propagate. Indeed assume $\rho \neq 0$ and that we may write

$$x_t(e_{j-1}) = \rho^{t/T}\left[t^{j-2}p_t^1 + t^{j-3}p_t^2 + \cdots + tp_t^{j-2} + p_t^{j-1}\right] \qquad (2.52)$$

where $\{p_t^k\}_t$ are periodic sequences. We shall show that we may write

$$x_t(e_j) = \rho^{t/T}\left[t^{j-1}q_t^1 + t^{j-2}q_t^2 + \cdots + tq_t^{j-1} + q_t^j\right] \qquad (2.53)$$

where $\{q_t^k\}_t$ are periodic sequences that can be defined step by step starting from the sequences $\{p_t^k\}_t$. We may write

$$
\begin{aligned}
x_{t+T}(e_j) &= \rho^{(t+T)/T}\left[(t+T)^{j-1}q_{t+T}^1 + (t+T)^{j-2}q_{t+T}^2 + \cdots \right.\\
&\qquad \left. \cdots + (t+T)q_{t+T}^{j-1} + q_{t+T}^j\right]\\
&= \rho x_t(e_j) + x_t(e_{j-1}) = \\
&= \rho^{(t+T)/T}\left[t^{j-1}q_t^1 + t^{j-2}q_t^2 + \cdots + tq_t^{j-1} + q_t^j\right] + \\
&\quad +\rho^{t/T}\left[t^{j-2}p_t^1 + t^{j-3}p_t^2 + \cdots + tp_t^{j-2} + p_t^{j-1}\right]
\end{aligned}
$$

From here it follows that

$$
\begin{aligned}
\rho\left[\left(t^{j-1} + C_{j-1}^1 t^{j-2}T + \cdots + C_{j-1}^k t^{j-1-k}T^k + \cdots + T^{j-1}\right)q_{t+T}^1 + \right.\\
+ \left(t^{j-2} + C_{j-2}^1 t^{j-3}T + \cdots + T^{j-2}\right)q_{t+T}^2 + \cdots \\
\left. \cdots + (t+T)q_{t+T}^{j-1} + q_{t+T}^j\right] = \\
= \rho\left(t^{j-1}q_t^1 + \cdots + tq_t^{j-1} + q_t^j\right) + t^{j-2}p_t^1 + t^{j-3}p_t^2 + \cdots + p_t^{j-1} .
\end{aligned}
$$

This equality may be reordered as follows:

$$
\begin{aligned}
\rho\left(q_{t+T}^1 - q_t^1\right)t^{j-1} + \rho\left(C_{j-1}^1 Tq_{t+T}^1 + q_{t+T}^2 - q_t^2\right)t^{j-2} + \\
+ \rho\left(C_{j-1}^2 T^2 q_{t+T}^1 + C_{j-2}^1 T^2 q_{t+T}^2 + q_{t+T}^3 - q_t^3\right)t^{j-3} + \cdots \\
\cdots + \rho\left(C_{j-1}^{j-2}T^{j-2}q_{t+T}^1 + \cdots + C_2^1 Tq_{t+T}^{j-2} + q_{t+T}^{j-1} - q_t^{j-1}\right)t + \\
+ \rho\left(T^{j-1}q_{t+T}^1 + T^{j-2}q_{t+T}^2 + \cdots + Tq_{t+T}^{j-1} + q_{t+T}^j - q_t^j\right) = \\
= t^{j-2}p_t^1 + t^{j-3}p_t^2 + \cdots + tp_t^{j-2} + p_t^{j-1}
\end{aligned}
$$

$$(2.54)$$

We make now the following choices:

$$\rho C^1_{j-1} T q^1_t = p^1_t$$
$$\rho \left(C^2_{j-1} T^2 q^1_t + C^1_{j-2} T^2 q^2_t \right) = p^2_t$$
$$\cdots\cdots\cdots\cdots\cdots\cdots\cdots\cdots\cdots$$
$$\rho \left(C^{j-2}_{j-1} T^{j-2} q^1_t + C^{j-3}_{j-1} T^{j-3} q^2_t + \cdots + C^1_2 T q^{j-2}_t \right) = p^{j-2}_t$$
$$\rho \left(T^{j-1} q^1_t + T^{j-2} q^2_t + \cdots + T q^{j-1}_t \right) = p^{j-1}_t$$

(2.55)

which allow determination step-by-step of the sequences q^k_t, $k = \overline{1, j-1}$ from the sequences p^k_t, $k = \overline{1, j-1}$. The sequence q^j_t is determined from

$$q^j_t = \rho^{-t/T} x_t(e_j) - \left(t^{j-1} q^1_t + t^{j-2} q^2_t + \cdots + t q^{j-1}_t \right) \qquad (2.56)$$

From the above choices it can be seen that T-periodicity of q^k_t follows from T-periodicity of p^k_t for $k = \overline{1, j-1}$. Also substitution of (2.55) in (2.54) will give

$$\rho \left(q^1_{t+T} - q^1_t \right) t^{j-1} + \left[\rho \left(q^2_{t+T} - q^2_t \right) + p^1_{t+T} - p^1_t \right] t^{j-2} +$$
$$+ \cdots + \left[\rho \left(q^{j-1}_{t+T} - q^{j-1}_t \right) + p^{j-2}_{t+T} - p^{j-2}_t \right] t +$$
$$+ \rho \left(q^j_{t+T} - q^j_t \right) + p^{j-1}_{t+T} - p^{j-1}_t \equiv 0$$

and T-periodicity of p^k_t and q^k_t, $k = \overline{1, j-1}$ will finally give $q^j_{t+T} = q^j_t$, i.e., $\{ q^k_t \}_t$ is also T-periodic.

We obtained in this way the structure of the solutions for linear periodic systems. For non-zero multipliers the system has a set of linearly independent solutions as follows: to each non-zero multiplier there are associated the solutions corresponding to the initial conditions defined by the vectors of Jordan canonical basis of U. These solutions have the form

$$x_t = \rho^{t/T}_i \left[t^{j-1} q^{i,1}_t + t^{j-2} q^{i,2}_t + \cdots + t q^{i,j-1}_t + q^{i,j}_t \right] \qquad (2.57)$$

with $\left\{ q^{i,k}_t \right\}_t$ being T-periodic.

For the case of a zero-multiplier the relation

$$x_{t+T}(e_j) = \rho x_t(e_j) + x_t(e_{j-1})$$

becomes $x_{t+T}(e_j) = x_t(e_{j-1})$ and iterating this equality we shall have $x_{t+jT}(e_j) = 0$ since $x_{t+T}(e_j) = 0$.

It follows that the solutions corresponding to a zero multiplier vanish after a finite number of steps.

B. The same structure of solutions may be obtained by applying to system (2.48) a nonsingular periodic mapping that associates a system with constant coefficients (*Liapunov reductibility*). *Assume that A_t is nonsingular for any t, i.e., for $t = \overline{0, T-1}$*. Therefore, U is nonsingular and the matrix $\ln U$ exists. Denoting $V = \dfrac{1}{T} \ln U$ we have $U = e^{VT} = (B)^T$ where we denoted $B = e^V$. We may write $X_t = Q_t (B)^t$ where Q_t is T-periodic, X_t being the above-defined state-transition matrix (the fundamental matrix of solutions). Indeed we have $Q_t = X_t B^{-t}$, hence, $Q_{t+T} = X_t U B^{-T} B^{-t} = X_t B^{-t} = Q_t$.

We may take now $T_t = Q_t^{-1}$ and introduce the new state vector $y_t = T_t x_t$. We shall have

$$y_{t+1} = T_{t+1} x_{t+1} = T_{t+1} A_t x_t = T_{t+1} A_t T_t^{-1} y_t$$

Now compute

$$T_{t+1} A_t T_t^{-1} = B^{t+1} X_{t+1}^{-1} A_t T_t^{-1} = BB^t X_t^{-1} A_t^{-1} A_t X_t B^{-t} = B$$

and we have that y_t satisfies:

$$y_{t+1} = By_t \tag{2.58}$$

which is a system with constant coefficients. If for this system we consider the solutions whose initial conditions are the vectors of the Jordan canonical basis related to B (see Section 2.1), then the solutions of (2.48) defined by $x_t = T_t^{-1} y_t = Q_t B^t y_0$ will have the structure described by (2.57). Remark that $y_0 = T_0 x_0 = B^0 X_0^{-1} = I$ which gives $X_t = Q_t B^t = Q_t Y_t$, a natural result, if we recall that the columns of the state transition matrix are solutions of the system.

2.4.2. Stability Results

The stability concepts for the linear periodic systems are those defined for systems with constant coefficients (Definition 2.1). The structure of the solutions allows one to obtain simple stability conditions in the form of *multiplier location in the complex plane*.

Theorem 2.12 *For the exponential stability of* (2.48) *it is necessary and sufficient that the multipliers, i.e., the eigenvalues of the monodromy matrix U are located inside the unit disk of the complex*

*plane. For the stability it is necessary and sufficient that the multi-
pliers are located inside the unit disk and on the unit circle, to the
ones on the unit circle corresponding one-dimensional Jordan cells.
The system is unstable if at least one multiplier on the unit circle is
located in a Jordan cell of dimension larger than 1. The instability is
exponential if at least one eigenvalue is located outside the unit disk.*

This result is valid without the nonsingularity assumption for A_t.
Indeed if we use the structure of the solutions a singular A_t means
that some multipliers are zero, i.e., inside the unit circle. The corresponding solutions in the canonical basis vanish after a finite number
of steps; hence, the asymptotic behavior holds. On the other hand
the asymptotic behavior of the state transition matrix X_t may be obtained without Floquet-Liapunov theory. Indeed since $X_{t+T} = X_t U$
we shall have $X_{t+kT} = X_t U^k$ and the asymptotic behavior of X_t for
$t \to \infty$ depends on the asymptotic behavior of U^k for $k \to \infty$.

The previous considerations allows one to obtain some stability
criteria for linear periodic systems expressed in the language of the
coefficients.

Theorem 2.13 *Consider (2.48) with A_t T -periodic; if there exists
a constant matrix C with the eigenvalues inside the unit disk such that
$\sum_{0}^{T-1} |A_t - C|$ is sufficiently small then (2.48) is exponentially stable.*

Proof. We may write

$$x_{t+1} = Cx_t + (A_t - C)x_t$$

and use the formula of variations of constants (2.29)

$$x_t = C^t x_0 + \sum_{0}^{t-1} C^{t-1-k}(A_k - C)x_k$$

which gives for the state transition matrix

$$X_t = C^t + \sum_{0}^{t-1} C^{t-1-k}(A_k - C)X_k .$$

The assumption on the eigenvalues of C gives that $|C^t| \le \beta_0 \rho^t$ with

$\beta_0 > 0$, $0 < \rho < 1$. Therefore,

$$|X_t| \le \beta_0 \rho^t + \beta_0 \sum_0^{t-1} \rho^{t-1-k} |A_k - C| |X_k| \ .$$

Denoting $u_t = \rho^{-t} |X_t|$ we shall have

$$u_t \le \beta_0 + (\beta_0/\rho) \sum_0^{t-1} |A_k - C| u_k$$

and a simple induction argument will give

$$|u_t| \le \beta_0 \prod_{k=0}^{t-1} (1 + (\beta_0/\rho) |A_k - C|) \ .$$

(Remark that this is a straightforward generalization of Proposition 2.9 - the discrete analogue of Gronwall lemma; see Lemma 0.6 of the book of Halanay (1966), for the continuous time version of the above inequality.)

From here we have

$$|X_t| \le \beta_0 \rho^t \prod_{k=0}^{t-1} (1 + (\beta_0/\rho) |A_k - C|)$$

and further

$$
\begin{aligned}
|X_{mT}| &\le \beta_0 \rho^{mT} \prod_{k=0}^{mT-1} (1 + (\beta_0/\rho) |A_k - C|) = \\
&= \beta_0 \rho^{mT} \left\{ \prod_{k=0}^{T-1} (1 + (\beta_0/\rho) |A_k - C|) \right\}^m
\end{aligned}
$$

the last equality being a consequence of the periodicity of A_t. On the other hand since $1 + q < e^q$ we have

$$1 + (\beta_0/\rho) |A_k - C| < \exp \left\{ \frac{\beta_0}{\rho} |A_k - C| \right\}$$

and from here

$$\prod_{k=0}^{T-1} (1 + (\beta_0/\rho) |A_k - C|)^m < \exp \left\{ (m\beta_0/\rho) \sum_0^{T-1} |A_k - C| \right\}$$

This will give the estimate

$$|X_{mT}| \leq \beta_0 \exp \left\{ m \left(-T \log \frac{1}{\rho} + (\beta_0/\rho) \sum_0^{T-1} |A_k - C| \right) \right\}$$

Assume that

$$(\beta_0/\rho) \sum_0^{T-1} |A_k - C| < T \log \frac{1}{\rho}$$

This will give

$$|X_{mT}| \leq \beta_0 e^{-\alpha mT}$$

for some $\alpha > 0$. Since $X_{mT} = (X_T)^m = U^m$, it follows that the eigenvalues of U are inside the unit disk; this gives the exponential stability that ends the proof.

Remarks. The condition on A_t may be written as follows:

$$\frac{1}{T} \sum_0^{T-1} |A_k - C| \leq (\rho/\beta_0) \log \frac{1}{\rho}$$

If A_t were not periodic we might have required that $\sum_0^{\infty} |A_k - C| < \infty$; in this case the infinite product $\prod_{k=0}^{\infty} (1 + (\beta_0/\rho) |A_k - C|)$ is convergent and X_t is subject to an exponential estimate that again gives exponential stability. The periodicity assumption allowed an important weakening of the stability condition; remark, for instance that if A_t is periodic then the series $\sum_0^{\infty} |A_k - C|$ *cannot be convergent* since $|A_k - C|$ tends to zero only if it is identically zero.

The next stability result has applications to a class of canonical systems.

Theorem 2.14 *Let* $x_{t+1} = A_t x_t$ *and* $y_{t+1} = B_t y_t$ *be two* T-*periodic systems. Assume that the multipliers of the first system satisfy a reciprocal equation and that the multipliers of the second one are located on the unit circle and are distinct; also* B_t *are invertible for any* t. *If* $\sum_0^{T-1} |A_k - B_k|$ *is sufficiently small then the solutions of*

$x_{t+1} = A_t x_t$ *are bounded for any integer* t.

Proof. Let X_t and Y_t be the state transition matrices associated to A_t and B_t respectively. Let U, V be the corresponding monodromy matrices $U = X_T$, $V = Y_T$. We may write

$$x_{t+1} = B_t x_t + (A_t - B_t) x_t$$

and apply the variations of constant formula

$$X_t = Y_t + \sum_{k=0}^{t-1} Y_{t,k+1} (A_k - B_k) X_k$$

Since B_t is invertible it follows that Y_t is invertible; we shall have then $Y_{t,k} = Y_t Y_k^{-1}$ and

$$X_t = Y_t \left[I + \sum_{k=0}^{t-1} Y_{k+1}^{-1} (A_k - B_k) X_k \right]$$

$$X_t - Y_t = \sum_{k=0}^{t-1} Y_t Y_{k+1}^{-1} (A_k - B_k) (X_k - Y_k) +$$

$$+ \sum_{k=0}^{t-1} Y_t Y_{k+1}^{-1} (A_k - B_k) Y_k$$

Since the eigenvalues of V are distinct and located on the unit circle it follows that Y_t and Y_t^{-1} are bounded. Therefore

$$|X_t - Y_t| \le \beta_1 \sum_{k=0}^{t-1} |A_k - B_k| + \beta_2 \sum_{k=0}^{t-1} |A_k - B_k| |X_k - Y_k|$$

As in the case of Theorem 2.13 we shall have

$$|X_t - Y_t| \le \beta_1 \left(\sum_{k=0}^{t-1} |A_k - B_k| \right) \prod_{k=0}^{t-1} (1 + \beta_2 |A_k - B_k|) \ .$$

Therefore,

$$|U - V| = |X_T - Y_T| \le$$

$$\le \beta_1 \left(\sum_{0}^{T-1} |A_k - B_k| \right) \prod_{k=0}^{T-1} (1 + \beta_2 |A_k - B_k|) \le$$

$$\leq \ \beta_1 \left(\sum_{k=0}^{T-1} |A_k - B_k| \right) \exp \left\{ \beta_2 \sum_{0}^{T-1} |A_k - B_k| \right\}$$

This shows that if $\sum_{0}^{T-1} |A_k - B_k|$ is sufficiently small, then U and V are arbitrarily close; hence their eigenvalues are arbitrarily close. The eigenvalues of V being distinct, if U and V are sufficiently close, in the neighborhood of each eigenvalue of V there is only one eigenvalue of U. Since the characteristic equation of U is reciprocal and since U is real, if ρ is an eigenvalue of U then $1/\bar{\rho}$ is also an eigenvalue of U. If ρ is not on the unit circle, then ρ and $1/\bar{\rho}$ are symmetric with respect to the unit circle; hence in a small neighborhood of an eigenvalue of V there were two eigenvalues of U which is a contradiction. We deduce that the eigenvalues of U are located on the unit circle and distinct which implies that the evolution defined by A_t is stable.

As an application of this result we shall consider the second order vector equation

$$y_{t+1} - (A + \varepsilon P_t) y_t + y_{t-1} = 0 \tag{2.59}$$

Proposition 2.15 *If in (2.59) $A = A^*$, $P_t = P_t^* = P_{t+T}$ and the eigenvalues of A are distinct and located inside the disk of radius 2, all solutions of the equation are bounded provided $|\varepsilon|$ is sufficiently small.*

Proof. Let $x_t = y_{t-1}$; we shall have the system

$$\begin{aligned} x_{t+1} &= y_t \\ y_{t+1} &= -x_t + (A + \varepsilon P_t) y_t \end{aligned}$$

This is a canonical system since its matrix is simplectic (by direct check). Moreover it is obtained as a small perturbation of the following canonical system with constant coefficients

$$\begin{aligned} x_{t+1} &= y_t \\ y_{t+1} &= -x_t + A y_t \end{aligned} \tag{2.60}$$

The characteristic equation of this system is

$$\det \begin{pmatrix} \lambda \mathbf{I} & -\mathbf{I} \\ \mathbf{I} & \lambda \mathbf{I} - A \end{pmatrix} = 0 \ ;$$

hence,

$$\det\begin{pmatrix} \mathbf{I} & -\frac{1}{\lambda}\mathbf{I} \\ \mathbf{I} & \lambda\mathbf{I} - A \end{pmatrix} = \det\begin{pmatrix} \mathbf{I} & -\frac{1}{\lambda}\mathbf{I} \\ 0 & \left(\lambda+\frac{1}{\lambda}\right)\mathbf{I} - A \end{pmatrix} =$$

$$= \det\left(\left(\lambda+\frac{1}{\lambda}\right)\mathbf{I} - A\right) = 0$$

Therefore, if λ is an eigenvalue of the canonical system (2.60) then $\mu = \lambda + \frac{1}{\lambda}$ is an eigenvalue of A. We have $\lambda^2 - \mu\lambda + 1 = 0$ with $|\mu| < 2$ and since μ_k are distinct, λ_k will be distinct and located on the unit circle. We may therefore try to apply Theorem 2.14. The unperturbed system is (2.60) whose eigenvalues λ_k are distinct and located on the unit circle; its matrix obviously is invertible. The perturbed system is also canonical; hence its matrix is simplectic; therefore the monodromy matrix is simplectic and consequently its characteristic equation is reciprocal. It follows that we are indeed in the conditions of Theorem 2.14 provided the perturbation is small and this follows if $|\varepsilon|$ is sufficiently small. Application of Theorem 2.14 ends the proof.

Remark. The above result holds true for the equation

$$y_{t+1} - (A + P_t)\, y_t + y_{t-1} = 0$$

if $\sum_{0}^{T-1} |P_t|$ is sufficiently small.

2.5. LIAPUNOV FUNCTIONS

2.5.1. Stability Concepts

Consider a system

$$y_{t+1} = g_t\,(y_t) \tag{2.61}$$

and let $\hat{y}_t$, $t > s$, where t and s are integers, be a solution of (2.61).

Definition 2.4 The solution $\hat{y}_t$ of (2.61) is called *Liapunov stable* if for each $\varepsilon > 0$ there exists $\delta\,(\varepsilon, s) > 0$ such that if $|y_s - \hat{y}_s| < \delta\,(\varepsilon, s)$

then $|y_t - \hat{y}_t| < \varepsilon$ for all $t > s$.
If δ does not depend on s, the stability is called *uniform*.

The general concept of stability of a solution can be reduced to the one of a special equilibrium, corresponding to a solution which identically equals zero. Indeed, with (2.61) we can associate a new system

$$x_{t+1} = f_t(x_t) \tag{2.62}$$

where $f_t(x) = g_t(x + \hat{y}_t) - g_t(\hat{y}_t)$.

We see that $f_t(0) \equiv 0$, hence the sequence $x_t \equiv 0$ is a solution of (2.62). We see next that if y_t is a solution to (2.61) then x_t defined by $x_t = y_t - \hat{y}_t$ is a solution of (2.62). Conversely, if x_t is a solution of (2.62), then y_t defined by $y_t = x_t + \hat{y}_t$ is a solution of (2.61). We see that stability of the solution $\hat{y}_t$ of (2.61) is equivalent to the stability of the solution $\hat{x}_t \equiv 0$ of (2.62). This is why in the future, without loss of generality, we can consider only the stability properties of the solution which is identically equal to zero.

Definition 2.5   The solution $\hat{x}_t \equiv 0$ of (2.62) is *asymptotically stable* if it is stable and, moreover, there exists $\delta_0 > 0$ such that if $|x_s| < \delta_0$ then $\lim\limits_{t \to \infty} x_t = 0$.

An important concept is uniform asymptotic stability.

Definition 2.6   The solution $\hat{x}_t \equiv 0$ of (2.62) is *uniformly asymptotically stable* if there exist $\delta_0 > 0$ and functions $\delta(\cdot), T(\cdot)$ such that $|x_s| < \delta(\varepsilon)$ implies $|x_t| < \varepsilon$ for all $t \geq s$ and if $|x_s| < \delta_0$ then $|x_t| < \varepsilon$ for $t \geq s + T(\varepsilon)$.

We note that uniform asymptotic stability means uniform Liapunov stability and an asymptotic behavior which is independent both of the initial moment and of the initial conditions of the perturbed motion provided these initial conditions are small enough.

We will finally consider the strongest stability property which corresponds not only to uniform asymptotic stability but gives also quantitative description of the behavior of solutions.

Definition 2.7   The solution $\hat{x}_t \equiv 0$ of (2.62) is *exponentially stable* if there exist $\delta_0 > 0$, $\beta \geq 1$ and $\rho \in (0, 1)$ such that for all x_s with $|x_s| < \delta_0$ we have $|x_t| \leq \beta \rho^{t-s} |x_s|$.

We would like to remark that Definition 2.7 fully agrees with the notion of exponential stability as used in the case of linear systems or in the case of stability by the first approximation. Remark also that in the case of linear systems $\delta_0 > 0$ could be arbitrarily large; this property is called *global exponential stability*.

2.5.2. Stability via Liapunov Functions

We have the following results:

Theorem 2.16   *Assume there exist a sequence of functions $V_t(x)$ and two continuous and strictly increasing functions $\alpha(\rho), \beta(\rho)$ with $\alpha(0) = \beta(0)$ having the properties:*
  i) $\alpha(|x|) \leq V_t(x) \leq \beta(|x|)$;
  ii) $V_{t+1}(x_{t+1}) \leq V_t(x_t)$ *for any solution x_t of system (2.62).*
Then the solution $\hat{x}_t \equiv 0$ of the system is uniformly stable.

Proof. From *ii)* we have $V_t(x_t) \leq V_s(x_s)$, $t \geq s$; from *i)* we have $\alpha(|x_t|) \leq V_t(x_t)$, $V_s(x_s) \leq \beta(|x_s|)$. Take now $\delta(\varepsilon) = \beta^{-1}(\alpha(\varepsilon))$ and $|x_s| < \delta(\varepsilon)$. Therefore

$$
\begin{aligned}
\alpha(|x_t|) &\leq& V_t(x_t) \leq V_s(x_s) \leq \beta(|x_s|) < \\
&<& \beta(\delta(\varepsilon)) = \beta\left(\beta^{-1}(\alpha(\varepsilon))\right) = \alpha(\varepsilon)
\end{aligned}
$$

hence $|x_t| < \varepsilon$ for $t \geq s$ and the proof ends.

Theorem 2.17   *Assume there exist a sequence of functions $V_t(x)$ with the properties:*
  i) $\alpha(|x|) \leq V_t(x) \leq \beta(|x|)$ *with $\alpha(\rho), \beta(\rho)$ as above,*
  ii) $V_{t+1}(x_{t+1}) - V_t(x_t) \leq -\gamma(|x_{t+1}|)$ *for any solution x_t of system (2.62) and $\gamma(0) = 0$, $\gamma(\rho)$ being continuous and strictly increasing. Then the solution $\hat{x}_t \equiv 0$ of the system is uniformly asymptotically stable.*

Proof. The two conditions of the theorem are sufficient for the fulfillment of the two conditions of Theorem 2.16 hence $\hat{x}_t \equiv 0$ is uniformly stable and $\delta(\varepsilon)$ from the definition of uniform stability exists. If $f_t(x)$ are defined in a ball $|x| < h$ we may take $\delta_0 = \delta(h)$ to deduce that solutions starting within the ball $|x| < \delta(h)$ are defined for all future values.

   Let $T(\varepsilon) = [\beta(\delta_0)/\gamma(\delta(\varepsilon))]_e + 2$, s be an arbitrary integer and let $|x_s| < \delta_0$. We shall show that there exists t, $s \leq t \leq s + T(\varepsilon)$

such that $|x_t| < \delta(\varepsilon)$. Indeed if this were not true it would follow that $|x_t| \geq \delta(\varepsilon)$ for all t between s and $s + T(\varepsilon)$. Consequently, $\gamma(|x_{t+1}|) \geq \gamma(\delta(\varepsilon))$ for $s \leq t \leq T(\varepsilon) + s - 1$. This would give

$$V_{t+1}(x_{t+1}) - V_t(x_t) \leq -\gamma(\delta(\varepsilon)) \quad \text{hence}$$
$$V_{T(\varepsilon)+s}(x_{T(\varepsilon)+s}) - V_s(x_0) \leq -(T(\varepsilon) - 1)\gamma(\delta(\varepsilon))$$

From here it follows that

$$V_{T(\varepsilon)+s}(x_{T(\varepsilon)+s}) \leq V_s(x_0) - (T(\varepsilon) - 1)\gamma(\delta(\varepsilon)) \leq$$
$$\leq \beta(|x_s|) - (T(\varepsilon) - 1)\gamma(\delta(\varepsilon)) \leq \beta(\delta_0) - (T(\varepsilon) - 1)\gamma(\delta(\varepsilon)) =$$
$$= \left[\frac{\beta(\delta_0)}{\gamma(\delta(\varepsilon))} - T(\varepsilon) + 1\right]\gamma(\delta(\varepsilon)) < 0, \quad \text{a contradiction.}$$

Consequently there exists $\hat{t} \leq T(\varepsilon) + s$ with $|x_{\hat{t}}| < \delta(\varepsilon)$ hence $|x_t| < \varepsilon$ for $t \geq \hat{t}$ which means that in any case $|x_t| < \varepsilon$ for $t \geq T(\varepsilon) + s$ and this ends the proof of uniform asymptotic stability.

It is worth mentioning that if $\gamma(|x_{t+1}|)$ is replaced in ii) by $\gamma(|x_t|)$ the proof goes unchanged and the same conclusion is obtained.

Remark. The functions α, β, γ with the properties mentioned above are called *Massera functions*.

We showed that existence of the sequences of Liapunov functions $V_t(x)$ with certain properties is sufficient for the two basic stability properties. The converse results are also true, i.e., fulfillment of the stability properties implies the existence of Liapunov functions having even some degree of smoothness.

Theorem 2.18 *If the zero solution $\hat{x}_t \equiv 0$ of (2.62) is uniformly stable, then there exists a sequence of functions $V_t(x)$ and Massera functions $\alpha(\rho), \beta(\rho)$ with the properties:*
   i) $\alpha(|x|) \leq V_t(x) \leq \beta(|x|)$,
   ii) $V_{t+1}(x_{t+1}) \leq V_t(x_t)$ for any solution of (2.62).

Proof. According to Massera the function $\delta(\varepsilon)$ from the definition of uniform stability may be chosen continuous and strictly increasing (see Halanay (1966), pp. 25-26, for the description of such a Massera type construction). Therefore, the inverse mapping $\varepsilon(\delta)$ exists. Let $x_k(t, x)$ be a solution of (2.62) defined for $k \geq t$ by the initial condition $x_t(t, x) = x$. Since the zero solution is uniformly stable we have

$|x_k(t,x)| < \varepsilon(|x|)$. Define now

$$V_t(x) = \sup_{k \geq t} |x_k(t,x)|$$

We obviously have $|x| \leq V(t,x) \leq \varepsilon(|x|)$ which is nothing else than i) with $\alpha(\rho) = \rho$, $\beta(\rho) = \varepsilon(\rho)$. Further

$$V_t(x_t) = \sup_{k \geq t} |x_k(t,x_t)| = \sup_{k \geq s} |x_k(t,x_s)| \quad \text{for any } s < t$$

From here we have

$$
\begin{aligned}
V_t(x_{t+1}) &= \sup_{k \geq t+1} |x_k(t+1,x_{t+1})| = \sup_{k \geq t+1} |x_k(t,x_t)| \leq \\
&\leq \sup_{k \geq t} |x_k(t,x_t)| = V_t(x_t)
\end{aligned}
$$

and the theorem is proved.

Theorem 2.19 *Consider system* (2.62) *under the assumption:*

$$|f_t(x) - f_t(\tilde{x})| \leq L_r |x - \tilde{x}| \;, \; |x| \leq r \;, \; |\tilde{x}| \leq r \qquad (2.63)$$

If the zero solution $\hat{x}_t \equiv 0$ of (2.62) *is uniformly asymptotically stable then there exists a sequence of functions $V_t(x)$ and Massera functions $\alpha(\rho)$, $\beta(\rho)$, $\gamma(\rho)$ with the properties:*
    i) $\alpha(|x|) \leq V_t(x) \leq \beta(|x|)$,
    ii) $V_{t+1}(x_{t+1}) - V_t(x_t) \leq -\gamma(|x_{t+1}|)$ *for any solution x_t of* (2.62),
    iii) $|V_t(x) - V_t(\tilde{x})| \leq M |x - \tilde{x}|$.

Proof. Consider a C^2 function $G : \mathbb{R}_+ \longrightarrow \mathbb{R}_+$ with the following properties: $G(0) = G'(0) = 0$, $G'(t) > 0$, $G''(r) > 0$. Under these assumptions if $\lambda > 1$ then $G(r/\lambda) < G(r)/\lambda$. Let

$$V_t(x) = \sup_{k \geq 0} \left[G(|x_{t+k}(t,x)|) \frac{1 + \lambda k}{1 + k} \right]$$

We shall have

$$G(|x|) \leq V_t(x) \leq \left(\lambda - \frac{\lambda - 1}{1 + k} \right) G(|x_{t+k}(t,x)|) \leq \lambda G(\varepsilon(|x|))$$

where $\varepsilon(r)$ is that obtained from the uniform stability using the construction of Massera; the above inequalities represent property *i)* since both the mappings G and $G \circ \varepsilon$ are of Massera type.

Let now $k \geq T(\varepsilon)$, $|x| < \delta_0$ where $T(\varepsilon)$ and δ_0 are those from the definition of uniform asymptotic stability. We shall have $|x_{t+k}(t,x)| < \varepsilon$. Consequently, for $k \geq T\left(\frac{1}{\lambda}|x|\right)$ we shall have $|x_{t+k}(t,x)| < \frac{1}{\lambda}|x|$. This will give

$$G\left(|x_{t+k}(t,x)|\right)\frac{1+\lambda k}{1+k} < \lambda G\left(\frac{1}{\lambda}|x|\right) < G(|x|) \leq V_t(x)$$

for $k \geq T\left(\frac{1}{\lambda}|x|\right)$. Therefore

$$V_t(x) = \sup_{0 \leq k \leq T\left(\frac{1}{\lambda}|x|\right)}\left[G\left(|x_{t+k}(t,x)|\right)\frac{1+\lambda k}{1+k}\right] =$$
$$= G\left(|x_{t+p}(t,x)|\right)\frac{1+\lambda p}{1+p}, \quad 0 \leq p \leq T\left(\frac{1}{\lambda}|x|\right)$$

From here we deduce

$$V_{t+1}(x_{t+1}) = G\left(|x_{t+1+p}(t+1,x_{t+1})|\right)\frac{1+\lambda p}{1+p} =$$
$$= G\left(|x_{t+1+p}(t,x_t)|\right)\frac{1+\lambda p}{1+p} =$$
$$= G\left(|x_{t+1+p}(t,x_t)|\right)\frac{1+\lambda(p+1)}{1+(p+1)}\left[1 - \frac{\lambda-1}{(1+p)(1+\lambda(p+1))}\right] \leq$$
$$\leq V_t(x_t)\left[1 - \frac{\lambda-1}{(1+p)(1+\lambda(p+1))}\right]$$

Since $V_t(x) \geq G(|x|)$ and $p \leq T\left(\frac{1}{\lambda}|x_{t+1}|\right)$ we shall have

$$V_{t+1}(x_{t+1}) - V_t(x_t) \leq \frac{-(\lambda-1)G(|x_t|)}{\left(1+T\left(\frac{1}{\lambda}|x_{t+1}|\right)\right)\left(1+\lambda+T\left(\frac{1}{\lambda}|x_{t+1}|\right)\right)}$$

We shall use now the additional assumption (2.63), i.e., that the right hand side of (2.62) is Lipschitz. Since $f_t(0) \equiv 0$ we have $|f_t(x_t)| \leq$

$L_{\delta_0} |x_t|$ provided $|x_t| < \delta_0$ hence $|x_{t+1}| \leq L_{\delta_0} |x_t|$. This will give

$$V_{t+1} (x_{t+1}) - V_t (x_t) \leq$$

$$\leq - \frac{(\lambda - 1) G \left(\dfrac{1}{L_{\delta_0}} |x_{t+1}| \right)}{\left(1 + T \left(\dfrac{1}{\lambda} |x_{t+1}| \right) \right) \left(1 + \lambda + T \left(\dfrac{1}{\lambda} |x_{t+1}| \right) \right)} = -\gamma (|x_{t+1}|)$$

The function $\gamma(\rho)$ is strictly increasing if $T(\varepsilon)$ is taken strictly decreasing. This is always possible by using, for instance, a construction of Massera type, like in the case of $\delta(\varepsilon)$ for the property of uniform stability (see Halanay (1966), pp. 25-26 for this construction). Property $ii)$ of the Liapunov function is thus proved.

We shall prove next property $iii)$ which ensures a certain smoothness of the Liapunov function. In this connection we take $q = 1/L_{\delta_0}$ and

$$G(r) = \alpha_0 \int_0^r q^{T\left(\frac{1}{\lambda} \delta(s)\right)} \mathrm{d}s \tag{2.64}$$

$\alpha_0 > 0$ being a constant. It is easy to see that this $G(r)$ satisfies the required properties for such functions.

Let $u_t = |x_t (s, x) - x_t (s, \tilde{x})|$; for those t for which the solutions $x_t (s, x)$ and $x_t (s, \tilde{x})$ remain in the ball $|x| \leq r$ we shall have $u_{t+1} \leq L_r u_t$. Therefore for such t we shall have $u_t \leq L_r^{t-s} |x - \tilde{x}|$. Now if $|x| \leq \delta(r)$, $|\tilde{x}| \leq \delta(r)$ both solutions will remain in the ball $|x| \leq r$ for all $t \geq s$.

We have $V_t (x) = G (|x_{t+p} (t, x)|) \dfrac{1 + \lambda p}{1 + p}$ where $p \leq T \left(\dfrac{1}{\lambda} |x| \right)$ has been defined above. If $|x_{t+p} (t, \tilde{x})| \geq |x_{t+p} (t, x)|$ then it follows that

$$V_t (\tilde{x}) \geq G (|x_{t+p} (t, \tilde{x})|) \frac{1 + \lambda p}{1 + p} \geq G (|x_{t+p} (t, x)|) \frac{1 + \lambda p}{1 + p} = V_t (x)$$

If $|x_{t+p} (t, \tilde{x})| \leq |x_{t+p} (t, x)|$ then, using the expression and the properties of $G(r)$ given by (2.64), it follows

$$0 \leq G (|x_{t+p} (t, x)|) - G (|x_{t+p} (t, \tilde{x})|) \leq$$
$$\leq G' (|x_{t+p} (t, x)|) (|x_{t+p} (t, x)| - |x_{t+p} (t, \tilde{x})|) \leq$$
$$\leq \alpha_0 q^{T\left(\frac{1}{\lambda} \delta(|x_{t+p}(t,x)|)\right)} L_r^{T\left(\frac{1}{\lambda}|x|\right)} |x - \tilde{x}|$$

If we had $|x| < \delta\left(\left|x_{t+p}\left(t,x\right)\right|\right)$ this would give $\left|x_{t+p}\left(t,x\right)\right| < \left|x_{t+p}\left(t,x\right)\right|$, a contradiction. Consequently $|x| > \delta\left(\left|x_{t+p}\left(t,x\right)\right|\right)$ and this implies

$$T\left(\frac{1}{\lambda}|x|\right) \leq T\left(\frac{1}{\lambda}\delta\left(\left|x_{t+p}\left(t,x\right)\right|\right)\right)$$

We deduce that for $|x| < \delta_0$, $|\tilde{x}| < \delta_0$ we have:

$$
\begin{aligned}
V_t\left(x\right) &\leq V_t\left(\tilde{x}\right) + \lambda\alpha_0 q^{T\left(\frac{1}{\lambda}\delta\left(\left|x_{t+p}\left(t,x\right)\right|\right)\right)} L_{\delta_0}^{T\left(\frac{1}{\lambda}\delta\left(\left|x_{t+p}\left(t,x\right)\right|\right)\right)} |x - \tilde{x}| = \\
&= V_t\left(\tilde{x}\right) + \lambda\alpha_0 |x - \tilde{x}|
\end{aligned}
$$

and this inequality holds regardless the relation between $\left|x_{t+p}\left(t,x\right)\right|$ and $\left|x_{t+p}\left(t,\tilde{x}\right)\right|$. Interchanging x and $\tilde{x}$ we shall have $V_t\left(\tilde{x}\right) \leq V_t\left(x\right) + \lambda\alpha_0 |x - \tilde{x}|$; hence

$$\left|V_t\left(x\right) - V_t\left(\tilde{x}\right)\right| \leq \lambda\alpha_0 |x - \tilde{x}|$$

for nonzero x and $\tilde{x}$. If $\tilde{x} = 0$ then $x_{t+p}\left(t,\tilde{x}\right) = 0$ since $\tilde{x} = 0$ is an equilibrium; obviously $\left|x_{t+p}\left(t,x\right)\right| > 0$ and we may proceed as above obtaining $V_t\left(x\right) \leq \lambda\alpha_0 |x|$. The theorem is thus completely proved.

2.5.3. Liapunov Functions for Linear Systems. Exponential Stability

A. Consider the linear system

$$x_{t+1} = A_t x_t \tag{2.65}$$

and assume that its zero solution $\hat{x}_t \equiv 0$ is uniformly asymptotically stable in the sense of Definition 2.6. Since uniform stability implies boundedness of all solutions it follows that

$$\left|x_t\left(s,x\right)\right| \leq M |x| \tag{2.66}$$

(Remember that in the linear case the solution is a linear mapping with respect to the initial condition. In fact inequality (2.66) is equivalent to $\left|X_{t,k}\right| \leq M$ where $X_{t,k}$ is the state transition matrix - see Section 2.2.)

Define now the following (candidate Liapunov) function

$$V_t\left(x\right) = \sum_{k=0}^{N} \left|x_{t+k}\left(t,x\right)\right| \tag{2.67}$$

From (2.66) it follows that $V_t(x) \leq (N+1) M |x|$ and from the definition it follows that $V_t(x) > |x|$. On the other hand we have

$$V_t(x_t) = \sum_{k=0}^{N} |x_{t+k}(t, x_t)| = \sum_{k=0}^{N} |x_{t+k}(s, x_s)|$$

for any $s \leq t$. Therefore

$$V_{t+1}(x_{t+1}) - V_t(x_t) = \sum_{k=0}^{N} |x_{t+1+k}(t+1, x_{t+1})| - \sum_{k=0}^{N} |x_{t+k}(t, x_t)| =$$

$$= \sum_{k=0}^{N} |x_{t+1+k}(t, x_t)| - \sum_{k=0}^{N} |x_{t+k}(t, x_t)| = |x_{N+1+k}(t, x_t)| - |x_t|$$

Uniform asymptotic stability implies that all solutions tend uniformly to 0; hence, $|X_{t,k}| \leq \Psi_{t-k}$ where the sequence Ψ_t tends to zero. We deduce that $|x_{N+1+k}(t, x_t)| \leq \Psi_{N+1} |x_t|$ which gives

$$V_{t+1}(x_{t+1}) - V_t(x_t) \leq -(1 - \Psi_{N+1}) |x_t|$$

By choosing N sufficiently large in order that $\Psi_{N+1} < 1/2$ we shall have

$$V_{t+1}(x_{t+1}) - V_t(x_t) \leq -\frac{1}{2} |x_t| \leq -\frac{1}{2(N+1) M} V_t(x_t)$$

In this special situation (of the linear systems) we have thus obtained a specific result concerning existence of a Liapunov function.

Proposition 2.20 *Assume that linear system (2.65) is uniformly asymptotically stable. Then there exists a sequence of functions $V_t(x)$ with the properties*
  i) $|x| \leq V_t(x) \leq c_0 |x|$,
  ii) $V_{t+1}(x_{t+1}) - V_t(x_t) \leq -c_1 V_t(x_t)$, $0 < c_1 < 1$ *for any solution of (2.65).*

This proposition allows to obtain the discrete version of the theorem of Persidskii:

Proposition 2.21 *For linear systems uniform asymptotic stability implies exponential stability.*

Proof. Uniform asymptotic stability implies existence of a Liapunov function $V_t(x)$ with the properties from Proposition 2.20. Therefore,

$$\frac{V_{t+1}(x_{t+1})}{V_t(x_t)} \leq 1 - c_1 \quad \text{which gives}$$

$$|x_t| \leq V_t(x_t) \leq (1-c_1)^{t-s} V_s(x_s) \leq c_0(1-c_1)^{t-s} |x_s|$$

and the proposition is proved.

B. The result on exponential stability of linear systems from the above proposition may be extended to a class of nonlinear systems. Indeed remark that the proof of Proposition 2.21 is a straightforward application of Proposition 2.20. On the other hand, the proof of Proposition 2.20 makes use essentially of two properties of the solutions of a uniformly asymptotically stable linear system: an upper estimate by a function that is linear in the initial conditions - see (2.66) - and an upper estimate of the norm of the state transition matrix by a sequence that tends to zero.

In the general nonlinear case uniform asymptotic stability means existence for $|x| < \delta_0$ of $\delta(\varepsilon)$ and $T(\varepsilon)$ that may be chosen monotonically increasing and monotonically decreasing respectively. Consequently we shall have

$$|x_t(s,x)| \leq \varphi(|x|)\sigma(t-s) \tag{2.68}$$

where $\varphi(\rho)$ is monotonically increasing with $\varphi(0) = 0$ and $\sigma(\lambda)$ is monotonically decreasing and $\lim_{\lambda \to \infty} \sigma(\lambda) = 0$.

Assume now that $\varphi(\rho) = \beta_0\rho$, $\beta_0 > 0$. We shall then have $|x_t(s,x)| \leq \beta_0\sigma(0)|x|$ - an upper estimate that is linear in the initial conditions - and by choosing $V_t(x)$ as defined by (2.67) we shall have the following:

$$V_{t+1}(x_{t+1}) - V_t(x_t) = |x_{N+1+t}(t,x_t)| - |x_t| \leq (\beta_0\sigma(N+1) - 1)|x_t|$$

Since $\sigma(\lambda)$ is decreasing we may choose N sufficiently large to have $\beta_0\sigma(N+1) < 1/2$ thus obtaining

$$V_{t+1}(x_{t+1}) - V_t(x_t) \leq -\frac{1}{2}|x_t| \leq -\frac{1}{2\beta_0\sigma(0)(N+1)}V_t(x_t)$$

Proceeding as previously we find

$$|x_t| \leq (N+1)\beta_0\sigma(0)\left(1 - \frac{1}{2\beta_0\sigma(0)(N+1)}\right)^{t-s}|x_s|$$

nothing else than exponential stability. We obtained in fact the following result:

Proposition 2.22   *If the zero solution $\hat{x}_t \equiv 0$ of (2.62) is uniformly asymptotically stable and the monotonically increasing function $\varphi(\rho)$ in (2.68) is linear then the stability is exponential.*

Remark. For systems described by differential equations this extension of the result of Persidskii had been obtained by Halanay (1960, 1966) with slightly different type of Liapunov function. The same function adapted to the discrete-time case may be used here too.
Consider the function

$$V_t(x) = \sum_{k=0}^{N} |x_{t+k}(t,x)|^2 \tag{2.69}$$

This function is subject to the inequalities

$$|x|^2 \leq V_t(x) \leq (N+1)\,\beta_0^2\sigma^2(0)\,|x|^2$$

since any solution satisfies (2.68) with $\varphi(\rho) = \beta_0\rho$, $\beta_0 > 0$. Also

$$V_{t+1}(x_{t+1}) - V_t(x_t) = |x_{N+1+t}(t,x_t)|^2 - |x_t|^2 \leq$$
$$\leq (\beta_0^2\sigma^2(N+1) - 1)\,|x_t|^2 \leq -\frac{1}{2}\,|x_t|^2 \leq -\frac{1}{2\beta_0^2\sigma^2(0)\,(N+1)}V_t(x_t)$$

provided N is chosen such that $\beta_0^2\sigma^2(0)(N+1) < 1/2$. Proceeding as previously we find

$$|x_t| \leq \beta_0\sigma(0)\sqrt{N+1}\left(1 - \frac{1}{2\beta_0^2\sigma^2(0)(N+1)}\right)^{\frac{t-s}{2}}|x_s| \tag{2.70}$$

which represents exponential stability. Remark that N has to be large enough in order that $2\beta_0^2\sigma^2(0)(N+1) < 1$. Therefore, N should be the largest of the two values prescribed by the two inequalities

$$2\beta_0^2\sigma^2(N+1) < 1 \,,\; 2\beta_0^2(N+1)\sigma^2(0) < 1 \,.$$

C. Exponential stability allows simpler constructions for the Liapunov functions $V_t(x)$. Indeed, assume that the zero solution $\hat{x}_t \equiv 0$ of (2.62) is exponentially stable, i.e., there exists $\beta \geq 1,\, 0 < \rho < 1$

such that $|x_t| \leq \beta \rho^{t-s} |x_s|$, $t \geq s$ provided $|x_s| < \delta_0$. Let us take

$$V_t(x) = \sup_{k \geq 0} \left(\frac{1}{\rho^k} |x_{t+k}(t, x)| \right) \tag{2.71}$$

We shall have

$$|x| \leq V_t(x) \leq \beta |x| \tag{2.72}$$

ρ and β being the ones in Definition 2.7 of exponential stability. Also

$$V_{t+1}(x_{t+1}) = \sup_{k \geq 0} |x_{t+1+k}(t+1, x_{t+1})| \frac{1}{\rho^k} = \sup_{k \geq 0} |x_{t+1+k}(t, x_t)| \frac{1}{\rho^k} =$$

$$= \sup_{k \geq 1} |x_{t+k}(t, x_t)| \frac{1}{\rho^{k-1}} = \rho V_t(x_t)$$

Therefore

$$V_{t+1}(x_{t+1}) - V_t(x_t) \leq -(1 - \rho) V_t(x_t) \tag{2.73}$$

Inequalities (2.72) and (2.73) show that the Liapunov function defined by (2.71) has the properties of Theorem 2.19 expressed in terms of the constants that characterize exponential stability.

If the system is linear then we shall have further

$$V_t(x) - V_t(\tilde{x}) = \sup_{k \geq 0} |x_{t+k}(t, x)| \frac{1}{\rho^k} - \sup_{k \geq 0} |x_{t+k}(t, \tilde{x})| \frac{1}{\rho^k} \leq$$

$$\leq \sup_{k \geq 0} |x_{t+k}(t, x) - x_{t+k}(t, \tilde{x})| \frac{1}{\rho^k} =$$

$$= \sup_{k \geq 0} |x_{t+k}(t, x - \tilde{x})| \frac{1}{\rho^k} = V_t(x - \tilde{x})$$

Taking into account (2.72) we obtain

$$|V_t(x) - V_t(\tilde{x})| \leq \beta |x - \tilde{x}| \tag{2.74}$$

that is regularity of the Liapunov function is also expressed in terms of the constants characterizing exponential stability.

D. We shall consider again linear systems that are exponentially stable and show that this allows construction of Liapunov functions that are quadratic forms. Let $x^* Q_t x$ be a sequence of quadratic forms satisfying

$$0 < \mu |x|^2 \leq x^* Q_t x \leq M |x|^2 \tag{2.75}$$

We define a sequence of quadratic forms $V_t(x)$ as follows

$$V_t(x) = x^* \left(\sum_{k=t}^{\infty} X_{k,t}^* Q_k X_{k,t} \right) x \qquad (2.76)$$

The convergence of the matrix series is ensured by exponential stability of system (2.65). Indeed we have the estimate $|X_{t,k}| \leq \beta_0 \rho^{t-k}$, $0 < \rho < 1$, a straightforward consequence of the exponential stability as formulated in Definition 2.7 and of the fact that the columns of the transition matrix are solutions of (2.65). Therefore,

$$\left| \sum_{k=t}^{\infty} X_{k,t}^* Q_k X_{k,t} \right| \leq M \beta_0^2 \sum_{k=t}^{\infty} \rho^{2(k-t)} = \frac{M \beta_0^2}{1 - \rho^2} ;$$

hence, $V_t(x) \leq \dfrac{M \beta_0^2}{1 - \rho^2} |x|^2$. Also $V_t(x) \geq x^* Q_t x \geq \mu |x|^2$ and

$$V_t(x_t) = \sum_{k=t}^{\infty} x_t^* X_{k,t}^* Q_k X_{k,t} x_t = \sum_{k=t}^{\infty} x_k^* Q_k x_k .$$

Therefore,

$$V_{t+1}(x_{t+1}) - V_t(x_t) = -x_t^* Q_t x_t$$

Let $\alpha \in (0,1)$; we have obviously $x_t^* Q_t x_t \geq \alpha x_t^* Q_t x_t$; hence,

$$V_{t+1}(x_{t+1}) - V_t(x_t) \leq -\alpha x_t^* Q_t x_t \leq -\alpha \mu |x_t|^2 \leq -\alpha \mu V_t(x_t)$$

The Liapunov function thus considered has the properties prescribed by Proposition 2.20 with quadratic upper and lower estimates for $V_t(x)$. Let us denote

$$P_t = \sum_{k=t}^{\infty} X_{k,t}^* Q_k X_{k,t} \qquad (2.77)$$

the matrix of the quadratic form $V_t(x)$. We shall have

$$\begin{aligned}
P_t &= \sum_{k=t}^{\infty} X_{k,t}^* Q_k X_{k,t} = \sum_{k=t+1}^{\infty} X_{k,t}^* Q_k X_{k,t} + Q_t = \\
&= \sum_{k=t+1}^{\infty} A_t^* X_{k,t+1}^* Q_k X_{k,t+1} A_t + Q_t =
\end{aligned}$$

$$= A_t^* \left(\sum_{k=t+1}^{\infty} X_{k,t+1}^* Q_k X_{k,t+1} \right) A_t + Q_t = A_t^* P_{t+1} A_t + Q_t \ .$$

We have shown that P_t is a solution of the equation

$$A_t^* X_{t+1} A_t - X_t = -Q_t \tag{2.78}$$

known as *Liapunov matrix discrete time equation*. Moreover, P_t is its unique bounded on the whole integer axis $\mathbb{Z}$ solution. Indeed since boundedness has been proved above, we have to prove uniqueness. Assume there is another solution $\tilde{P}_t$ satisfying (2.78). We shall have

$$\begin{aligned} P_t &= \sum_{k=t}^{\infty} X_{k,t}^* Q_k X_{k,t} = \sum_{k=t}^{\infty} X_{k,t}^* \left(-A_k^* \tilde{P}_{k+1} A_k + \tilde{P}_k \right) X_{k,t} = \\ &= \sum_{k=t}^{\infty} X_{k,t}^* \tilde{P}_k X_{k,t} - \sum_{k=t}^{\infty} X_{k+1,t}^* \tilde{P}_{k+1} X_{k+1,t} = \\ &= \sum_{k=t}^{\infty} X_{k,t}^* \tilde{P}_k X_{k,t} - \sum_{k=t+1}^{\infty} X_{k,t}^* \tilde{P}_k X_{k,t} = \tilde{P}_t \end{aligned}$$

We proved in fact the following

Proposition 2.23 *If the sequence of matrices A_t, $t \in \mathbb{Z}$, defines an exponentially stable evolution (the zero solution of the associated linear system is exponentially stable) then the Liapunov matrix discrete-time equation (2.78) has for each sequence of matrices Q_t a unique solution that is bounded on the whole integer axis $\mathbb{Z}$. Moreover, if Q_t are uniformly positively definite, P_t has the same property of uniform positive definiteness.*

The converse of this result is an usual stability result for a linear system using a quadratic Liapunov function.

Proposition 2.24 *Assume that for a sequence of uniformly positively definite matrices Q_t (i.e., satisfying $Q_t \geq \mu\mathbf{I} > 0$) the inequation (2.78) admits a solution P_t satisfying $0 < \lambda\mathbf{I} \leq P_t \leq \Lambda\mathbf{I}$. Then A_t defines an exponentially stable evolution.*

Proof. We consider the sequence of Liapunov functions $V_t(x) =$

$x^* P_t x$. We shall have $0 < \lambda |x|^2 \leq V_t(x) \leq \Lambda |x|^2$. We have also

$$V_{t+1}(x_{t+1}) - V_t(x_t) = x_{t+1}^* P_{t+1} x_{t+1} - x_t^* P_t x_t =$$
$$= x_t^* (A_t^* P_{t+1} A_t - P_t) x_t = -x_t^* Q_t x_t \leq -\mu |x_t|^2 \leq -\frac{\mu}{\Lambda} V_t(x_t)$$

From here we have

$$\frac{V_{t+1}(x_{t+1})}{V_t(x_t)} \leq 1 - \frac{\mu}{\Lambda} \; ; \quad \text{hence,} \quad V_t(x_t) \leq \left(1 - \frac{\mu}{\Lambda}\right)^{t-s} V_s(x_s)$$

Using the upper and lower estimates for $V_t(x)$ will give

$$\lambda |x_t|^2 \leq \left(1 - \frac{\mu}{\Lambda}\right)^{t-s} \Lambda |x_s|^2 \; , \quad |x_t| \leq \sqrt{\frac{\Lambda}{\lambda}} \left(1 - \frac{\mu}{\Lambda}\right)^{\frac{t-s}{2}} |x_s|$$

which ends the proof of the proposition.

2.5.4. Stability under Perturbations

It is well known that Liapunov stability means stability with respect to short-time perturbations whose effect is incorporated in the initial conditions. There exist nevertheless practical problems where perturbations cannot be considered any longer short-time - the best known case is that of the mechanical systems that are subject, as pointed by Četaev (1936), one of the pioneers of the stability studies, to a field of weak, usually (but not always) negligible forces. Another case, belonging to control systems, is that of the uncertainty incorporated in the mathematical model, due to neglected factors.

Such considerations requires one to consider together with the basic (unperturbed) system (2.62) the following perturbed system:

$$z_{t+1} = f_t(z_t) + h_t(z_t) \tag{2.79}$$

with $h_t(z)$ being "small" with respect to $f_t(z)$. Stability under perturbations allows one to obtain stability properties for the perturbed system (2.79) from stability properties of the basic system (2.62) under certain assumptions about the perturbations.

A. A typical case of such results is *stability by the first approximation*; in the previous sections we obtained such results for the case when $f_t(x) = A_t x$, i.e., in the linear case. Now we shall consider a slightly more general result.

Theorem 2.25 *If* $|f_t(y') - f_t(y'')| < L_r |y' - y''|$ *for* $|y'| < r$, $|y''| < r$ *and* $|h_t(y)| < L_r' |y|$ *for* $|y| < r$ *then from the exponential stability* *of the zero solution of* (2.62) *will follow the exponential stability of the zero solution of* (2.79) *provided* L'_r *is sufficiently small.*

Proof. We shall consider again the Liapunov function defined using the solutions of system (2.62) whose zero solution is exponentially stable. We shall have

$$|V_t(x') - V_t(x'')| \le \sum_{k=0}^{N} (|x_{t+k}(t, x')| + |x_{t+k}(t, x'')|) \times$$
$$\times |x_{t+k}(t, x') - x_{t+k}(t, x'')|$$

since $|x - y| \ge ||x| - |y||$.

From the exponential stability of the zero solution of (2.62) we shall have $|x_{t+k}(t, x)| \le \beta \rho^k |x|$; hence

$$|V_t(x') - V_t(x'')| \le \beta \sum_{k=0}^{N} \rho^k (|x'| + |x''|) |x_{t+k}(t, x') - x_{t+k}(t, x'')| .$$

On the other hand, we shall have

$$x_p(t, x') - x_p(t, x'') = x' - x'' + \sum_{i=t}^{p-1} [f_i(x_i(t, x')) - f_i(x_i(t, x''))] .$$

If $|x'| < \delta(r)$, $|x''| < \delta(r)$ where $\delta(\cdot)$ is the one from the definition of the uniform stability then $|x_i(t, x')| < r$, $|x_i(t, x'')| < r$. Therefore

$$|x_p(t, x') - x_p(t, x'')| \le |x' - x''| + L_r \sum_{i=t}^{p-1} |x_i(t, x') - x_i(t, x'')| .$$

Using Proposition 2.9 we find

$$|x_p(t, x') - x_p(t, x'')| \le (1 + L_r)^{p-t} |x' - x''|$$

and this will give

$$|V_t(x') - V_t(x'')| \le \beta (|x'| + |x''|) |x' - x''| \sum_{k=0}^{N} (\rho(1 + L_r))^k =$$
$$= K_r(N) (|x'| + |x''|) |x' - x''| ,$$

where $K_r(N) = \beta \dfrac{(\rho(1 + L_r))^{N+1} - 1}{\rho(1 + L_r) - 1}$.

($|V_t(x') - V_t(x'')| \leq 2\delta(r)K_r|x' - x''| = \tilde{L}_r|x' - x''|$ for $|x'| \leq$
$\delta(r)$, $|x''| \leq \delta(r)$ and any t.)

We have further

$$|x|^2 \leq V_t(x) \leq \sum_{k=0}^{N} \beta^2 \rho^{2k} |x|^2 \leq \frac{\beta^2}{1-\rho^2} |x|^2$$

Let us consider the behavior of the Liapunov function (2.69) along the solutions of the unperturbed system (2.62)

$$V_{t+1}(x_{t+1}) - V_t(x_t) = |x_{N+1+t}(t,x)|^2 - |x_t|^2 \leq$$
$$\leq (\beta^2 \rho^{2(N+1)} - 1)|x_t|^2 \leq -(1 - \beta^2 \rho^{2(N+1)}) V_t(x_t)$$

Consider now z_t a solution of (2.79). We shall have

$$V_{t+1}(z_{t+1}) - V_t(z_t) = V_{t+1}(x_{t+1}(t,z_t)) - V_t(z_t) +$$
$$+V_{t+1}(z_{t+1}) - V_{t+1}(x_{t+1}(t,z_t)) \leq -(1 - \beta^2 \rho^{2(N+1)}) V_t(z_t) +$$
$$+K_r(N)(|z_{t+1}| + |x_{t+1}(t,z_t)|)|z_{t+1} - x_{t+1}(t,z_t)| \leq$$
$$\leq -(1 - \beta^2 \rho^{2(N+1)}) V_t(z_t) + K_r(N)(2|f_t(z_t)| + |h_t(z_t)|)|h_t(z_t)|$$

Let $|z_t| < r$. It will follow that

$$V_{t+1}(z_{t+1}) - V_t(z_t) \leq -(1 - \beta^2 \rho^{2(N+1)}) V_t(z_t) +$$
$$+K_r(N)(2L_r + L'_r)|z_t|^2 \leq$$
$$\leq -(1 - \beta^2 \rho^{2(N+1)} - K_r(N)(2L_r + L'_r)L'_r) V_t(z_t)$$

This inequality ensures exponential stability of the zero solution of (2.79) if

$$1 - \beta^2 \rho^{2(N+1)} - K_r(N)(2L_r + L'_r)L'_r > 0,$$

i.e., provided L'_r is sufficiently small. If this condition holds then we shall have

$$V_{t+1}(z_{t+1}) - V_t(z_t) \leq -\alpha V_t(z_t), \quad 0 < \alpha < 1$$

This will give $V_t(z_t) \leq V_s(z_s)(1-\alpha)^{t-s}$, $t \geq 0$ and therefore

$$|z_t|^2 \leq \frac{\beta^2}{1-\rho^2}(1-\alpha)^{t-s}|z_s|^2$$

as long as $|z_t| < r$. If $|z_s| < \dfrac{r\sqrt{1-\rho^2}}{\beta}$ then $|z_t| < r$ for all $t \geq s$

and the above inequalities hold for all $t \geq s$. The inequality $|z_t| \leq \hat{\beta}(1-\alpha)^{t-s}|z_s|$ proves the exponential stability of the zero solution of the perturbed system (2.79).

B. We shall introduce now another stability concept.

Definition 2.8 The zero solution of (2.62) is called *stable with respect to permanent perturbation (totally stable)* if for every $\varepsilon > 0$ there exist $\delta_1(\varepsilon) > 0$ and $\delta_2(\varepsilon) > 0$ such that for any z with $|z| < \delta_1(\varepsilon)$ and any $h_t(x)$ with $|h_t(x)| < \delta_2(\varepsilon)$ for $|x| \leq \varepsilon$ the solution $z_t(s,z)$ of (2.79) verifies $|z_t(s,z)| < \varepsilon$ for $t > s$.

Theorem 2.26 *Assume that $f_t(x)$ is uniformly locally Lipschitz i.e. it satisfies $|f_t(x) - f_t(\tilde{x})| \leq L_r|x - \tilde{x}|$ for all $|x| \leq r, |\tilde{x}| \leq r$. If the zero solution of* (2.62) *is uniformly asymptotically stable then it is also stable under permanent perturbations.*

Proof. Since the zero solution of (2.62) is uniformly asymptotically stable and $f_t(x)$ is uniformly locally Lipschitz, the assumptions of Theorem 2.19 are fulfilled and there exist a sequence of Liapunov functions $V_t(x)$ and Massera functions $\alpha(\rho), \beta(\rho), \gamma(\rho)$ satisfying

$i)$ $\alpha(|x|) \leq V_t(x) \leq \beta(|x|)$,

$ii)$ $V_{t+1}(x_{t+1}) - V_t(x_t) \leq -\gamma(|x_{t+1}|)$ for any solution x_t of (2.62),

$iii)$ $|V_t(x) - V_t(\tilde{x})| \leq M|x - \tilde{x}|$.

Let $\varepsilon > 0$ and consider the following choices for $\delta_1(\varepsilon)$ and $\delta_2(\varepsilon)$:
$\delta_1(\varepsilon) < (1/2)\beta^{-1}(\alpha(\varepsilon))$, e.g., $\delta_1(\varepsilon) = (1/3)\beta^{-1}(\alpha(\varepsilon))$, and

$\delta_2(\varepsilon) < \min\left\{\delta_1(\varepsilon), \dfrac{1}{M}\gamma(\delta_1(\varepsilon))\right\}$. Assume also that $\varepsilon > 0$ is small

enough to fulfil $L_\varepsilon \varepsilon + \delta_2(\varepsilon) < \delta(\delta_0)$ where $\delta(\cdot)$ is the one from the definition of uniform stability and $\delta_0 > 0$ is the one from the definition of asymptotic stability.

Consider now $z_t(s,z)$ the solution of the perturbed system (2.79) which coincides with z at $t = s$ and assume that $|z| < \varepsilon$; consider also $x_k(t,z)$ the solution of the unperturbed system (2.62) which coincides with z at $k = t$. We shall have

$$x_{t+1}(t,z) = f_t(x_t(t,z)) = f_t(z)$$

and since $|z| < \varepsilon$ we deduce $|x_{t+1}(t,z)| \leq L_\varepsilon \varepsilon$. On the other hand

$$z_{t+1}(t,z) = f_t(z_t(t,z)) + h_t(z_t(t,z)) = f_t(z) + h_t(z) ;$$

hence

$$|z_{t+1}(t, z) - x_{t+1}(t, z)| = |h_t(z)| < \delta_2(\varepsilon) \quad \text{since} \quad |z| < \varepsilon.$$

This will give

$$|z_{t+1}(t, z)| \leq \delta_2(\varepsilon) + L_\varepsilon \varepsilon < \delta(\delta_0) \quad \text{since} \quad |z| < \varepsilon$$

We deduce

$$V_{t+1}(z_{t+1}(t, z)) - V_t(z) = V_{t+1}(x_{t+1}(t, z)) - V_t(z) +$$
$$+V_{t+1}(z_{t+1}(t, z)) - V_{t+1}(x_{t+1}(t, z)) < -\gamma(|x_{t+1}(t, z)|) +$$
$$+M|z_{t+1}(t, z) - x_{t+1}(t, z)| < -\gamma(|x_{t+1}(t, z)|) + M\delta_2(\varepsilon) <$$
$$< -\gamma(|x_{t+1}(t, z)|) + \gamma(\delta_1(\varepsilon))$$

Consider now $\tilde{z}$ such that $|\tilde{z}| < \delta_1(\varepsilon)$ and the solution $z_t(s, \tilde{z})$ of the perturbed system (2.79). We claim that $|z_t(s, \tilde{z})| < \varepsilon$; if this were not true then it would exist some $p > s$ such that $|z_t(s, \tilde{z})| \geq \varepsilon$ and $|z_t(s, \tilde{z})| < \varepsilon$ for $s \leq t \leq p-1$. Therefore,

$$V_p(z_p(s, \tilde{z})) \geq \alpha(\varepsilon) > \beta(2\delta_1(\varepsilon)) \geq \beta(\delta_1(\varepsilon) + \delta_2(\varepsilon))$$

and $V_s(\tilde{z}) \leq \beta(\delta_1(\varepsilon))$. We deduce existence of some index m, $s \leq m < p$ such that

$$V_m(z_m(s, \tilde{z})) \leq \beta(\delta_1(\varepsilon) + \delta_2(\varepsilon)),$$
$$V_{m+1}(z_{m+1}(s, \tilde{z})) \geq \beta(\delta_1(\varepsilon) + \delta_2(\varepsilon))$$

and since $V_{m+1}(z_{m+1}(s, \tilde{z})) \leq \beta(|z_{m+1}(s, \tilde{z})|)$ we deduce that $|z_{m+1}(s, \tilde{z})| \geq \delta_1(\varepsilon) + \delta_2(\varepsilon)$. From here it follows

$$|x_{m+1}(m, z_m(s, \tilde{z}))| =$$
$$= |z_{m+1}(s, \tilde{z}) - z_{m+1}(m, z_m(s, \tilde{z})) + x_{m+1}(m, z_m(s, \tilde{z}))| \geq$$
$$\geq |z_{m+1}(s, \tilde{z})| - |z_{m+1}(m, z_m(s, \tilde{z})) - x_{m+1}(m, z_m(s, \tilde{z}))| \geq$$
$$\geq \delta_1(\varepsilon) + \delta_2(\varepsilon) - \delta_2(\varepsilon) = \delta_1(\varepsilon)$$

We deduce further

$$0 \leq V_{m+1}(z_{m+1}(s, \tilde{z})) - V_m(z_m(s, \tilde{z})) <$$
$$< -\gamma(|x_{m+1}(s, \tilde{z})|) + \gamma(\delta_1(\varepsilon)) =$$
$$= -\gamma(|x_{m+1}(m, z_m(s, \tilde{z}))|) + \gamma(\delta_1(\varepsilon)) \leq 0$$

and this contradiction proves the theorem.

2.6. INVARIANCE PRINCIPLE.
BARBASHIN-KRASOVSKII-LA SALLE THEOREM

The invariance principle of Barbashin-Krasovskii-La Salle is the main tool allowing one to obtain stability conclusions via Liapunov functions which have weaker properties than those from the theorems of Liapunov type. This is particularly true for those cases when the Liapunov function is not strictly decreasing but only nonincreasing along solutions. A long experience in applications of the Liapunov method shows that most of the Liapunov functions that are naturally associated to some practical problems (e.g., the energy-like functions) are only nonincreasing along solutions. In the following, we shall give the basic results which make it possible to obtain stability results with such Liapunov functions with weaker properties in the discrete-time case.

2.6.1. Discrete Dynamical Systems

We start with the following

Definition 2.9 A *discrete dynamical system* on some set $S \subset \mathbb{R}^n$ is a sequence of mappings $\pi_t : S \longrightarrow S$ defined for nonnegative integers $t \in \mathbb{N}$ and satisfying the following properties:

    i) $\pi_0(x) = x$ for all $x \in S$;

    ii) $\pi_t(\pi_s(x)) = \pi_{t+s}(x)$ for all $t, s \in \mathbb{N}$ and all $x \in S$;

    iii) the mappings are continuous.

Every discrete system of the form

$$x_{t+1} = f(x_t) \tag{2.80}$$

defines a dynamical system π by the sequence of mappings of the iterates of f : $\pi_t(x) = f^t(x)$. Here $f^t(x)$ is the t-iterate of f : $f^{t+1}(x) = f(f^t(x))$, $f^0(x) = x$. Conversely, every discrete dynamical system has associated with it the discrete-time system $x_{t+1} = \pi_1(x_t)$.

Condition *ii)* is the semigroup property and expresses the uniqueness of the solution in the forward direction of time; π is often called a *semiflow* or a *semi-dynamical system*, the term *dynamical system* being used when $\mathbb{N}$ can be replaced by $\mathbb{Z}$ (f has an inverse).

Definition 2.10 A point $\bar{x}$ is a *limit point* of $f^t(x)$ - or of a trajectory $x_t(x)$ - if there exists a sequence $\{t_i\}_i$ such that $\lim_{i \to \infty} t_i = \infty$

and $\lim_{i \to \infty} x_{t_i}(x) = \bar{x}$.

The ω-limit set $\Omega(x)$ of the trajectory $x_t(x)$ is the set of all limit points of it.

Definition 2.11   A set H is said to be *positively invariant* if $f(H) \subset H$ and *negatively invariant* if $H \subset f(H)$. If $f(H) = H$ the set is called *invariant*.

Definition 2.12   A closed invariant set H is said to be *invariantly connected* if it is not the union of two nonempty disjoint closed invariant sets.

We are now in position to state and prove some basic properties of limit sets.

Theorem 2.27   *Every ω -limit set is closed and positively invariant.*

Proof. Let $\Omega(x)$ be a limit set and $\bar{\Omega}$ its closure; let $q \in \bar{\Omega}$; hence, there exists a sequence $\{q_k\}$ such that $q_k \in \Omega$ for any k and $\lim_{k \to \infty} q_k = q$. Consider some $\varepsilon > 0$; there exists k sufficiently large such that $|q_k - q| < \varepsilon/2$ and since $q_k \in \Omega$, there exists t_i sufficiently large such that $|x_{t_i}(x) - q_k| < \varepsilon/2$. Therefore

$$|x_{t_i}(x) - q| \le |x_{t_i}(x) - q_k| + |q_k - q| < \varepsilon$$

It follows that q is an ω -limit point hence $\bar{\Omega} \subset \Omega$ i.e. $\Omega(x)$ is closed.

To prove positive invariance consider some limit point $q \in \Omega(x)$; there will exist a sequence of integers $\{t_i\}_i$, $t_i \to \infty$ and $\lim_{i \to \infty} x_{t_i}(x) = \lim_{i \to \infty} \pi_{t_i}(x) = q$. Since $\pi_t(\cdot)$ is continuous it follows that $f(\pi_{t_i}(x)) = \pi_1(\pi_{t_i}(x)) = \pi_{t_i+1}(x)$ tends to $f(q)$ hence $f(q) \in \Omega(x)$, i.e., $f(\Omega) \subset \Omega$ what ends the proof.

Theorem 2.28   *Assume that $\pi_t(x)$ is bounded for $t > 0$ (such property is called positive Lagrange stability). Then $\Omega(x)$ is nonempty, compact, invariant, invariantly connected and is the smallest closed set that $\pi_t(x)$ approaches as $t \to \infty$.*

Proof. Since $\{\pi_t(x)\}_t$ is a bounded sequence it contains a convergent subsequence; hence, $\Omega(x)$ is *nonempty*. The limit points of a bounded sequence being contained in a bounded set, $\Omega(x)$ is *bounded*. We already know from the previous theorem that $\Omega(x)$ is closed. Being closed and bounded in $\mathbb{R}^n$, $\Omega(x)$ is *compact*.

We also know from the previous theorem that $\Omega(x)$ is positively invariant. Let us show that it is in fact invariant. Let $\bar{x} \in \Omega(x)$ and select $\{t_i\}_i$ such that $x_{t_i}(x) \to \bar{x}$. Since $x_t(x)$ is bounded we may assume that $x_{t_i-1}(x)$ also converges (by selecting a subsequence if necessary). Let z be the limit of $x_{t_i-1}(x)$; hence, $z \in \Omega(x)$. On the other hand $f(x_{t_i-1}(x))$ converges to $f(z)$, from continuity, but $f(x_{t_i-1}(x)) = x_{t_i}(x)$ which converges to $\bar{x}$ hence $f(z) = \bar{x}$. We found for a given $\bar{x} \in \Omega(x)$ a point $z \in \Omega(x)$ such that $f(z) = \bar{x}$, i.e., we obtained $\Omega(x) \subset f(\Omega(x))$. Since we already have $f(\Omega(x)) \subset \Omega(x)$ from the previous theorem, it follows that $f(\Omega(x)) = \Omega(x)$, i.e., $\Omega(x)$ is *invariant*.

We show next that $x_t(x)$ tends to $\Omega(x)$ when $x_t(x)$ is bounded. We have to show that $\lim_{t\to\infty} d(x_t(x), \Omega(x)) = 0$, i.e., for any $\varepsilon > 0$ there exists $T(\varepsilon) > 0$ such that $d(x_t(x), \Omega(x)) < \varepsilon$ for $t > T(\varepsilon)$. Here $d(x, S) = \inf_{z \in S} |x - z|$ represents the distance from the point x to the set S. Assume this is not true; hence, there exists some $\varepsilon_0 > 0$ such that for any $T > 0$ there exists $t > T$ with the property $d(x_t(x), \Omega(x)) \geq \varepsilon_0$. Consider now an arbitrary sequence $\{T_k\}_k$, $T_k \to \infty$; therefore, for any k it is possible to find $t_k > T_k$ such that $d(x_{t_k}(x), \Omega(x)) \geq \varepsilon_0$. Since $x_t(x)$ is bounded the sequence $x_{t_k}(x)$ introduced above is bounded. By selecting (possibly) a subsequence we obtain that $x_{t_k}(x)$ is convergent to some $\bar{x} \in \Omega(x)$ because $t_k \to \infty$. Taking the limit in $d(x_{t_k}(x), \Omega(x)) \geq \varepsilon_0$ we shall have $d(\bar{x}, \Omega(x)) \geq \varepsilon_0$ hence $\bar{x} \notin \Omega(x)$ a contradiction. We have thus proved that *a bounded trajectory approaches its ω-limit set*.

Assume now that $x_t(x)$ tends to a closed set E. Therefore, for any $\varepsilon > 0$ there exists $T(\varepsilon)$ such that $d(x_t(x), E) < \varepsilon$ for $t > T(\varepsilon)$. This means that $\inf_{z \in E} |x_t(x) - z| < \varepsilon$ for $t > T(\varepsilon)$. We deduce existence for each $t > T(\varepsilon)$ of some $z_t \in E$ such that $|x_t(x) - z_t| < \varepsilon$.

Let $\bar{x} \in \Omega(x)$; there exists a sequence $\{t_k\}_k$, $t_k \to \infty$ with the property that $x_{t_k}(x) \to \bar{x}$; by selecting (possibly) a subsequence we shall find $\{z_{t_k}\}_k$ such that $|x_{t_k}(x) - z_{t_k}| < \varepsilon$, $t_k > T(\varepsilon)$. If the sequence $\{z_{t_k}\}_k$ is bounded then it contains a convergent subsequence. Let $z_{t_k} \to z$. Since E is closed we have $z \in E$. We have also $\bar{x} = z$ what shows that the ω-limit set $\Omega(x)$ is contained in the closed set E. But this closed set was arbitrary; hence, any closed set E to which a trajectory tends will contain its ω-limit set; being itself a closed set approached by the trajectory, *the ω-limit set is the smallest set that the trajectory approaches as $t \to \infty$.*

It remains to show that $\Omega(x)$ is invariantly connected. If this were not true, then $\Omega(x)$ would be the union of two disjoint closed nonempty invariant sets Ω_1 and Ω_2. Since $\Omega(x)$ is compact, so are Ω_1 and Ω_2. There exist then disjoint open sets U_1 and U_2 such that $\Omega_1 \subset U_1$ and $\Omega_2 \subset U_2$. Let $\bar{x} \in \Omega_1$; from invariance we have $f(\bar{x}) \in \Omega_1$; hence, $f(\bar{x}) \in U_1$. From the continuity of f it follows that there exists a neighborhood $N(\bar{x})$ such that $f(N(\bar{x})) \subset U_1$. The open set $V_1 = \bigcup_{\bar{x} \in \Omega(x)} N(\bar{x})$ has the property that $\Omega \subset V_1$ and $f(V_1) \subset U_1$.

Since $\Omega(x)$ is the smallest closed set that $x_t = f^t(x)$ approaches, it has to intersect both V_1 and U_2 an infinite number of times. There exist then two sequences of indices k'_j and k''_j such that $\left\{f^{k'_j}(x)\right\}_j \subset V_1, \left\{f^{k''_j}(x)\right\}_j \subset U_2$. Now, if $f^{k'_j}(x) \in V_1$, then $f^{k'_j+1}(x) \in U_1$ hence not to U_2; if it is neither in V_1 we denote $f^{k'_j+1}(x) = f^{k'''_j}(x)$. If $f^{k'_j+1}(x) \in V_1$ then $f^{k'_j+2}(x) \in U_1$, hence not to U_2; if it is neither in V_1 we denote $f^{k'_j+2}(x) = f^{k'''_j}(x)$. If $f^{k'_j+2}(x) \in V_1$ then $f^{k'_j+3}(x) \in U_1$ hence not to U_2 etc. In any case we shall be able to find some index such that $f^{k'''_j}(x)$ is neither in V_1 nor in U_2 since otherwise we would reach some index of the sequence k''_j which would send the sequence from V_1 to U_2. The above described selection of an index k'''_j may be performed for any j; the sequence $\left\{f^{k'''_j}(x)\right\}$ is neither in V_1 nor in U_2 and contains a convergent subsequence (being bounded). This convergent subsequence approaches $\Omega(x)$ but $\Omega(x)$ is included in the union of V_1 and U_2 while the sequence is not. This contradiction shows that $\Omega(x)$ is *invariantly connected* and the proof ends.

2.6.2. The Invariance Principle

As pointed out by La Salle (1976), information about the location of the limit set of a trajectory (motion) is information about its asymptotic behavior. In the following it will be shown that, suitably defined, Liapunov functions give information about the location of limit sets. This is done using, in particular, the invariance property of limit sets what justifies the name of *invariance principle* for the idea behind the approach we shall present below.

Let $V : G \subset \mathbb{R}^n \longrightarrow \mathbb{R}$. We associate to it the following function

$$W(x) = V(f(x)) - V(x) \qquad (2.81)$$

where f defines the discrete system (2.80).

Definition 2.13 The function $V : G \subset \mathbb{R}^n \longrightarrow \mathbb{R}$ is called a *Liapunov function* of (2.80) on G if: *i)* V is continuous; *ii)* $W(x) \leq 0$ on G; *iii)* for any $x \in G$ it can be found a neighborhood where V is bounded from below.

Some remarks are necessary. First of all, if x_t is a trajectory of (2.80) then we shall have $V(x_{t+1}) - V(x_t) = V(f(x_t)) - V(x_t) = W(x_t) \leq 0$ that is the Liapunov function thus defined is nonincreasing along the trajectories. Boundedness from below associated to nonincreasing will give convergence of the sequence $\{V(x_t)\}_t$ which is important in the study of asymptotic behavior. From the usual definition of the Liapunov function the constant sign property has been eliminated. This fact will allow one to obtain several types of asymptotic behavior among which asymptotic stability of an equilibrium point is a rather peculiar one.

We shall introduce the following set

$$E = \left\{ x \in \bar{G}; \ W(x) = 0 \right\} \tag{2.82}$$

where $\bar{G}$ is the closure of G and let M denote the largest invariant set in E. Denote $V^{-1}(c) = \{ x \in G; \ V(x) = c \}$.

Theorem 2.29 (Invariance Principle) *Consider the system* (2.80) *and let* $V : G \subset \mathbb{R}^n \longrightarrow \mathbb{R}$ *be a Liapunov function of it. For any bounded solution of* (2.80) *that is contained in G for all $t \geq 0$ there exists a number c such that* $x_t \to M \cap V^{-1}(c)$ *as* $t \to \infty$.

Proof. Consider some bounded trajectory $x_t(x)$ contained in G for $t \geq 0$. Since $V(x_t)$ is a nonincreasing and bounded from below sequence, $V(x_t) \to c$ as $t \to \infty$. Let $\bar{x} \in \Omega(x)$ (the ω-limit set of a bounded trajectory is nonempty); there will exist a sequence t_k such that $t_k \to \infty$ and $x_{t_k} \to \bar{x}$ as $k \to \infty$. Since V is continuous, $V(x_{t_k}(x)) \to V(\bar{x}) = c$. It follows that $\bar{x} \in V^{-1}(c)$ and since $\bar{x}$ is arbitrary we have $\Omega(x) \subset V^{-1}(c)$. From the invariance of $\Omega(x)$ it follows that $V(f(\bar{x})) = c$ hence $W(\bar{x}) = 0$. We obtained that $\Omega(x) \subset E$. Being invariant, $\Omega(x)$ *is* included in M. We obtained that $\Omega(x) \subset M \cap V^{-1}(c)$. Since $x_t(x) \to \Omega(x)$ it follows that $x_t(x) \to M \cap V^{-1}(c)$ what ends the proof.

The following consequences of Theorem 2.29 are important for applications.

Corollary 2.30   *If $\tilde{G}$ is bounded, open and positively invariant, and if $M \subset \tilde{G}$ then M is an attractor and $\tilde{G}$ is contained in the basin of attraction of M.*

Proof.   For $x \in \tilde{G}$ it follows that $x_t(x) \in \tilde{G}$ for all $t \geq 0$ since $\tilde{G}$ is positively invariant. From the boundedness of $\tilde{G}$ we deduce boundedness of $x_t(x)$. Applying the theorem just proved we deduce that $x_t(x) \to M$ hence M is an attractor and $\tilde{G}$ lies in its basin of attraction.

In applications $\tilde{G}$ is chosen as a connected component of a set of the form $\{x;\ V(x) < c\}$; such set will be open and invariant (from the properties of V); to apply Corollary 2.30 one has to check the boundedness of this set.

Corollary 2.31   (Barbashin-Krasovskii theorem for discrete-time systems)
If $M = \{\hat{x}\}$ (is a singleton), $V(x) > 0$ for x in a neighborhood of $\hat{x}$ then $\hat{x}$ is asymptotically stable. If moreover $V(x) \geq \alpha(|x|)$ with $\alpha(\rho)$ increasing, continuous, $\alpha(0) = 0$ and $\lim_{\rho \to \infty} \alpha(\rho) = \infty$ then $\hat{x}$ is globally asymptotically stable.

Proof.   We have Liapunov stability of $\hat{x}$ since $V(\hat{x}) = 0$, $V(x) > 0$ in a neighborhood of $\hat{x}$ and $W(x) \leq 0$ for all x in this neighborhood. That means that solutions starting in a neighborhood of $\hat{x}$ are bounded. From Theorem 2.29 we deduce that they tend to M, i.e., to $\hat{x}$; $\hat{x}$ is asymptotically stable. For the global asymptotic stability we shall have

$$\alpha(|x_t(x)|) \leq V(x_t(x)) \leq V(x) \quad \text{for any } x.$$

Since $\lim_{\rho \to \infty} \alpha(\rho) = \infty$ we may find some r in the range of the mapping defined by α such that $V(x) \leq r$ hence $|x_t(x)| \leq \alpha^{-1}(r)$. We may apply now Theorem 2.29 to obtain asymptotic stability. The proof is complete.

2.7. STABILITY IN DISCRETE MODELS OF CHEMICAL KINETICS

The information used in chemical kinetics models is obtained mainly from the steady state theories and experiments; the mass action law which is the basis of reaction velocity description is formulated at steady state. Once the reaction velocity is defined we may choose between two ways: either to consider the deduced at steady-state reaction velocity as valid during dynamic processes and describe the dynamics by the differential equations of the concentrations or to consider a discrete dynamics generated by the finite differences of the concentrations, assuming that the reaction velocity is constant on some small (but finite) intervals of time. This second approach seems quite realistic and the theoretical computations may be verified experimentally by using discretely recorded data. In the following we shall consider this approach which leads to discrete-time systems in chemical kinetics.

In the considered model the elements of the state vector are defined by the concentrations of the reacting substances. The reaction velocities are defined by a vector valued C^1 function $G : Q_M \subset \mathbb{R}^N \longrightarrow \mathbb{R}^N$ with $Q_M = \{x \in \mathbb{R}^N, x^i \geq 0, i = \overline{1, N}, \sum_1^N x^i \leq M\}$. The assumptions on $G(x)$ given below are motivated by chemical kinetics properties (see Frank-Kamenetskii (1987)):

i) $G(0) = 0$; ii) $\dfrac{\partial G^i}{\partial x^j} \geq 0, i \neq j$; iii) $\sum_1^N G^i(x) \equiv 0$.

If the concentrations are measured at discrete times t_k the evolution may be described by

$$x(t_{k+1}) - x(t_k) = (t_{k+1} - t_k) G(x(t_k)) .$$

If $t_{k+1} - t_k = \tau_0$ for any k, denoting $F(x) = \tau_0 G(x)$ and (by an abuse of notation) $x(t_k) = x_k$ we obtain the model

$$x_{k+1} - x_k = F(x_k) \tag{2.83}$$

where obviously $F(x)$ satisfies the same assumptions i) - iii) as $G(x)$.

Proposition 2.32 *If τ is sufficiently small then for $x_0 \in Q_M$ it follows that $x_k \in Q_M$ for all $k > 0$.*

Proof. Remark that Q_M is a compact set hence $\partial G^i/\partial x^i, i = \overline{1, N}$, are bounded on Q_M. It follows that for τ_0 small enough (depending on M) we shall have $\left| \dfrac{\partial F^i}{\partial x^i}(x) \right| < 1$ for all $x \in Q_M$ hence $1 + \dfrac{\partial F^i}{\partial x^i}(x) > 0$.

Let now x_k be a solution of (2.83) with initial value $x_0 \in Q_M$. We have

$$x_{k+1}^i \;=\; x_k^i + F^i(x_k) = x_k^i + F^i(x_k) - F^i(0) =$$

$$= x_k^i + \sum_{j=1}^{N} \frac{\partial F^i}{\partial x^j} (\theta x_k) x_k^j \, , \, 0 < \theta < 1 \, .$$

Assuming that $x_k \in Q_M$, $\theta x_k \in Q_M$ and the above formula is consistent. We may write further

$$x_{k+1}^i = \left(1 + \frac{\partial F^i}{\partial x^i} (\theta x_k) \right) x_k^i + \sum_{j \neq i} \frac{\partial F^i}{\partial x^j} (\theta x_k) x_k^j$$

and since $x_k \in Q_M$ we deduce, using the properties of F and its partial derivatives, that $x_{k+1}^i \geq 0, i = \overline{1, N}$.

Let us calculate now

$$\sum_{i=1}^{N} \left(x_{k+1}^i - x_k^i \right) = \sum_{i=1}^{N} F^i (x_k) = 0; \quad \text{hence}$$

$$\sum_{i=1}^{N} x_{k+1}^i = \sum_{i=1}^{N} x_k^i \leq M \quad \text{if } x_k \in Q_M \, .$$

We deduce from the above that if $x_k \in Q_M$ then $x_{k+1} \in Q_M$ and the evolution is defined for all k. This ends the proof.

Proposition 2.33 *If x_k , y_k are solutions of (2.83) in Q_M and $x_0^i \geq y_0^i$ for $i = \overline{1, N}$, then $x_k^i \geq y_k^i$, $i = \overline{1, N}$, $k > 0$.*

Proof. Let $z_k = x_k - y_k$; therefore,

$$z_{k+1}^i - z_k^i = F^i (x_k) - F^i (y_k) = \sum_{j=1}^{N} \frac{\partial F^i}{\partial x^j} (\tilde{x}_k) z_k^i$$

$$z_{k+1}^i = \left(1 + \frac{\partial F^i}{\partial x^i} (\tilde{x}_k) \right) z_k^i + \sum_{j \neq i} \frac{\partial F^i}{\partial x^j} (\tilde{x}_k) z_k^j \qquad (2.84)$$

where $\tilde{x}_k = \theta x_k + (1 - \theta) y_k$, $0 < \theta < 1$. We shall have $\tilde{x}_k \in Q_M$, the assumptions on the partial derivatives are valid, $z_0^i \geq 0$, $i = \overline{1, N}$, hence $z_k^i \geq 0$ what ends the proof.

Corollary 2.34 *All solutions are bounded.*

Proof. From (2.83) we deduce

$$\sum_{i=1}^{N} \left(x_{k+1}^i - x_k^i \right) = \sum_{i=1}^{N} F^i \left(x_k \right) \equiv 0 ; \quad \text{hence}$$

$$0 \le \sum_{i=1}^{N} x_k^i = \sum_{i=1}^{N} x_0^i , \quad 0 \le x_k^i \le \sum_{j=1}^{N} x_0^j , \; i = \overline{1, N} .$$

Remark. The equality

$$\sum_{i=1}^{N} x_k^i = \sum_{i=1}^{N} x_0^i , \; k > 0$$

shows that *the model admits a conservation law.*

Proposition 2.35 *Denote, for* $x \in \mathbb{R}^N$, $\|x\| = \sum_{i=1}^{N} |x^i|$. *If* x_k, y_k *are solutions of* (2.83) *then*

$$\|x_k - y_k\| \le \|x_0 - y_0\| , \; k > 0 \qquad (2.85)$$

Proof. Let I_1 be the set of indices such that $x_0^i > y_0^i$ and I_2 be the set of indices such that $x_0^i \le y_0^i$. Let $z_0^i = \max \left\{ x_0^i, y_0^i \right\}, i = \overline{1, N}$. We have $z_0^i = x_0^i, i \in I_1$, $z_0^i = y_0^i, i \in I_2$. This will give

$$\|z_0 - x_0\| + \|z_0 - y_0\| = \sum_{i=1}^{N} |z_0^i - x_0^i| + \sum_{i=1}^{N} |z_0^i - y_0^i| =$$

$$= \sum_{i \in I_2} |y_0^i - x_0^i| + \sum_{i \in I_1} |x_0^i - y_0^i| = \|x_0 - y_0\| .$$

We have also $x_0^i \le z_0^i, i = \overline{1, N}$; let z_k be the solution defined by z_0. From Proposition 2.33 we shall have $x_k^i \le z_k^i, i = \overline{1, N}, k > 0$. We deduce

$$
\begin{aligned}
\|z_k - x_k\| &= \sum_{i=1}^{N} |z_k^i - x_k^i| = \sum_{i=1}^{N} \left(z_k^i - x_k^i \right) = \\
&= \sum_{i=1}^{N} \left(z_0^i - x_0^i \right) = \|z_0 - x_0\|
\end{aligned}
$$

In the same way we shall have $\|z_k - y_k\| = \|z_0 - y_0\|$. It follows that

$$
\begin{aligned}
\|x_k - y_k\| &\leq \|x_k - z_k\| + \|z_k - y_k\| = \\
&= \|x_0 - z_0\| + \|z_0 - y_0\| = \|x_0 - y_0\|
\end{aligned}
$$

Remark. We may take instead of 0 any other starting point and the above result reads

$$\|x_k - y_k\| \leq \|x_j - y_j\| \ , \ k > j \ . \tag{2.86}$$

Corollary 2.36 *If x_k is a solution then $\xi_k = \|x_{k+1} - x_k\|$ is a decreasing sequence.*

Proof. Since system (2.83) is time invariant both x_{k+1} and x_k are solutions. Applying the above proposition we obtain

$$\xi_k = \|x_{k+1} - x_k\| \leq \|x_{j+1} - x_j\| = \xi_j \ , \ k \geq j \ .$$

Remark. The function $V(x) = \|F(x)\| = \sum_{i=1}^{N} |F^i(x)|$ is a *Liapunov function* since for all solutions x_k the sequence $V(x_k) = \|F(x_k)\| = \|x_{k+1} - x_k\|$ is decreasing.

In the following we shall deduce some useful properties connected with this Liapunov function.

, Let x_k be a solution of (2.83) and define $f_k^i = F^i(x_k)$. We shall have

$$
\begin{aligned}
\sum_{i=1}^{N} f_k^i &= \sum_{i=1}^{N} F^i(x_k) = 0 \,, \\
f_{k+1}^i - f_k^i &= F^i(x_{k+1}) - F^i(x_k) = \sum_{j=1}^{N} \frac{\partial F^i}{\partial x^j}(\tilde{x}_k)\left(x_{k+1}^j - x_k^j\right) = \\
&= \sum_{j=1}^{N} \frac{\partial F^i}{\partial x^j}(\tilde{x}_k) F^j(x_k) = \sum_{j=1}^{N} \frac{\partial F^i}{\partial x^j}(\tilde{x}_k) f_k^j
\end{aligned}
$$

We obtained that f_k^i are solutions of the following linear time varying discrete system

$$f_{k+1}^i = \left(1 + \frac{\partial F^i}{\partial x^i}(\tilde{x}_k)\right) f_k^i + \sum_{j \neq i}^{N} \frac{\partial F^i}{\partial x^j}(\tilde{x}_k) f_k^j \tag{2.87}$$

which is like (2.84). Let l be an integer and define $\left(f_l^i\right)_+ = \max\left(0, f_l^i\right)$, $\left(f_l^i\right)_- = \min\left(0, f_l^i\right)$; we shall have $\left(f_l^i\right)_+ \geq 0$ for all i and $\left(f_l^i\right)_- \leq 0$ for all i. Also $f_l^i = \left(f_l^i\right)_+ + \left(f_l^i\right)_-$. For $k > l$ we define g_k^i, $i = \overline{1, N}$ as the solution of (2.87) with the initial condition $g_l^i = \left(f_l^i\right)_+$ and h_k^i, $i = \overline{1, N}$ as the solution of (2.87) with the initial condition $h_l^i = \left(f_l^i\right)_-$. Since (2.87) is a linear system we shall have $f_k^i = g_k^i + h_k^i$, $i = \overline{1, N}$, $k \geq l$. Also from the properties of (2.87), i.e., of (2.84), we deduce that $g_k^i \geq 0$, $i = \overline{1, N}$, $k \geq l$, $h_k^i \leq 0$, $i = \overline{1, N}$, $k \geq l$.

On the other hand we have for any solution z_k of (2.87)

$$
\begin{aligned}
\sum_{i=1}^{N} \left(z_{k+1}^i - z_k^i\right) &= \sum_{i=1}^{N}\sum_{j=1}^{N} \frac{\partial F^i}{\partial x^j}(\tilde{x}_k) z_k^j = \sum_{j=1}^{N}\left(\sum_{i=1}^{N} \frac{\partial F^i}{\partial x^j}(\tilde{x}_k)\right) z_k^j = \\
&= \sum_{j=1}^{N}\left(\frac{\partial}{\partial x^j}\left(\sum_{i=1}^{N} F^i(x)\right)(\tilde{x}_k)\right) z_k^j \equiv 0,
\end{aligned}
$$

that is, $\displaystyle\sum_{i=1}^{N} z_k^i$ is constant for any k. Since g_k^i and h_k^i are solutions we deduce that $\displaystyle\sum_{i=1}^{N} g_k^i$ and $\displaystyle\sum_{i=1}^{N} h_k^i$ are constant for any k.

We deduce further from the assumptions on the partial derivatives of $F^i(x)$ and the already proved facts that $g_k^i \geq 0$, $h_k^i \leq 0$, $i = \overline{1, N}$, $k > l$:

$$
g_{k+1}^i - g_k^i = \sum_{j=1}^{N} \frac{\partial F^i}{\partial x^j}(\tilde{x}_k) g_k^j \geq \frac{\partial F^i}{\partial x^i}(\tilde{x}_k) g_k^i
$$

$$
g_k^i \geq \left(\prod_{j=l}^{k-1}\left(1 + \frac{\partial F^i}{\partial x^i}(\tilde{x}_j)\right)\right)\left(f_l^i\right)_+, \quad i = \overline{1, N}, \ k > l
$$

and, in the same way,

$$
h_k^i \leq \left(\prod_{j=l}^{k-1}\left(1 + \frac{\partial F^i}{\partial x^i}(\tilde{x}_j)\right)\right)\left(f_l^i\right)_-, \quad i = \overline{1, N}, \ k > l.
$$

We deduce that either $g_k^i \equiv 0$, $k > l$ (if $g_l^i = 0$) or $g_k^i > 0$, $k > l$ (if $g_l^i > 0$); also either $h_k^i \equiv 0$, $k > l$ (if $h_l^i = 0$) or $h_k^i < 0$, $k > l$ (if

$h_l^i < 0$).

We are now in position to state the following

Lemma 2.37 *If there exists some integer l such that*
$$V(x_l) = \sum_{i=1}^{N} |F^i(x_l)| > 0 \text{ then it will also exist } k > l \text{ such that}$$
$V(x_k) < V(x_l)$ *i.e. if sequence $\{V(x_k)\}_k$ is not identically zero it cannot be constant.*

Proof. According to previous definitions and notations $V(x_l) = \sum_{i=1}^{N} |f_l^i| > 0$ hence there exists at least a nonzero f_l^i. We shall assume first that there exists a nonvoid set of indices I_1 such that $(f_l^i)_+ > 0$, $i \in I_1$. Therefore, $g_k^i > 0$ for $i \in I_1$, $k > l$ and also $g_k^i \equiv 0$, $i \notin I_1$, $k > l$. It is obvious that for $i \in I_1$ we shall have $(f_l^i)_- = 0$; hence, $h_k^i \equiv 0$, $i \in I_1$, $k > l$. Therefore,

$$V(x_k) = \sum_{i=1}^{N} |f_k^i| = \sum_{i=1}^{N} |g_k^i + h_k^i| = \sum_{i \in I_1} g_k^i + \sum_{i \notin I_1} |h_k^i| = \sum_{i \in I_1} g_k^i - \sum_{i \notin I_1} h_k^i.$$

But we already know that

$$0 \equiv \sum_{i=1}^{N} f_k^i = \sum_{i=1}^{N} (g_k^i + h_k^i) = \sum_{i \in I_1} g_k^i + \sum_{i \notin I_1} h_k^i.$$

Summing the two equalities we shall have

$$V(x_k) = 2 \sum_{i \in I_1} g_k^i = 2 \sum_{i \in I_1} f_k^i = 2 \left[\sum_{i \in I_1} x_{k+1}^i - \sum_{i \in I_1} x_k^i \right].$$

It follows that if $\{V(x_k)\}_k$ is a constant nonzero sequence then $\left\{ \sum_{i \in I_1} x_k \right\}_k$ is an increasing arithmetic progression what contradicts boundedness of solutions.

Assume now that the set I_1 of indices defined above is void. Since there exists at least a nonzero f_l^i there will exist a nonvoid set of indices I_2 such that $(f_l^i)_- < 0$, $i \in I_2$. Therefore $h_k^i < 0$ for $i \in I_2$, $k > l$ and also $h_k^i \equiv 0$, $i \notin I_2$, $k > l$. Obviously for $i \in I_2$ we shall have $\left(f_l^i\right)_+ = 0$ hence $g_k^i \equiv 0$, $i \in I_2$, $k > l$. Therefore, in this case

$$V\left(x_k\right) = \sum_{i=1}^{N} \left|f_k^i\right| = \sum_{i=1}^{N} \left|g_k^i + h_k^i\right| = -\sum_{i \in I_2} h_k^i + \sum_{i \notin I_2} g_k^i = -\sum_{i \in I_2} h_k^i .$$

(The last equality is a consequence of the assumption that the set of indices where $g_k^i \neq 0$, $k > l$ is void.) We have also

$$0 \equiv \sum_{i=1}^{N} f_k^i = \sum_{i \in I_2} h_k^i + \sum_{i \notin I_2} g_k^i = \sum_{i \in I_2} h_k^i .$$

Therefore, $V\left(x_k\right) = 0 \neq V\left(x_l\right)$. The lemma is proved.

We may state now the main stability result.

Theorem 2.38 *For any $L < M$ there exist equilibria $\hat{x}$ such that $\sum_{i=1}^{N} \hat{x}^i = L$ and every solution $\{x_k\}_k$ with $\sum_{i=1}^{N} x_0^i = L$ tends asymptotically to such equilibrium.*

Proof. Let x_0 be such that $\sum_{i=1}^{N} x_0^i = L$ what gives for the corresponding solution $\sum_{i=1}^{N} x_k^i = L$. Since this solution is bounded its ω -limit set is nonempty. For $\hat{x}$ in the ω -limit set there exists a subsequence $\{k_m\}_m$ with $k_m \to \infty$ as $m \to \infty$, such that $\lim_{m \to \infty} x_{k_m} = \hat{x}$. From the continuity of V we have $\lim_{m \to \infty} V\left(x_{k_m}\right) = V\left(\hat{x}\right)$; the positive sequence $V\left(x_k\right)$ is nonincreasing; hence, it has a limit. The same limit will be obtained for the subsequence $V\left(x_{k_m}\right)$ hence $\lim_{k \to \infty} V\left(x_k\right) = V\left(\hat{x}\right)$.

Consider now the solution $\hat{x}_k$ with $\hat{x}_0 = \hat{x}$; since $\hat{x}$ is an ω -limit point and the ω -limit set is invariant, $\hat{x}_k$ are ω -limit points for all k. We deduce from the above that $V\left(\hat{x}_k\right) = V\left(\hat{x}\right)$, i.e., the sequence $V\left(\hat{x}_k\right)$ is constant. From the lemma we deduce that this sequence is zero hence $V\left(\hat{x}\right) = 0$. This will give

$$\lim_{k \to \infty} V\left(x_k\right) = \lim_{k \to \infty} \sum_{i=1}^{N} \left|F^i\left(x_k\right)\right| = 0 ;$$

hence, $\lim\limits_{k \to \infty} F^i(x_k) = 0$. We deduce

$$\lim_{m \to \infty} F^i(x_{k_m}) = F^i\left(\lim_{m \to \infty} x_{k_m}\right) = F^i(\hat{x}) = 0,$$

that is, $\hat{x}$ is an equilibrium. The sequence $\sum\limits_{i=1}^{N} x_k^i$ being constant, it follows that

$$\sum_{i=1}^{N} x_{k_m}^i = \sum_{i=1}^{N} x_0^i = L.$$

This will give $\lim\limits_{m \to \infty} \sum\limits_{i=1}^{N} x_{k_{m_i}}^i = L = \sum\limits_{i=1}^{N} \hat{x}^i$. Moreover $\|x_k - \hat{x}\| \leq \|x_{k_m} - \hat{x}\|$ for $k > k_m$ (according to (2.86)) and this shows that $\lim\limits_{k \to \infty} \|x_k - \hat{x}\| = 0$ since $\hat{x}$ is an ω -limit point. This ends the proof.

2.8. STABILITY RESULTS IN NEURODYNAMICS

Neural networks are computing devices for Artificial Intelligence, belonging to the class of *learning machines*. The basic feature of the neural networks is the interconnection of some simple computing elements in a very dense network of connections. The simple computing element is the *neuron* or, more precisely, the *artificial neuron*, a model of the biological neuron. We shall give below some elements of neuron modeling.

Each neuron is a device with several inputs u_i, $i = \overline{1,m}$ and a single output y. To each neuron there are associated the weights of the inputs, some real numbers w_i, $i = \overline{1,m}$, called synaptic weights; if $w_i > 0$ the weight is of *excitatory type* and if $w_i < 0$ the weight is of *inhibitory type*. The weighted sum of the inputs represents the *overall activation function* of the neuron (called also internal state)

$$s = \sum_{i=1}^{m} w_i u_i. \tag{2.88}$$

The output of the neuron is given by

$$y = f(s), \tag{2.89}$$

where the function f is called neuronal function, output function or activation function. The activation function is nonlinear.

We do not intend to develop here the discussion of neural networks. We shall only give one example concerning stability. As pointed out in various reference books (Kosko (1992), Haykin (1994), De Wilde (1996)) satisfactory operation of a neural network requires its evolution (motion) towards some equilibria which are significant in the application. This requirement is obviously a stability property of the equilibrium set of the neural network.

Our study will be concerned with the so called *Bidirectional Associative Memory* (BAM) designed by Kosko (1987). This device is composed of two neuron layers that are fully interconnected but in the same layer two neurons are not connected.

Denote by x^i, $i = \overline{1,n}$, the neuron states in one layer and by y^j, $j = \overline{1,p}$, the neuron states in the other layer; let $S^i\left(x^i\right)$, $\hat{S}\left(y^i\right)$ be the activation functions associated to the neurons. As already pointed out the inputs of a neuron come from the neurons of the other layer; each neuron has possibly an input coming from the environment; the inputs to neuron x^i that come from the neurons y^j are weighted by w_{ij}, $j = \overline{1,p}$ as well as the inputs to neuron y^j that come from the neurons x^i are weighted by w_{ij}, $i = \overline{1,n}$ (this shows that all neuron connections are bidirectional). These considerations lead to the following BAM model:

$$x_{t+1}^i = \sum_{j=1}^{p} w_{ij}\hat{S}^j\left(y_t^j\right) + I^i \ , \ i = \overline{1,n}$$

$$y_{t+1}^j = \sum_{i=1}^{n} w_{ij}S^i\left(x_t^i\right) + J^j \ , \ j = \overline{1,p} \ ,$$

(2.90)

where I^i, $i = \overline{1,n}$ and J^j, $j = \overline{1,p}$ denote constant inputs from the environment.

As activation functions we shall consider a significant case in neural network theory and applications - *threshold activation function* defined by

$$S^i\left(x\right) = \begin{cases} 1 & \text{if} \quad x \geq U^i \\ 0 & \text{if} \quad x < U^i \end{cases} \ , \ i = \overline{1,n};$$

$$\hat{S}^j\left(y\right) = \begin{cases} 1 & \text{if} \quad y \geq V^j \\ 0 & \text{if} \quad y < V^j \end{cases} \ , \ j = \overline{1,p}$$

(2.91)

The threshold functions are associated with the oldest neuron model - the McCulloch-Pitts model. There are other activation functions used in neuron models (sign functions, sigmoidal functions, etc.) but we shall not consider them here.

A. We shall analyze first some features of the model. First, it is always possible to adjust the thresholds in order that the arguments of the threshold functions never equal the threshold values. Indeed, any expression of the form $\sum_{j=1}^{p} w_{ij} \hat{S}^j \left(y^j\right) + I^i$ or of the form $\sum_{i=1}^{n} w_{ij} S^i \left(x^i\right) + J^j$ can have only a finite number of values (2^p and 2^n respectively) since the activation functions are binary-valued. We may compute

$$
\delta_x \;=\; \min_{\substack{y,i \\ \sum_{j=1}^{p} w_{ij}\hat{S}^j\left(y^j\right)+I^i-U^i\neq 0}} \left\{ \left| \sum_{j=1}^{p} w_{ij}\hat{S}^j\left(y^j\right) + I^i - U^i \right| \right\}
$$

$$
\delta_y \;=\; \min_{\substack{x,j \\ \sum_{i=1}^{n} w_{ij}S^i\left(x^i\right)+J^j-V^j\neq 0}} \left\{ \left| \sum_{i=1}^{n} w_{ij}S^i\left(x^i\right) + J^j - V^j \right| \right\} .
$$

We then subtract $\delta_x/2$ from U^i, $i = \overline{1,n}$ and $\delta_y/2$ from V^j, $j = \overline{1,p}$. In this way those state values satisfying equality with the thresholds (leading to the value 1 for the activation function) will be larger than the threshold value with a small nonzero quantity (leading again to the value 1 for the activation function). Former inequalities remain unmodified; hence, the dynamics are not affected. In the following we shall consider that U^i and V^j denote the already modified thresholds; hence, we may assume that neuron states never equal the threshold values of the activation functions.

Next we shall define an equilibrium state of (2.90) as usually as a constant solution satisfying

$$
\begin{aligned}
x^i &= \sum_{j=1}^{p} w_{ij}\hat{S}^j \left(y^j\right) + I^i \;,\; i = \overline{1,n} \\
y^j &= \sum_{i=1}^{n} w_{ij}S^i \left(x^i\right) + J^j \;,\; j = \overline{1,p}.
\end{aligned}
\tag{2.92}
$$

In the following we shall show that the equilibria defined by the above equations have the property that *they might be attained in a finite number of steps.*

To this end a Liapunov function is associated under the name of *energy function.* Following De Wilde (1995, p. 100) we choose the energy function expressed along the solutions of (2.91) as

$$E_t = -\sum_{i=1}^{n}\sum_{j=1}^{p} w_{ij} S^i \left(x_t^i\right) \hat{S}^j \left(y_{t-1}^j\right) - \sum_{j=1}^{p}\sum_{i=1}^{n} w_{ij} \hat{S}^j \left(y_t^j\right) S^i \left(x_{t-1}^i\right) -$$

$$- \sum_{i=1}^{n} \left(I^i - U^i\right) \left(S^i \left(x_t^i\right) + S^i \left(x_{t-1}^i\right)\right) -$$

$$- \sum_{j=1}^{p} \left(J^j - V^j\right) \left(\hat{S}^j \left(y_t^j\right) + \hat{S}^j \left(y_{t-1}^j\right)\right) .$$

(2.93)

This function has the following properties:

   i) It is *bounded*: this follows from the fact that functions $S^i(x)$, $\hat{S}^j(y)$ take only the values 0 and 1; it is clear that

$$|E_t| \leq 2 \sum_{i=1}^{n}\sum_{j=1}^{p} |w_{ij}| + \sum_{i=1}^{n} \left(|I^i| + |U^i|\right) + \sum_{j=1}^{p} \left(|J^j| + |V^j|\right)$$

   ii) It is *decreasing along any solution* of (2.94)

$$E_{t+1} - E_t = -\sum_{i=1}^{n}\sum_{j=1}^{p} w_{ij} S^i \left(x_{t+1}^i\right) \hat{S}^j \left(y_t^j\right) -$$

$$- \sum_{j=1}^{p}\sum_{i=1}^{n} w_{ij} \hat{S}^j \left(y_{t+1}^j\right) S^i \left(x_t^i\right) -$$

$$- \sum_{i=1}^{n} \left(I^i - U^i\right) \left(S^i \left(x_{t+1}^i\right) + S^i \left(x_t^i\right)\right) -$$

$$- \sum_{j=1}^{p} \left(J^j - V^j\right) \left(\hat{S}^j \left(y_{t+1}^j\right) + \hat{S}^j \left(y_t^j\right)\right) +$$

$$+ \sum_{i=1}^{n}\sum_{j=1}^{p} w_{ij} S^i \left(x_t^i\right) \hat{S}^j \left(y_{t-1}^j\right) + \sum_{j=1}^{p}\sum_{i=1}^{n} w_{ij} \hat{S}^j \left(y_t^j\right) S^i \left(x_{t-1}^i\right) +$$

$$+ \sum_{i=1}^{n} \left(I^i - U^i \right) \left(S^i \left(x_t^i \right) + S^i \left(x_{t-1}^i \right) \right) +$$

$$+ \sum_{j=1}^{p} \left(J^j - V^j \right) \left(\hat{S}^j \left(y_t^j \right) + \hat{S}^j \left(y_{t-1}^j \right) \right) =$$

$$= - \sum_{i=1}^{n} \left(S^i \left(x_{t+1}^i \right) - S^i \left(x_{t-1}^i \right) \right) \left(I^i - U^i + \sum_{j=1}^{p} w_{ij} \hat{S}^j \left(y_t^j \right) \right) -$$

$$- \sum_{j=1}^{p} \left(\hat{S}^j \left(y_{t+1}^j \right) - \hat{S}^j \left(y_{t-1}^j \right) \right) \left(J^j - V^j + \sum_{i=1}^{n} w_{ij} S^i \left(x_t^i \right) \right) =$$

$$= - \sum_{i=1}^{n} \left(S^i \left(x_{t+1}^i \right) - S^i \left(x_{t-1}^i \right) \right) \left(x_{t+1}^i - U^i \right) - $$
$$\tag{2.94}$$
$$- \sum_{j=1}^{p} \left(\hat{S}^j \left(y_{t+1}^j \right) - \hat{S}^j \left(y_{t-1}^j \right) \right) \left(y_{t+1}^j - V^j \right) .$$

Now, if $x_{t+1}^i > U^i$ then $S^i \left(x_{t+1}^i \right) = 1$; hence $S^i \left(x_{t+1}^i \right) - S^i \left(x_{t-1}^i \right) \geq 0$ and the product $\left(S^i \left(x_{t+1}^i \right) - S^i \left(x_{t-1}^i \right) \right) \left(x_{t+1}^i - U^i \right) \geq 0$. If $x_{t+1}^i < U^i$ then $S^i \left(x_{t+1}^i \right) = 0$; hence $S^i \left(x_{t+1}^i \right) - S^i \left(x_{t-1}^i \right) \leq 0$ and $\left(S^i \left(x_{t+1}^i \right) - S^i \left(x_{t-1}^i \right) \right) \left(x_{t+1}^i - U^i \right) \geq 0$. The same justification is valid for the other sum in (2.94). We deduce that $E_{t+1} - E_t \leq 0$ for any t and regardless of neuron transition (synchronous, asynchronous).

iii) Let us consider the case when $E_{t+1} = E_t$, i.e., when E is constant along some trajectory. More precisely, consider a fixed t such that $E_{t+1} = E_t$. Since the thresholds cannot be attained (see above) this equality is true when

$$S^i \left(x_{t-1}^i \right) = S^i \left(x_{t+1}^i \right), \ i = \overline{1, n}; \quad \hat{S}^j \left(y_{t-1}^j \right) = \hat{S}^j \left(y_{t+1}^j \right), \ j = \overline{1, p}$$
$$\tag{2.95}$$

Equalities (2.95) hold in the following cases:

1. $x_{t-1}^i = x_{t+1}^i$, $y_{t-1}^j = y_{t+1}^j$, $i = \overline{1,n}, j = \overline{1,p}$; this corresponds to a cycle of period 2. Indeed we have

$$x_{t+2}^i = \sum_{j=1}^p w_{ij} \hat{S}^j \left(y_{t+1}^j\right) + I^i = \sum_{j=1}^p w_{ij} \hat{S}^j \left(y_{t-1}^j\right) + I^i = x_t^i \quad \text{etc.}$$

Such cycles are called in neurodynamics *reverberations* (Caianiello (1966), (1967)). Remark that the above equalities hold also for cycles of period lower than 2, i.e., on *equilibria*.

2. The equalities $x_{t-1}^i = x_{t+1}^i$ and $y_{t-1}^j = y_{t+1}^j$ may not hold but $x_{t-1}^i - U^i$ and $x_{t+1}^i - U^i$ just have the same sign this being true also for $y_{t-1}^j - V^j$ and $y_{t+1}^j - V^j$. Assume that

$$x_{t-1}^i - U^i < 0 \,, \; x_{t+1}^i - U^i < 0 \,, \; i = \overline{1, n_1}$$
$$x_{t-1}^i - U^i > 0 \,, \; x_{t+1}^i - U^i > 0 \,, \; i = \overline{n_1 + 1, n}$$
$$y_{t-1}^j - V^j < 0 \,, \; y_{t+1}^j - V^j < 0 \,, \; j = \overline{1, p_1}$$
$$y_{t-1}^j - V^j > 0 \,, \; y_{t+1}^j - V^j > 0 \,, \; j = \overline{p_1 + 1, p}$$

We deduce from here

$$S^i \left(x_{t-1}^i\right) = 0 \,, \; S^i \left(x_{t+1}^i\right) = 0 \,, \; i = \overline{1, n_1}$$
$$S^i \left(x_{t-1}^i\right) = 1 \,, \; S^i \left(x_{t+1}^i\right) = 1 \,, \; i = \overline{n_1 + 1, n}$$
$$\hat{S}^j \left(y_{t-1}^j\right) = 0 \,, \; \hat{S}^j \left(y_{t+1}^j\right) = 0 \,, \; j = \overline{1, p_1}$$
$$\hat{S}^j \left(y_{t-1}^j\right) = 1 \,, \; \hat{S}^j \left(y_{t+1}^j\right) = 1 \,, \; j = \overline{p_1 + 1, p}$$

and

$$x_t^i = \sum_{j=p_1+1}^p w_{ij} + I^i \,, \; i = \overline{1, n} \,,$$

$$y_t^j = \sum_{i=n_1+1}^n w_{ij} + J^j \,, \; j = \overline{1, p} \,,$$

$$x_{t+2}^i = \sum_{j=p_1+1}^p w_{ij} + I^i = x_t^i \,, \; i = \overline{1, n} \,,$$

$$y_{t+2}^j = \sum_{i=n_1+1}^n w_{ij} + J^j = y_t^j \,, \; j = \overline{1, p}$$

which again shows a 2 -cycle.

We deduce that $E_{t+1} - E_t = 0$ *(i.e., E is constant) on equilibria or on 2 -cycles (reverberations of 2nd order)*.

iv) If the system is not on some equilibrium or in a 2 -cycle *the difference* $E_t - E_{t+1}$ *is bounded away from* 0. Indeed we know that $x_{t+1}^i - U^i$ and $y_{t+1}^j - V^j$ are bounded away from zero by $\delta_x/2$ and $\delta_y/2$ respectively (see above). The differences $S^i\left(x_{t+1}^i\right) - S^i\left(x_{t-1}^i\right)$ and $\hat{S}^j\left(y_{t+1}^j\right) - \hat{S}^j\left(y_{t-1}^j\right)$ which are nonzero in this case may equal only 1 or -1. It follows that $E_t - E_{t+1} > \frac{1}{2}\min\{\delta_x, \delta_y\}$ which shows that if the system is not at rest (equilibrium) or in a 2-cycle, the difference $E_t - E_{t+1}$ is bounded from below by a certain strictly positive number, independent of t and of the solution.

Summarizing, we obtained the following properties of the energy function defined by (2.93):

(a) it is bounded;

(b) it is non-increasing along any solution;

(c) it is constant on equilibria and 2-cycles;

(d) it is strictly decreasing by a bounded away from zero quantity which is independent of t and of the solution unless this solution is an equilibrium or a 2-cycle.

From the above properties of the energy function we may deduce now that any trajectory ends in an equilibrium or on a 2-cycle after a finite number of steps. This conclusion follows from the fact that at each step the energy decreases by at least a fixed amount and if this evolution would not end the energy would be unbounded.

It follows that after a finite number of steps the energy function is constant, i.e., the ω-limit set is reached in finite number of steps. Our result is thus the following one:

Theorem 2.39 *Consider the BAM neural network described by (2.90) with threshold activation functions (2.91). After a finite number of steps any solution of (2.90) ends in an equilibrium point or in a 2-cycle.*

Let us remark that the fourth property of the energy function cannot be present if the activation functions (hence also the energy function) are continuous. We may write indeed

$$z_{t+1} = F\left(z_t\right)$$

with $z = \mathrm{col}\left(x, y\right)$, $F\left(z\right)$ being continuous and $\tilde{E}\left(z_t, z_{t-1}\right) = E\left(x_t, x_{t-1}, y_t, y_{t-1}\right)$ being also continuous. Assume for awhile that

$$E_{t+1} - E_t = \tilde{E}\left(F\left(z_t\right), F\left(z_{t-1}\right)\right) - \tilde{E}\left(z_t, z_{t-1}\right) \leq -\mu$$

with $\mu > 0$ unless z_t is in the ω-limit set.

Let now $\hat{z}$ be an ω -limit point and let $\left\{z_0^k\right\}_k$ be a sequence tending to $\hat{z}$ where z_0^k do not belong to the ω -limit set. Let z_t^k be the solution corresponding to z_0^k. We shall have

$$
\begin{aligned}
E_2 - E_1 &= \tilde{E}\left(F\left(z_1^k\right), F\left(z_0^k\right)\right) - \tilde{E}\left(z_1^k, z_0^k\right) = \\
&= \tilde{E}\left(F\left(z_1^k\right), F\left(z_0^k\right)\right) - \tilde{E}\left(F^2\left(\hat{z}\right), F\left(\hat{z}\right)\right) + \\
+E\left(F\left(\hat{z}\right), \hat{z}\right) &- \tilde{E}\left(F\left(z_0^k\right), z_0^k\right) = \tilde{E}\left(F^2\left(z_0^k\right), F\left(z_0^k\right)\right) - \\
-\tilde{E}\left(F^2\left(\hat{z}\right), F\left(\hat{z}\right)\right) &+ E\left(F\left(\hat{z}\right), \hat{z}\right) - \tilde{E}\left(F\left(z_0^k\right), z_0^k\right) .
\end{aligned}
$$

Since both $\tilde{E}\left(v, w\right)$ and $F\left(z\right)$ are continuous we shall have

$$
\lim_{k \to \infty}\left(\tilde{E}\left(F\left(z_1^k\right), F\left(z_0^k\right)\right) - \tilde{E}\left(z_1^k, z_0^k\right)\right) = 0
$$

contradicting $\tilde{E}\left(F\left(z_1^k\right), F\left(z_0^k\right)\right) - \tilde{E}\left(z_1^k, z_0^k\right) \leq \mu$. (We used the fact that on the ω -limit set the Liapunov function is constant and that this set is invariant.)

Summarizing, we see a type of evolution that is quite different from those presented in this book. We nevertheless introduced it here in order to show how the ideas connected to Liapunov functions may be useful in another context.

2.9. STABILITY VIA INPUT/OUTPUT PROPERTIES

In this section we shall consider the following type of problems: for systems of the form

$$
\begin{aligned}
x_{t+1} &= f_t\left(x_t, u_t\right) \\
y_t &= h_t\left(x_t, u_t\right),
\end{aligned}
\tag{2.96}
$$

where $\left\{u_t\right\}_t$ is an input sequence belonging to a certain class, x_t is the solution corresponding to the initial condition x_s and to the input sequence $\left\{u_t\right\}_t$ and y_t is constructed using x_t and u_t, we shall try to obtain stability and boundedness properties for the state x_t based on some properties of the input/output pairs $\left(u_t, y_t\right)$ defined above.

2.9.1. The Perron Condition

We give the following

Definition 2.14 It is said that the linear system $x_{t+1} = A_t x_t$ *satisfies the Perron condition if for any bounded sequence* $\left\{f_t\right\}_t$ *the solu-*

tion with zero initial conditions of the forced system $x_{t+1} = A_t x_t + f_t$ is bounded.

A. In the following we shall prove

Theorem 2.40   *If the system $x_{t+1} = A_t x_t$ satisfies the Perron condition then the zero solution is exponentially stable.*

Proof. The variation of constants formula will give

$$x_t = \sum_{k=\tau}^{t-1} X_{t,k} f_k , \qquad \tau \geq 0 . \tag{2.97}$$

Since $\{x_t\}_t$ is bounded for any bounded sequence $\{f_t\}_t$ it follows that the linear operator U_τ defined by $(U_\tau f)_t = \sum_{k=\tau}^{t-1} X_{t,k} f_k$ applies the Banach space of bounded sequences with the *sup* norm into itself. We can apply Banach-Steinhaus lemma and deduce existence of a constant $M > 0$ such that $\| U_\tau f \| \leq M \| f \|$ that is $\sup_t |(U_\tau f)_t| \leq M \sup_k |f_k|$.

In a more explicit form we have $\sup_t \left| \sum_{k=\tau}^{t-1} X_{t,k} f_k \right| \leq M \sup_k |f_k|$.

Denote by $x_{t,k}^{ij}$ the entries of the matrix $X_{t,k}$ and by f_k^j the entries of f_k; let $\xi_{t,k}^i = \sum_{j=1}^{n} x_{t,k}^{ij} f_k^j$ and choose $f_k^j = \operatorname{sgn} x_{t,k}^{ij}$ what gives $\xi_{t,k}^i = \sum_{j=1}^{n} \left| x_{t,k}^{ij} \right|$. From here it follows that:

$$\left| \sum_{k=\tau}^{t-1} X_{t,k} f_k \right| \leq \sum_{i=1}^{n} \sum_{k=\tau}^{t-1} \xi_{t,k}^i = \sum_{k=\tau}^{t-1} \sum_{i=1}^{n} \sum_{j=1}^{n} \left| x_{t,k}^{ij} \right| \leq Mn .$$

(We have used for a vector $v \in \mathbb{R}^n$ with coordinates v_i the norm $|v| = \sum_{1}^{n} |v_i| .$)

Since all finite dimensional norms are equivalent we obtained in fact that $\sum_{k=\tau}^{t-1} |X_{t,k}| \leq C$ where C may depend on the chosen matrix norm but not on t. From here we have that $|X_{t,k}| \leq C + 1$ for any $t > \tau$ and $\tau \leq k \leq t$.

Consider a solution of the free system. We have $x_t = X_{t,s} x_s$; hence $|x_t| \leq (C + 1) |x_s|$ and this is nothing more than uniform stability.

We have further $X_{t,k} = X_{t,l} X_{l,k}$ which holds for $t \geq l \geq k$ and since the matrix sequence $\{X_{l,k}\}_l$ is bounded the Perron condition implies that $\sum_{k}^{t-1} X_{t,k} = (t - k) X_{t,k} = \sum_{l=k}^{t-1} X_{t,l} X_{l,k}$ is bounded by a constant $M_1(k)$. This gives $|X_{t,k}| \leq M_1(k)(t-k)^{-1}$ and asymptotic stability holds. Moreover

$$\left| \sum_{l=k}^{t-1} X_{t,l} X_{l,k} \right| \leq \sum_{l=k}^{t-1} |X_{t,l}| |X_{l,k}| \leq (C+1) \sum_{0}^{t-1} |X_{t,l}| \leq C(C+1) \, ;$$

hence, M_1 may be taken independent of k and asymptotic stability is uniform. Since the system is linear the stability is exponential (Proposition 2.21).

B. We shall consider here the linear system

$$\begin{aligned} x_{t+1} &= A_t x_t + B_t u_t \\ y_t &= C_t x_t \end{aligned} \tag{2.98}$$

under the assumptions of uniform stabilizability and uniform detectability. There exist a bounded matrix sequence $\{F_t\}_t$ and a bounded matrix sequence $\{H_t\}_t$ such that the matrix sequence $\{A_t + B_t F_t\}_t$ and $\{A_t + H_t C_t\}_t$ define exponentially stable evolutions, i.e., the systems

$$x_{t+1} = (A_t + B_t F_t) x_t \tag{2.99}$$

and

$$x_{t+1} = (A_t + H_t C_t) x_t \tag{2.100}$$

are exponentially stable.

Theorem 2.41 *Assume that uniform stabilizability and uniform detectability hold for system (2.98). If for any bounded input sequence $\{u_t\}$ the output sequence $\{y_t\}$ corresponding to a zero initial state is bounded, then the free system $x_{t+1} = A_t x_t$ is exponentially stable.*

Proof. Consider the system

$$
\begin{aligned}
x^1_{t+1} &= (A_t + B_t F_t)\, x^1_t + u_t \\
x^2_{t+1} &= B_t F_t x^1_t + A_t x^2_t \\
x^3_{t+1} &= -H_t C_t x^1_t + H_t C_t x^2_t + (A_t + H_t C_t)\, x^3_t + u_t
\end{aligned}
\qquad (2.101)
$$

where F_t and H_t are bounded and such that $(A_t + B_t F_t)$ and $(A_t + H_t C_t)$ define exponentially stable evolutions. Let $\{u_t\}_t$ be a bounded sequence. Since $(A_t + B_t F_t)$ defines an exponentially stable evolution it follows that x^1_t with zero initial condition is bounded, uniformly with respect to the initial moment. We deduce that $F_t x^1_t$ is bounded. If we use bounded input/bounded output stability of (2.98) we obtain that $C_t x^2_t$ is bounded. Therefore, $u_t - H_t C_t x^1_t + H_t C_t x^2_t$ is a bounded input sequence for the equation of x^3_t whose matrix $A_t + H_t C_t$ defines an exponentially stable evolution. It follows that $\{x^3_t\}_t$ is also bounded.

Define further $z_t = x^1_t - x^2_t$; we may write

$$
\begin{aligned}
z_{t+1} &= A_t z_t + u_t \\
x^3_{t+1} &= -H_t C_t z_t + (A_t + H_t C_t)\, x^3_t + u_t \\
x^3_{t+1} - z_{t+1} &= (A_t + H_t C_t)\left(x^3_t - z_t\right).
\end{aligned}
$$

Since the initial conditions are 0 for x^1, x^2, x^3 we have a zero initial condition for z which gives $x^3_t \equiv z_t,\ \forall t > \tau$. But $\{x^3_t\}_t$ is bounded for any bounded $\{u_t\}_t$; hence, $\{z_t\}$ is bounded for any bounded $\{u_t\}_t$. We deduce that the system $z_{t+1} = A_t z_t + u_t$ satisfies the condition of Perron. Applying Theorem 2.40 we obtain that A_t defines an exponentially stable evolution what ends the proof.

Remark. The above theorem expresses exponential stability as a consequence of bounded input/bounded output (BIBO) stability under usual assumptions of uniform stabilizability and uniform detectability. The result is known (Anderson and Moore (1969), but the above proof due to V. Drăgan is much simpler than the previous ones).

2.9.2. Dissipativity and Stability

We shall consider here systems of the form

$$
\begin{aligned}
x_{t+1} &= f(x_t) + g(x_t)\, u_t \\
y_t &= h(x_t),
\end{aligned}
\qquad (2.102)
$$

where the functions f, g, h are continuous in some neighborhood D

of the equilibrium point $\hat{x}$ of the free system (with $u_t \equiv 0$); we have $\hat{x} = f(\hat{x})$. We shall assume also that $h(\hat{x}) = 0$.

A. We state first the following:

Theorem 2.42 *Assume there exist a constant $\gamma > 0$ and a function $\varphi : D \subset \mathbb{R}^n \longrightarrow \mathbb{R}_+$ such that for any $x_0 \in D$, any input sequence $\{u_t\}_t$ and for any $T > 0$ the following holds:*

$$\sum_{1}^{T} |y_t|^2 \leq \gamma^2 \sum_{0}^{T-1} |u_t|^2 + \varphi(x_0) , \qquad (2.103)$$

where $\{x_t\}_t$ is the state sequence corresponding to x_0 and to the input sequence and $y_t = h(x_t)$. Assume further that if $h(x_t) = 0$, $t = 0, 1, 2, ...$ for some solution x_t of the free system $x_{t+1} = f(x_t)$, then $x_t \equiv \hat{x}$. Then the equilibrium at $\hat{x}$ is a asymptotically stable.

Remark. The property in the first assumption is called ℓ^2-*gain lower than* γ (van der Schaft (1993)) and that of the second assumption is called *observability of* (h, f) *at* $\hat{x}$ (e.g., Byrnes and Lin (1994), Hill and Moylan (1974)).

In order to prove the above theorem we shall need first a preliminary result.

Proposition 2.43 *If (h, f) is observable at $\hat{x}$ then if x_t is a solution of the free system $x_{t+1} = f(x_t)$ and $y_t = h(x_t)$ then $\lim_{t \to \infty} y_t = 0$ implies $\lim_{t \to \infty} x_t = \hat{x}$.*

Proof. Let x_t be a solution of the free system located in a compact neighborhood of $\hat{x}$ such that $\lim_{t \to \infty} h(x_t) = 0$. Since $\{x_t\}_t$ belongs to a compact set there exist ω-limit points of this solution. If $\tilde{x}$ is such a limit point then there exists a sequence $t_k \to \infty$ such that $x_{t_k} \to \tilde{x}$. From the continuity of h it follows that $h(\tilde{x}) = 0$, i.e., $h(x)$ vanishes on the ω-limit points. Define the solution $\tilde{x}_t$ by $\tilde{x}_{t+1} = f(\tilde{x}_t)$, $\tilde{x}_0 = \tilde{x}$. From the invariance of the ω-limit sets it follows that $\tilde{x}_t$ are ω-limit points hence $h(\tilde{x}_t) \equiv 0$. Observability will then give $\tilde{x}_t \equiv \hat{x}$, hence $\tilde{x} = \hat{x}$. Since every limit point of the sequence x_t equals $\hat{x}$ it follows that $\lim_{t \to \infty} x_t = \hat{x}$ and the proof ends.

Proof of Theorem 2.42. Let $u_t = 0$, $t = \overline{0, T-1}$; from finite

ℓ^2-gain inequality we deduce

$$\sum_{0}^{T} |y_t|^2 \leq \varphi(x_0) .$$

Since the sum is bounded by a constant that does not depend on T, the series $\sum_{0}^{\infty} |y_t|^2$ is convergent; hence, $\lim_{t \to \infty} y_t = 0$ and this will give $\lim_{t \to \infty} x_t = \hat{x}$, i.e., attractivity of the equilibrium (we used observability and Proposition 2.43).

In order to prove Liapunov stability we shall define

$$V(x) = \sup_{T,u} \left\{ \sum_{0}^{T} |y_t|^2 - \gamma^2 \sum_{0}^{T-1} |u_t|^2 \right\} , \qquad (2.104)$$

where $y_t = h(x_t)$, $x_{t+1} = f(x_t) + g(x_t) u_t$, $x_0 = x$, the supremum being taken for all $T > 0$ and all finite sequences u_t, $t = \overline{0, T-1}$. The property of ℓ^2-gain lower than γ shows that $V(x)$ thus defined is bounded.

Let now $t > 0$; we define x_t using $x_0 = x$ and some input sequence $u_0, u_1, ..., u_{t-1}$. We may now construct $\tilde{x}_s$ by using $\tilde{x}_0 = x_t$ and an input sequence $\tilde{u}_0, \tilde{u}_1, ..., \tilde{u}_{\tilde{T}-1}$, the input sequence and $\tilde{T} > 0$ being arbitrary. We can now define

$$V(x_t) = V(\tilde{x}_0) = \sup_{\tilde{T},\tilde{u}} \left\{ \sum_{0}^{\tilde{T}} |\tilde{y}_s|^2 - \gamma^2 \sum_{0}^{\tilde{T}-1} |\tilde{u}_s|^2 \right\} .$$

On the other hand, given any $T > t$ we may construct x_i on $[0, T]$ with $x_0 = x$ and with the input sequence $u_0, u_1, ..., u_{t-1}, u_t, ..., u_{T-1}$ where $u_0, u_1, ..., u_t$ are those defining x_t. We consider now

$$\sum_{i=t}^{T} |y_i|^2 - \gamma^2 \sum_{i=t}^{T-1} |u_i|^2 = \sum_{s=0}^{T-t} |y_{t+s}|^2 - \gamma^2 \sum_{s=0}^{T-1-t} |u_{t+s}|^2 =$$

$$= \sum_{0}^{\tilde{T}} |\tilde{y}_s|^2 - \gamma^2 \sum_{0}^{\tilde{T}-1} |\tilde{u}_s|^2 ,$$

where we denoted $\tilde{T} = T - t$, $\tilde{u}_s = u_{t+s}$, $\tilde{x}_s = x_{t+s}$, $\tilde{y}_s = y_{t+s}$. It will follow that

$$V(x_t) = \sup_{\tilde{T}, \tilde{u}} \left\{ \sum_0^{\tilde{T}} |\tilde{y}_s|^2 - \gamma^2 \sum_0^{\tilde{T}-1} |\tilde{u}_s|^2 \right\} =$$

$$= \sup_{T > t, u} \left\{ \sum_t^T |y_s|^2 - \gamma^2 \sum_t^{T-1} |u_s|^2 \right\}$$

because of the coincidence of the two sets where *supremum* is taken.

From the definition of the *supremum* we deduce that for any $\varepsilon > 0$ there will exist $T(\varepsilon) > t + 1$ and a sequence $u_{t+1}^\varepsilon, u_{t+2}^\varepsilon, ..., u_{T-1}^\varepsilon$ such that

$$V(x_{t+1}) \le \sum_{t+1}^{T(\varepsilon)} |y_s^\varepsilon|^2 - \gamma^2 \sum_{t+1}^{T(\varepsilon)-1} |u_s^\varepsilon|^2 + \varepsilon$$

and also

$$V(x_t) \ge \sum_{t+1}^{T(\varepsilon)} |y_s^\varepsilon|^2 - \gamma^2 \sum_{t+1}^{T(\varepsilon)-1} |u_s^\varepsilon|^2 + |y_t|^2 - \gamma^2 |u_t|^2$$

for an arbitrary u_t. Therefore,

$$V(x_{t+1}) - V(x_t) + |y_t|^2 - \gamma^2 |u_t|^2 \le \varepsilon$$

and since $\varepsilon > 0$ is arbitrary we shall have

$$V(x_{t+1}) - V(x_t) + |y_t|^2 - \gamma^2 |u_t|^2 \le 0 \qquad (2.105)$$

for an arbitrary u_t. By taking $u_t = 0$ we obtain

$$V(x_{t+1}) - V(x_t) \le -|y_t|^2 = -|h(x_t)|^2 \le 0;$$

hence the sequence $\{V(x_t)\}_t$ is decreasing along the solution sequence of $x_{t+1} = f(x_t)$.

Take now $u_t = 0$, $t = \overline{0, T-1}$. We shall have

$$V(x) \ge \sum_0^T |h(x_t)|^2 = h^*(x) h(x) + h^*(f(x)) h(f(x)) +$$

$$+ \cdots + h^*(f^T(x)) h(f^T(x))$$

for every $T > 0$ where $f^t(x)$ means, as in Section 2.6, the t-iterate of $f(x)$. Obviously $V(x) \ge 0$. Assume there exists some $\hat{t}$ such that

$h(x_{\hat{t}}) \neq 0$, $x_0 = x$. Then $V(x) > 0$ since we may take in the above estimate $T > \hat{t}$. It follows that if $V(x) = 0$ then $h(x_t) \equiv 0$ for $x_0 = x$. Observability implies now $x_t \equiv \hat{x}$.

We obtained that $V(x) > 0$ for all $x \neq \hat{x}$ in the neighborhood D containing $\hat{x}$. We may now apply the theorem of Liapunov on stability (Theorem 2.16) and obtain stability of the equilibrium $\hat{x}$. The proof is complete.

Remark. The inequality

$$V(x_{t+1}) - V(x_t) \leq \gamma^2 |u_t|^2 - |y_t|^2$$

obtained above and denoted by (2.105) is called *dissipation inequality* (Willems (1972)). Its form is called *local form* while the following form obtained by summing from 0 to $T - 1$ is called "integral form"

$$V(x_T) \leq V(x_0) + \sum_{t=0}^{T-1} \left(\gamma^2 |u_t|^2 - |y_t|^2 \right). \tag{2.106}$$

In terms of dissipative systems theory the state function $V(x)$ is called *storage function* while $w(u, x) = \gamma^2 |u|^2 - |h(x)|^2$ is called *supply rate*. The dissipation inequality is usually written as

$$V(x_T) \leq V(x_0) + \sum_{t=0}^{T-1} w(u_t, x_t).$$

The above proved theorem shows in fact that a sufficient condition for Liapunov stability is the existence of a strictly positive storage function and of a supply rate that is nonpositive along the trajectories of the free system.

B. The results contained in Theorem 2.40 will allow one to extend the concept of *stability radius* to the nonlinear case.

Consider the system

$$x_{t+1} = f(x_t) \tag{2.107}$$

with the equilibrium $\hat{x}$ and assume this equilibrium is asymptotically stable. Associate the perturbed system

$$x_{t+1} = f(x_t) + g(x_t) \Delta(x_t) h(x_t) \tag{2.108}$$

with $h(\hat{x}) = 0$. We look for an estimate of the disturbance Δ for which the equilibrium $\hat{x}$ remains asymptotically stable. Such an estimate may be obtained if the associated input/output system (2.102) satisfies the assumptions in the previous theorem (Theorem 2.42). We may state

Theorem 2.44 *Consider system (2.108) and assume that the associated input/output system (2.102) satisfies the assumptions of Theorem 2.42. If $\hat{x}$ is an equilibrium of the free system (2.107) then it is an asymptotically stable equilibrium for the perturbed system (2.108) for all perturbations $\Delta(x)$ satisfying $|\Delta(x)|^2 < \delta^2$ with $\delta < 1/\gamma$.*

Proof. Indeed let $\{u_t\}_t$ be some input sequence , let x_t be the corresponding solution of

$$x_{t+1} = f(x_t) + g(x_t)[\Delta(x_t)h(x_t) + u_t], \quad x_0 = x$$

and let $y_t = h(x_t)$. The ℓ^2-gain assumption will give

$$\sum_0^T |y_t|^2 \leq \gamma^2 \sum_0^{T-1} |\Delta(x_t)h(x_t) + u_t|^2 + \varphi(x).$$

This inequality may be written as follows:

$$\sum_0^T y_t^* y_t - \gamma^2 \sum_0^{T-1} [y_t^* \Delta^*(x_t)\Delta(x_t)h(x_t) + y_t^* \Delta^*(x_t)u_t + \\ + u_t^* \Delta(x_t)h(x_t) + u_t^* u_t] - \varphi(x) \leq 0.$$

Taking into account the inequality for $\Delta(x)$ we deduce

$$\sum_0^T |y_t|^2 - \gamma^2 \sum_0^{T-1}\left(\delta^2 |y_t|^2 + 2\delta |y_t| |u_t| + |u_t|^2\right) - \varphi(x) \leq 0$$

and denoting $\lambda^2 = \sum_0^T |y_t|^2$, $\mu^2 = \sum_0^{T-1} |u_t|^2$ it follows that

$$\left(1 - \gamma^2 \delta^2\right)\lambda^2 - 2\gamma^2\delta\mu\lambda - \left(\gamma^2\mu^2 + \varphi(x)\right) \leq 0$$

for $\lambda \geq 0$. We deduce, since $\delta\gamma < 1$:

$$0 \leq \lambda \leq \frac{\gamma^2\delta\mu + \sqrt{\gamma^2\mu^2 + \left(1 - \gamma^2\delta^2\right)\varphi(x)}}{1 - \gamma^2\delta^2}$$

and from here

$$\lambda^2 \;\leq\; 2\frac{\gamma^4\delta^2\mu^2 + \gamma^2\mu^2 + \left(1 - \gamma^2\delta^2\right)\varphi\left(x\right)}{\left(1 - \gamma^2\delta^2\right)^2} =$$

$$= \; 2\frac{\gamma^2\left(1 + \gamma^2\delta^2\right)}{\left(1 - \gamma^2\delta^2\right)^2}\mu^2 + \frac{2}{1 - \gamma^2\delta^2}\varphi\left(x\right).$$

The above inequality reads

$$\sum_0^T |y_t|^2 \leq 2\frac{\gamma^2\left(1 + \gamma^2\delta^2\right)}{\left(1 - \gamma^2\delta^2\right)^2}\sum_0^{T-1}|u_t|^2 + \frac{2}{1 - \gamma^2\delta^2}\varphi\left(x\right)$$

which shows that the perturbed system satisfies the ℓ^2-gain assumption. Since (h, f) is observable at $\hat{x}$ it obviously follows that $(h, f + g\Delta h)$ is observable at $\hat{x}$. The assumptions of Theorem 2.42 are thus fulfilled. It follows that $\hat{x}$ is an asymptotically stable equilibrium for the perturbed system. (The fact that $\hat{x}$ is an equilibrium of the perturbed system follows from $h\left(\hat{x}\right) = 0$.) The proof is complete.

C. For the next result we have to recall some useful notions about inverse systems. If we denote by $(\sum)$ the basic system (2.102) and by (I) the identity system $y = u$ the system $(I - \sum)$ is defined by

$$\begin{aligned} x_{t+1} &= f\left(x_t\right) + g\left(x_t\right)u_t \\ y_t &= u_t - h\left(x_t\right). \end{aligned} \tag{2.109}$$

The inverse of (2.109) denoted $(I - \sum)^{-1}$ is

$$\begin{aligned} z_{t+1} &= f\left(z_t\right) + g\left(z_t\right)h\left(z_t\right) + g\left(z_t\right)w_t \\ v_t &= h\left(z_t\right) + w_t. \end{aligned} \tag{2.110}$$

The two systems are indeed one the inverse of the other. Take in (2.109) $u_t = v_t = h\left(z_t\right) + w_t$, i.e., apply to the input of (2.109) the output of (2.110); assume also that the initial states coincide. We deduce

$$z_{t+1} - x_{t+1} = f\left(z_t\right) - f\left(x_t\right) + \left(g\left(z_t\right) - g\left(x_t\right)\right)\left(h\left(z_t\right) + w_t\right)$$

and since $z_0 = x_0$ a simple induction argument shows that $z_t \equiv x_t$. The output of this tandem will be

$$y_t = u_t - h\left(x_t\right) = h\left(z_t\right) + w_t - h\left(x_t\right) \equiv w_t,$$

i.e., the resulting system is (I). The same result is obtained if output of (2.109) is applied to the input of (2.110), i.e., if $w_t = y_t = u_t - h(x_t)$ provided $z_0 = x_0$.

We may now state

Theorem 2.45 *If the assumptions of Theorem 2.42 hold with $\gamma < 1$ and $\hat{x}$ is an equilibrium of the free system $(\sum)$ then it is an asymptotically stable equilibrium of the free system $(I - \sum)^{-1}$*

Proof. The free system $(I - \sum)^{-1}$ is given by

$$z_{t+1} = f(z_t) + g(z_t) h(z_t)$$

being a perturbed version of the free system $(\sum)$: $x_{t+1} = f(x_t)$ by a structured perturbation $\Delta(x) \equiv 1$ - see (2.108). We may apply Theorem 2.44 and obtain the required result since $\gamma^{-1} > 1$ and $|\Delta(x)| \equiv 1 < \gamma^{-1}$.

D. The framework of this section allows to consider a *small gain theorem.*

Let

$$\begin{aligned} x_{t+1}^i &= f^i(x_t^i) + g^i(x_t^i) u_t^i \\ y_t^i &= h^i(x_t^i) , \quad i = 1,2 \end{aligned} \tag{2.111}$$

be two systems that have ℓ^2-gain lower that γ_i, $i = 1,2$ respectively. We assume satisfied all basic assumptions of this section for both systems: continuity of f^i, g^i, h^i in some neighborhoods D_i of the equilibria $\hat{x}^i = \hat{f}^i(\hat{x}^i)$ and $h^i(\hat{x}^i) = 0$, $i = 1,2$.

The feedback connection of the two systems means taking $u^1 = y^2$, $u^2 = y^1$, i.e., the output of each system is fed to the input of the other. The resulting system is an autonomous one

$$\begin{aligned} x_{t+1}^1 &= f^1(x_t^1) + g^1(x_t^1) h^2(x_t^2) \\ x_{t+1}^2 &= f^2(x_t^2) + g^2(x_t^2) h^1(x_t^1) \end{aligned} \tag{2.112}$$

This system has $(\hat{x}^1, \hat{x}^2)$ as equilibrium. We may state

Theorem 2.46 *If each of the systems (2.111) satisfies the assumptions of Theorem 2.42, i.e., it has ℓ^2-gain lower than γ_i and (h^i, f^i) are observable at $\hat{x}^i$, then the equilibrium $(\hat{x}^1, \hat{x}^2)$ of (2.112) is asymptotically stable provided the small gain condition $\gamma_1 \gamma_2 < 1$ holds.*

Proof. Consider the following system with 2 inputs and 2 outputs:

$$x_{t+1}^1 = f^1\left(x_t^1\right) + g^1\left(x_t^1\right) h^2\left(x_t^2\right) + g^1\left(x_t^1\right) u_t^1$$
$$x_{t+1}^2 = f^2\left(x_t^2\right) + g^2\left(x_t^2\right) h^1\left(x_t^1\right) + g^2\left(x_t^2\right) u_t^2 \qquad (2.113)$$
$$y_t^1 = h^1\left(x_t^1\right) , \ y_t^2 = h^2\left(x_t^2\right)$$

This system has the ℓ^2-gain property in the sense that there exist $\gamma > 0$ and $\Psi\left(x_0^1, x_0^2\right)$ such that

$$\sum_1^T \left(\left|y_t^1\right|^2 + \left|y_t^2\right|^2\right) \leq \gamma^2 \sum_0^{T-1} \left(\left|u_t^1\right|^2 + \left|u_t^2\right|^2\right) + \Psi\left(x_0^1, x_0^2\right)$$

for any $T > 0$. This property will be obtained by using the assumption that $\gamma_1 \gamma_2 < 1$. We shall have, using the ℓ^2-gain property of systems (2.112)

$$\sum_1^T \left(\left|y_t^1\right|^2 + \left|y_t^2\right|^2\right) \leq \gamma_1^2 \sum_0^{T-1} \left|h^2\left(x_t^2\right) + u_t^1\right|^2 + \varphi^1\left(x_0^1\right) +$$

$$+\gamma_2^2 \sum_0^{T-1} \left|h^1\left(x_t^1\right) + u_t^2\right|^2 + \varphi^2\left(x_0^2\right) \leq \gamma_1^2 \sum_1^T \left|y_t^2\right|^2 + 2\gamma_1^2 \sum_0^{T-1} y_t^2 u_t^1 +$$

$$+\gamma_1^2 \sum_0^{T-1} \left|u_t^1\right|^2 + \gamma_1^2 \left|h^2\left(x_0^2\right)\right|^2 + \varphi^2\left(x_0^2\right) + \gamma_2^2 \sum_1^T \left|y_t^1\right|^2 +$$

$$+2\gamma_2^2 \sum_0^{T-1} y_t^1 u_t^2 + \gamma_2^2 \sum_0^{T-1} \left|u_t^2\right|^2 + \gamma_2^2 \left|h^1\left(x_0^1\right)\right|^2 + \varphi^1\left(x_0^1\right) \leq$$

$$\leq \gamma_1^2 \gamma_2^2 \sum_1^T \left(\left|y_t^1\right|^2 + \left|y_t^2\right|^2\right) + 2\gamma_2^2\left(1 + \gamma_1^2\right) \sum_0^{T-1} y_t^1 u_t^2 +$$

$$+2\gamma_1^2\left(1 + \gamma_2^2\right) \sum_0^{T-1} y_t^2 u_t^1 + \gamma_1^2\left(1 + \gamma_2^2\right) \sum_0^{T-1} \left|u_t^1\right|^2 +$$

$$+\gamma_2^2\left(1 + \gamma_1^2\right) \sum_0^{T-1} \left|u_t^2\right|^2 + \gamma_2^2\left(1 + \gamma_1^2\right) \left|h^1\left(x_0^1\right)\right|^2 +$$

$$+ \left(1 + \gamma_2^2\right)^2 \varphi^1 \left(x_0^1\right) + \gamma_1^2 \left(1 + \gamma_2^2\right) \left|h^2 \left(x_0^2\right)\right|^2 + \left(1 + \gamma_1^2\right) \varphi^2 \left(x_0^2\right)$$

Using Cauchy-Schwarz inequality it follows that

$$\sum_0^{T-1} y_t^1 u_t^2 \leq \left(\sum_1^T \left|y_t^1\right|^2\right)^{1/2} \left(\sum_0^{T-1} \left|u_t^2\right|^2\right)^{1/2} +$$

$$+ \left(\sum_0^{T-1} \left|u_t^2\right|^2\right)^{1/2} \left|h^1 \left(x_0^1\right)\right| \leq \left(\sum_1^T \left|y_t^1\right|^2\right)^{1/2} \left(\sum_0^{T-1} \left|u_t^2\right|^2\right)^{1/2} +$$

$$+ \frac{1}{2} \left(\sum_0^{T-1} \left|u_t^2\right|^2 + \left|h^1 \left(x_0^1\right)\right|^2\right) ,$$

$$\sum_0^{T-1} y_t^2 u_t^1 \leq \left(\sum_1^T \left|y_t^2\right|^2\right)^{1/2} \left(\sum_0^{T-1} \left|u_t^1\right|^2\right)^{1/2} +$$

$$+ \frac{1}{2} \left(\sum_0^{T-1} \left|u_t^1\right|^2 + \left|h^2 \left(x_0^2\right)\right|^2\right)$$

Substituting in the above inequality will give

$$\left(1 - \gamma_1^2 \gamma_2^2\right) \sum_1^T \left(\left|y_t^1\right|^2 + \left|y_t^2\right|^2\right) \leq$$

$$\leq 2\gamma_2^2 \left(1 + \gamma_1^2\right) \left(\sum_1^T \left|y_t^1\right|^2\right)^{1/2} \left(\sum_0^{T-1} \left|u_t^2\right|^2\right)^{1/2} +$$

$$+ 2\gamma_1^2 \left(1 + \gamma_2^2\right) \left(\sum_1^T \left|y_t^2\right|^2\right)^{1/2} \left(\sum_0^{T-1} \left|u_t^1\right|^2\right)^{1/2} +$$

$$+ 2\gamma_1^2 \left(1 + \gamma_2^2\right) \left(\sum_0^{T-1} \left|u_t^1\right|^2\right)^{1/2} + 2\gamma_2^2 \left(1 + \gamma_1^2\right) \sum_0^{T-1} \left|u_t^2\right|^2 +$$

$$+ 2\gamma_2^2 \left(1 + \gamma_1^2\right) \left|h^1 \left(x_0^1\right)\right|^2 + \left(1 + \gamma_2^2\right) \varphi^1 \left(x_0^1\right) +$$

$$+2\gamma_1^2\left(1+\gamma_2^2\right)\left|h^2\left(x_0^2\right)\right|^2+\left(1+\gamma_1^2\right)\varphi^2\left(x_0^2\right).$$

In the following some notations are useful:

$$\|u^i\|^2 \;=\; \sum_0^{T-1}|u_t^i|^2,\; \|y^i\|^2=\sum_0^{T-1}|y_t^i|^2,$$

$$\Psi^i\left(x_0^i\right) \;=\; 2\left|h^i\left(x_0^i\right)\right|^2+\varphi^i\left(x_0^i\right),\; i=1,2;$$

$$\|u\|^2 \;=\; \|u^1\|^2+\|u^2\|^2,\; \|y\|^2=\|y^1\|^2+\|y^2\|^2.$$

With these notations the last inequality becomes

$$\left(1-\gamma_1^2\gamma_2^2\right)\|y\|^2\le 2\left(1+\gamma_1^2+\gamma_2^2\right)\left(\|y^1\|\,\|u^2\|+\|y^2\|\,\|u^1\|\right)+$$
$$+2\left(1+\gamma_1^2+\gamma_2^2\right)\|u\|^2+\left(1+\gamma_1^2+\gamma_2^2\right)\left(\Psi^1\left(x_0^1\right)+\Psi^2\left(x_0^2\right)\right).$$

We use again Cauchy-Schwarz-Buniakowski inequality

$$\|y^1\|\,\|u^2\|+\|y^2\|\,\|u^1\| \;\le\; \left(\|y^1\|^2+\|y^2\|^2\right)^{1/2}\left(\|u^1\|^2+\|u^2\|^2\right)^{1/2}$$
$$=\; \|y\|\,\|u\|.$$

This will give

$$\frac{1-\gamma_1^2\gamma_2^2}{1+\gamma_1^2+\gamma_2^2}\|y\|^2-2\|u\|\,\|y\|-2\|u\|^2-\left(\Psi^1\left(x_0^1\right)+\Psi^2\left(x_0^2\right)\right)\le 0$$

and since $\gamma_1^2\gamma_2^2<1$ we obtain

$$\|y\| \le \frac{\|u\|+\sqrt{\|u\|^2+\dfrac{\left(1-\gamma_1^2\gamma_2^2\right)\left(2\|u\|^2+\left(\Psi^1\left(x_0^1\right)+\Psi^2\left(x_0^2\right)\right)\right)}{\left(1+\gamma_1^2+\gamma_2^2\right)}}}{\left(1-\gamma_1^2\gamma_2^2\right)\left(1+\gamma_1^2+\gamma_2^2\right)^{-1}}$$

and from here

$$\|y\| \;\le\; \frac{1+\sqrt{1+2\left(1-\gamma_1^2\gamma_2^2\right)\left(1+\gamma_1^2+\gamma_2^2\right)^{-1}}}{\left(1-\gamma_1^2\gamma_2^2\right)\left(1+\gamma_1^2+\gamma_2^2\right)^{-1}}\|u\|+$$
$$+\sqrt{\frac{\Psi^1\left(x_0^1\right)+\Psi^2\left(x_0^2\right)}{\left(1-\gamma_1^2\gamma_2^2\right)\left(1+\gamma_1^2+\gamma_2^2\right)^{-1}}};$$

hence,

$$\|y\|^2\le\gamma^2\|u\|^2+\Psi\left(x_0^1,x_0^2\right).$$

We show next that for system (2.113) we have observability at $(\hat{x}^1, \hat{x}^2)$. Let $u_t^i \equiv 0$ and assume that for some solution (x_t^1, x_t^2) of the free system (2.113) we have $h^1(x_t^1) \equiv 0$, $h^2(x_t^2) \equiv 0$. It follows that x_t^i are solutions of $x_{t+1}^i = f^i(x_t^i)$, $i = 1, 2$. From observability of (h^i, f^i) at $\hat{x}^i$, $i = 1, 2$, we deduce that $x_t^i \equiv \hat{x}^i$, $i = 1, 2$; hence, $(x_t^1, x_t^2) \equiv (\hat{x}^1, \hat{x}^2)$. Observability of (2.113) is proved. Further, if $u_t^1 \equiv 0$, $u_t^2 \equiv 0$ the ℓ^2-gain inequality for system (2.113) will give the convergence of the series $\sum_0^\infty \left(|y_t^1|^2 + |y_t^2|^2\right)$ hence $\lim\limits_{t \to \infty} y_t^1 = \lim\limits_{t \to \infty} y_t^2 = 0$. Since system (2.113) is observable at $(\hat{x}^1, \hat{x}^2)$ it follows from Proposition 2.43 that $\lim\limits_{t \to \infty} (x_t^1, x_t^2) = (\hat{x}^1, \hat{x}^2)$ and attractivity is proved.

In order to prove Liapunov stability we consider the Liapunov function

$$V\left(x^1, x^2\right) = \sup_{T, u} \left\{ \sum_0^T \left(|y_t^1|^2 + |y_t^2|^2\right) - \gamma^2 \sum_0^{T-1} \left(|u_t^1|^2 + |u_t^2|^2\right) \right\}.$$

Doing exactly as in the proof of Theorem 2.42 we shall obtain

$$V\left(x_{t+1}^1, x_{t+1}^2\right) - V\left(x_t^1, x_t^2\right) \leq -\left(|y_t^1|^2 + |y_t^2|^2\right) + \gamma^2 \left(|u_t^1|^2 + |u_t^2|^2\right)$$

and the Liapunov function is decreasing along the solutions of the free system (2.113). Now if $u_t^i = 0$, $t = \overline{0, T-1}$ then we shall have

$$V\left(x^1, x^2\right) \geq \sum_0^T \left(|h^1(x_t^1)|^2 + |h^2(x_t^2)|^2\right).$$

As in the case of Theorem 2.42 if $V\left(x^1, x^2\right) = 0$ then $h^i(x_t^i) \equiv 0$, $i = 1, 2$ and observability will give $x_t^i \equiv \hat{x}^i$, $i = 1, 2$; hence, $V\left(x^1, x^2\right) > 0$ for all $x^i \neq \hat{x}^i$, $i = 1, 2$ in the neighborhood of the equilibrium $(\hat{x}^1, \hat{x}^2)$. Application of the theorem of Liapunov on stability (Theorem 2.16) ends the proof.

2.9.3. Stabilization of Bilinear Systems

Consider the discrete-time bilinear control system described by

$$x_{t+1} = Ax_t + \sum_1^m \left(d^i + B^i x_t\right) u_t^i \qquad (2.114)$$

where $x \in \mathbb{R}^n$, $A, B^i, i = \overline{1, m}$ are constant $n \times n$ matrices and $d^i, i =$

$\overline{1,m}$ are constant n -vectors.

We assume that the free system

$$x_{t+1} = A x_t$$

is Liapunov stable but not asymptotically stable. Since in applications asymptotic stability is important, the following problem is naturally stated: find a state feedback law

$$u^i = f^i(x) , \quad f^i(0) = 0 , \ i = \overline{1,m}$$

such that the resulting closed loop system

$$x_{t+1} = A x_t + \sum_1^m \left(d^i + B^i x_t \right) f^i(x_t) \tag{2.115}$$

has the zero solution globally asymptotically stable.

A. Following Lin and Byrnes (1994, 1995) but also some earlier work in the field (e.g., Slemrod (1978), Halanay and Răsvan (1980)) we remark that since the free linear system is stable it follows as a consequence of Propositions 2.4 and 2.5 that there exists a positive definite matrix P such that the matrix $A^* P A - P$ is negative semi-definite.

We consider the Liapunov function $V(x) = x^* P x$ of the free system and compute its variation along the solutions of the controlled bilinear system (2.114). To do this some notations are useful. If we denote by u the control vector with the entries u^i, $i = \overline{1,m}$, with D the matrix having as columns the vectors d^i, $i = \overline{1,m}$ and with $B(x)$ the matrix having as columns the vectors $B^i x$, $i = \overline{1,m}$, then system (2.114) may be written as

$$x_{t+1} = A x_t + (D + B(x_t)) u_t . \tag{2.116}$$

We have further

$$V(x_{t+1}) - V(x_t) =$$
$$= (A x_t + (D + B(x_t)) u_t)^* P (A x_t + (D + B(x_t)) u_t) - x_t^* P x_t =$$
$$= x_t^* (A^* P A - P) x_t + u_t^* (D + B(x_t))^* P A x_t +$$
$$+ x_t^* A^* P (D + B(x_t)) u_t + u_t^* (D + B(x_t))^* P (D + B(x_t)) u_t \tag{2.117}$$

We chose u from the following equality

$$u = - (D + B(x))^* P A x - \frac{1}{2} (D + B(x))^* P (D + B(x)) u$$

which becomes

$$\left[\mathbf{I} + \frac{1}{2} \left(D + B\left(x\right)\right)^* P \left(D + B\left(x\right)\right) \right] u = - \left(D + B\left(x\right)\right)^* P A x \, .$$

Remark that matrix $\mathbf{I} + \frac{1}{2} \left(D + B\left(x\right)\right)^* P \left(D + B\left(x\right)\right)$ is invertible due to the strict positivity of both $\mathbf{I}$ - the identity matrix - and P. We deduce the state feedback control law

$$u\left(x\right) = - \left[\mathbf{I} + \frac{1}{2} \left(D + B\left(x\right)\right)^* P \left(D + B\left(x\right)\right) \right]^{-1} \left(D + B\left(x\right)\right)^* P A x \, .$$
(2.118)

With this choice of the control we obtain

$$V\left(x_{t+1}\right) - V\left(x_t\right) = x_t^* \left(A^* P A - P\right) x_t -$$
$$-2 \left| \left[\mathbf{I} + \frac{1}{2} \left(D + B\left(x_t\right)\right)^* P \left(D + B\left(x_t\right)\right) \right]^{-1} \left(D + B\left(x_t\right)\right)^* P A x_t \right|^2 \leq 0$$
(2.119)

which gives Liapunov stability of the zero solution of the closed loop system

$$x_{t+1} = A x_t - \left(D + B\left(x_t\right)\right) \left[\mathbf{I} + \frac{1}{2} \left(D + B\left(x_t\right)\right)^* P \left(D + B\left(x_t\right)\right) \right]^{-1} \times$$
$$\times \left(D + B\left(x_t\right)\right) P A x_t \, .$$
(2.120)

In order to obtain global asymptotic stability of the zero solution we shall use Barbashin-Krasovskii theorem for discrete-time systems (Corollary 2.31). The set where $V\left(x_t\right)$ is constant for any $t \in \mathbb{N}$ is defined by

$$x_t^* \left(A^* P A - P\right) x_t = 0 \, , \ \left(D + B\left(x_t\right)\right) P A x_t = 0 \, , \ t = 0, 1, 2, \dots \, ,$$

where x_t is a solution of the free system. This set clearly coincides with the set $\Omega \cap S$ where we denoted

$$\Omega = \left\{ x \in \mathbb{R}^n : \left(A^t x\right)^* \left(A^* P A - P\right) \left(A^t x\right) = 0, \ t = 0, 1, 2, \dots \right\}$$
$$S = \left\{ x \in \mathbb{R}^n : \left(A^{t+1} x\right)^* P \left(D + B\left(A^t x\right)\right) = 0, \ t = 0, 1, 2, \dots \right\}$$
(2.121)

Remark that both Ω and S hence $\Omega \cap S$ are invariant with respect to the free system $x_{t+1} = A x_t$. Since $P > 0$, $V\left(x\right) > \lambda_{\min} \left| x \right|^2$ where $\lambda_{\min} > 0$ is the smallest eigenvalue of P. If $\Omega \cap S = \{0\}$ then the closed loop system (2.120) has the zero solution globally asymptotically stable. We are in position to state now

Theorem 2.47   *Consider the bilinear control system* (2.116); *assume the free system is stable. Then the state feedback control law* (2.118) *makes the zero solution of the closed loop system* (2.120) *globally asymptotically stable provided* $\Omega \cap S = \{0\}$ *where* Ω *and* S *are defined by* (2.121).

B. This result has an interesting significance in terms of dissipative systems. We may consider the following bilinear controlled system with output

$$x_{t+1} = Ax_t + (D + B(x_t)) u_t$$
$$y_t = (D + B(x_t))^* PAx_t + \tfrac{1}{2}(D + B(x_t))^* P(D + B(x_t)) u_t$$
$$\text{(2.122)}$$

If $V(x) = x^* Px$ is associated then equality (2.117) and the fact that $A^* PA - P \leq 0$ will give

$$V(x_{t+1}) - V(x_t) \leq u_t^* y_t + y_t^* u_t \tag{2.123}$$

and this is *a dissipation inequality with $V(x)$ as a storage function and $w(u, y) = u^* y + y^* u$ as a supply rate.* The result of Theorem 2.47 may be viewed as follows: the control u is chosen as $u = -y$ and this will give

$$V(x_{t+1}) - V(x_t) \leq -2 |y_t|^2 \; ;$$

hence, system (2.122) with unitary "negative" feedback has the zero solution Liapunov stable.

Function $V(x)$ is constant along those solutions for which output y vanishes and this implies that they are solutions of the free linear system. For the asymptotic stability of the zero solution these solutions have to approach zero asymptotically. In fact, we need $\lim_{t \to \infty} x_t = 0$ for those solutions of $x_{t+1} = Ax_t$ such that

$$h(x_t) = (D + B(x_t))^* PAx_t = 0 \, , \, t = 0, 1, 2, \dots .$$

This property is called *detectability* at 0 for the couple $((D + B(x_t))^* PAx, Ax)$ - see also previous section 2.9.2.

Summarizing, the above result represents an application of a fairly more general result which states that a dissipative system with unitary "negative" feedback has the zero solution asymptotically stable provided it is detectable at zero with respect to its output (Lin and Byrnes (1995)). The special structure of the system - linear in state and bilinear in state and control - allows one to express detectability in a "checkable" form (Lin and Byrnes (1994)).

CHAPTER 3

ABSOLUTE STABILITY OF CONTROL SYSTEMS

The problem of absolute stability of nonlinear control systems has been formulated more than 50 years ago. If some physical system is described by the evolution of the deviations with respect to a steady state (equilibrium), its equations have the origin of the coordinates as equilibrium. If this zero equilibrium is stable, then the considered steady state (with respect to which the deviations are written) is also stable. If the zero equilibrium is globally asymptotically stable, there are no other stable equilibria and from the stability point of view the considered system while nonlinear is much like the linear one.

If the information about the nonlinearity is quite poor then we may consider an entire class of nonlinear functions and try to obtain global asymptotic stability conditions which are valid for this class of nonlinearities (and associated systems). This is called absolute stability and defines some robustness of the stability property with respect to the nonlinearity. This robustness property has been seen to depend upon the linear part and this fact become clear only when V. M. Popov stated his frequency domain condition for absolute stability (Popov (1959), (1961)). The problem was stated first for continuous-time systems of the form

$$\dot{x} = Ax - b\varphi\left(c^*x\right) \tag{3.1}$$

with $\varphi(0) = 0$. Clearly $x = 0$ is an equilibrium point. The only available information about $\varphi : \mathbb{R} \to \mathbb{R}$ was that it was restricted to a sector of the plane (σ, φ)

$$0 \leq \frac{\varphi(\sigma)}{\sigma} \leq \bar{\varphi} \tag{3.2}$$

119

If we think about (3.1) as defined for each $\varphi(\sigma)$ satisfying (3.2), then we associate with each nonlinear function of the sector the corresponding nonlinear system, defining in this way a class of systems (3.1) with $\varphi(\sigma)$ in the class defined by (3.2). Such a situation arises if we start with a linear control system:

$$\dot{x} = Ax + b\mu$$

and apply a feedback control $\mu = -\varphi(\sigma)$ where σ is a measurable output. Since the control device is not always precisely modelled mathematically, in order to analyze the behavior of the system, we may only use the available information on the feedback control which may be reduced, for instance, to the sector condition above.

Let $x(t, x_0)$ be a solution of (3.1) - for fixed φ. Consider the linear controlled system:

$$\begin{aligned} \dot{z} &= Az + b\mu(t) \\ \sigma &= c^* z \end{aligned} \qquad (3.3)$$

with the control defined by $\mu(t) = -\varphi(c^* x(t, x_0))$. It is easy to prove that $z(t)$, the solution of (3.3) with this control function and with $z(0) = x_0$ coincides with $x(t, x_0)$. For this reason we can consider (3.1) as composed of the linear part defined by (3.3) and the nonlinear control $\mu = -\varphi(\sigma)$. From the uncertainty/robustness point of view the equations of the linear part are considered certain while the nonlinear control defined by the class (3.2) is considered uncertain. It is now obvious that robustness of the stability property has to be understood as the validity of the property for the entire class, the stability conditions being expressed in the language of certain elements - the parameters of the linear part and of the sector to which the nonlinear functions belong.

For instance, if we define the following rational function of a complex variable

$$\gamma(s) = c^* (sI - A)^{-1} b \qquad (3.4)$$

called the *transfer function* of (3.3), then the frequency domain inequality of Popov (1959, 1961) reads

$$\frac{1}{\overline{\varphi}} + \mathrm{Re}\,(1 + i\omega q)\,\gamma(i\omega) > 0\,, \quad \omega \in \mathbb{R}. \qquad (3.5)$$

If A in (3.1) is a Hurwitz matrix then (3.5) is a sufficient condition for the absolute stability of (3.1) in the class defined by (3.2). Obviously

(3.5) depends on the certain elements $A, b, c, \bar{\varphi}$ only.

During the last half-century the absolute stability problem has known various developments among which is the version on discrete time systems. This version has several motivations but we cite only such applications as impulse systems with amplitude modulation (see e.g., Gelig (1982), (1993)) and computer controlled systems (e.g., Mahmoud and Singh (1984)). We shall describe briefly both these applications.

In the case of the amplitude modulation the control signal $\mu(t)$ in (3.3) is obtained as the output of a modulator described as follows (see e.g., Gelig (1993), p. 10)

$$\mu(t) = \begin{cases} \dfrac{1}{\tau} \varphi(\sigma(nT)) & , \quad nT \le t < nT + \tau \\ 0 & , \quad nT + \tau \le t < (n+1)T \end{cases} \tag{3.6}$$

where $\tau \in (0, T)$ is the pulse width, $\varphi(\sigma)$ is a continuous bounded function, $\varphi(0) = 0$, $\varphi(\sigma) > 0$ for $\sigma > 0$; by $T > 0$ it is denoted the sampling period. The input signal of the modulator is defined only for $t = nT$, $n \in \mathbb{Z}$, being a sampled signal. If this sampled signal is the output signal of the linear continuous time system (3.3) but measured at $t = nT$, then we have a feedback system with amplitude modulation of the pulses. In the sequel we shall show that only the sampling instances nT are of interest and we may associate with the system defined by (3.3) and (3.6) a discrete time system.

Indeed, consider (3.3) with the input defined by (3.6). The variation of constants formula will give

$$z((n+1)T) = e^{AT} z(nT) + \frac{1}{\tau} \int\limits_{nT}^{nT+\tau} e^{A((n+1)T-\theta)} b\varphi(\sigma(nT))\, d\theta$$

where $\sigma(nT)$ is some sampled input signal. A straightforward computation gives

$$z((n+1)T) = e^{AT} z(nT) + \frac{1}{\tau} e^{AT} A^{-1} \left(\mathbf{I} - e^{-A\tau} \right) b\varphi(\sigma(nT))$$

If $\sigma(t)$ - the input to the modulator - is the output of (3.3) - $\sigma = c^* z$ - then denoting $z(nT) = z_n$, $\sigma(nT) = \sigma_n$, the above equation becomes

$$z_{n+1} = A_d z_n - b_d \varphi(c^* z_n) \tag{3.7}$$

where we denoted $A_d = e^{AT}$, $b_d = \dfrac{1}{\tau}\left(e^{A(T-\tau)} - e^{AT}\right)A^{-1}b$.

b. Consider now the problem of controlling (3.3) by using a digital computer and a nonlinear actuator:

$$\begin{aligned} \dot{z} &= Az + b\psi\left(\mu_c\right) \\ \sigma &= c^* z \end{aligned} \tag{3.8}$$

The control signal μ_c is generated by the digital computer together with a digital - to - analogue converter modelled as a Zero - Order - Hold as follows: the output signal of (3.8) measured at the sampling instants $n\delta$ is compared to a digital reference (which for convenience is considered constant); the generated error is processed in the computer according to an algorithm that defines the digital compensator. The output of the compensator is obtained synchronously with the measurements, i.e., at the sampling instants $n\delta$ and kept constant between the sampling instants, generating in this way the piecewise constant control signal μ_c. All this computer control structure is described as follows:

$$\begin{aligned} \varepsilon\left(k\delta\right) &= \sigma_r - \sigma\left(k\delta\right) \\ z_c\left((k+1)\,\delta\right) &= A_c z_c\left(k\delta\right) + b_c\varepsilon\left(k\delta\right) \\ \sigma_c\left(k\delta\right) &= f_c^* z_c\left(k\delta\right) + \gamma_c\varepsilon\left(k\delta\right) \\ \mu_c\left(t\right) &= \sigma_c\left(k\delta\right), \quad k\delta \le t < (k+1)\,\delta \end{aligned} \tag{3.9}$$

Using as previously the variation of constants formula to (3.8) we have

$$z\left((n+1)\,\delta\right) = e^{AT}z\left(n\delta\right) + \frac{1}{\tau}e^{AT}A^{-1}\left(\mathbf{I} - e^{-A\tau}\right)b\varphi\left(\sigma\left(n\delta\right)\right) \tag{3.10}$$

The equations (3.9) and (3.10) show that only the sampling instants define the dynamic behavior. Therefore we denote

$$z\left(k\delta\right) = z_k, \quad z_c\left(k\delta\right) = z_k^c, \quad \sigma\left(k\delta\right) = \sigma_k, \quad \varepsilon\left(k\delta\right) = \varepsilon_k$$

and find the following discrete time system

$$\begin{aligned} \left(\begin{array}{c} z_{k+1} \\ z_{k+1}^c \end{array}\right) &= \left(\begin{array}{cc} e^{AT} & 0 \\ -b_c c^* & A_c \end{array}\right)\left(\begin{array}{c} z_k \\ z_k^c \end{array}\right) + \\ + \left(\begin{array}{c} \left(e^{AT} - \mathbf{I}\right)A^{-1}b \\ 0 \end{array}\right) &\psi\left(-\gamma_c c^* z_k + f_c^* z_k^c + \gamma_c\sigma_r\right) + \left(\begin{array}{c} 0 \\ b_c \end{array}\right)\sigma_r \end{aligned}$$

which has the form

$$X_{k+1} = A_d X_k + b_d \psi \left(c_d^* X_k + \gamma_c \sigma_r \right) + q \sigma_r$$

with $X = \mathrm{col}(z, z^c)$ and appropriate notations. This system has a steady state solution, imposed by the constant reference signal σ_r and given by

$$\hat{X} = A_d \hat{X} + b_d \psi \left(c_d^* \hat{X} + \gamma_c \sigma_r \right) + q \sigma_r$$

The aim of the feedback control is to maintain the system in a neighborhood of the steady state, i.e., to maintain small the deviation with respect to this steady state. Therefore we are more interested in the deviations than in the states themselves. Consequently we write down the system in deviations. Denoting $x = X - \hat{X}$ it follows that:

$$x_{k+1} = A_d x_k - b_d \left[\psi \left(c_d^* \hat{X}_k + \gamma_c \sigma_r \right) - \psi \left(c_d^* x_k + c_d^* \hat{X}_k + \gamma_c \sigma_r \right) \right]$$

With the final notation

$$\varphi(\sigma) = \psi \left(c_d^* \hat{X}_k + \gamma_c \sigma_r \right) - \psi \left(\sigma + c_d^* \hat{X}_k + \gamma_c \sigma_r \right)$$

we find

$$x_{k+1} = A_d x_k - b_d \varphi \left(c_k^* x_k \right) \tag{3.11}$$

which is like (3.7).

3.1. THE SIMPLEST ABSOLUTE STABILITY CRITERION OF TSYPKIN

This seems to be the oldest extension of absolute stability ideas to the case of discrete systems, being published more than three decades ago (Tsypkin (1962)). The model problem is a feedback system containing a linear continuous time block and an impulse - amplitude modulator - see above. The discrete time equations describing this system are

$$x_{t+1} = A x_t - b \varphi \left(c^* x_t \right) , \tag{3.12}$$

where $\varphi(\sigma)$ is subject to the following sector condition

$$0 \leq \varphi(\sigma) \sigma \leq \bar{\varphi} \sigma^2 \tag{3.13}$$

and A has the eigenvalues inside the unit disk $D = \{z \in \mathbb{C} \mid |z| < 1\}$ i.e., the linear controlled system

$$x_{t+1} = Ax_t + b\mu_t , \quad \sigma = c^*x \tag{3.14}$$

is exponentially stable. We already mentioned that (3.12) may be viewed as (3.14) with the control $\mu = -\varphi(\sigma)$. It can be seen from (3.13) that the control thus defined satisfies the quadratic inequality

$$\mu(\mu + \bar{\varphi}\sigma) \le 0 \tag{3.15}$$

Following Tsypkin (1962) - in fact Popov (1959) - consider some $T > 0$ and define μ_t^T as follows:

$$\mu_t^T = \begin{cases} -\varphi(\sigma_t) = -\varphi(c^*x_t) & , \quad 0 \le t \le T-1 \\ 0 & , \quad \text{elsewhere} \end{cases} \tag{3.16}$$

where x_t is a solution of (3.12). Let the input to (3.14) be μ_t^T thus defined. The representation (variation of constants) formula will give for the corresponding solution of (3.14)

$$x_t^T = A^t x_0 + \sum_0^{t-1} A^{t-1-k} b\mu_k^T , \quad t \ge 1 \tag{3.17}$$

$$\sigma_t^T = c^* x_t^T = c^* A^t x_0 + \sum_0^{t-1} \left(c^* A^{t-1-k} b \right) \mu_k^T , \quad t \ge 1 \tag{3.18}$$

Because of the assumptions about the eigenvalues of A we have

$$\left| c^* A^t x_0 \right| \le \beta |c| \rho^t |x_0| \; ; \; \left| c^* A^t b \right| \le \beta |c| |b| \rho^t \, , \, 0 < \rho < 1$$

and, taking into account (3.16) we have the following estimate for σ_t^T, provided $t > T$

$$\left| \sigma_t^T \right| \le \beta |c| \left(|x_0| + |b| \rho^{-(T+2)} \sup_{t \le T} \left| \varphi\left(\sigma_t^T\right) \right| \right) \rho^t = \beta_1(T) \rho^t \tag{3.19}$$

Consider the following index

$$\eta(T) = \sum_0^{T-1} \mu_t \left(\mu_t/\bar{\varphi} + \sigma_t - \delta\mu_t \right) , \tag{3.20}$$

where (μ_t, σ_t) might be any input/output pair of (3.14) and $\delta > 0$ will be chosen later.

Let μ_t^T be defined by (3.16) and σ_t^T by (3.17), (3.18). Denoting

$$w_t^T = \sum_0^{t-1} c^* A^{t-1-k} b \mu_k^T \ , \ t \geq 1 \tag{3.21}$$

we have $\sigma_t^T = c^* A^t x_0 + w_t^T$. We define also the so called *sequence of weighting patterns* for the linear system (3.14)

$$h_t = \begin{cases} c^* A^{t-1} b & , & t \geq 1 \\ 0 & , & t < 1. \end{cases} \tag{3.22}$$

Because $\mu_t^T = 0$ for $t < 0$ we may express w_t^T as a discrete convolution

$$w_t^T = \sum_{-\infty}^{\infty} h_{t-k} \mu_k^T . \tag{3.23}$$

Consider now $\eta(T)$ for μ_t^T, σ_t^T as defined above

$$\eta(T) = \sum_0^{T-1} \mu_t^T \left(\mu_t^T / \bar\varphi + \sigma_t^T - \delta\mu_t^T \right) =$$
$$= \sum_{-\infty}^{\infty} \mu_t^T \left(\mu_t^T / \bar\varphi + w_t^T - \delta\mu_t^T \right) + \sum_0^{T-1} \mu_t^T \left(c^* A^t x_0 \right) \tag{3.24}$$

Take now the discrete Fourier transforms of the sequences $\{\mu_t^T\}_t$, $\{\sigma_t^T\}_t$:

$$\tilde\mu_T(\theta) = \sum_{-\infty}^{\infty} \mu_t^T e^{-i\theta t} , \quad \tilde\sigma_T(\theta) = \sum_{-\infty}^{\infty} \sigma_t^T e^{-i\theta t} , \quad -\pi \leq \theta < \pi ,$$

where the convergence follows from (3.16) and (3.19). In fact these series have a finite number of terms. The electronics engineers call this Fourier transform DCFT (discrete-to-continuous Fourier transform) because it associates with the discrete range of time ($t \in \mathbb{Z}$) the continuous range of frequencies, θ being real between $-\pi$ and π (Kwakernaak and Sivan (1991)). Conversely, the terms of the sequence may be viewed as the Fourier coefficients of a continuous function defined on $[-\pi, \pi)$ and extended by periodicity.

For DCFT the convolution theorems and Plancherel - Parseval

equality are valid. Therefore,

$$\tilde{w}_T(\theta) = \sum_{-\infty}^{\infty} w_t^T e^{-i\theta t} = \tilde{h}(\theta)\,\tilde{\mu}_T(\theta)\ ,$$

where $\tilde{h}(\theta) = \sum_{-\infty}^{\infty} h_t e^{-i\theta t} = c^*\left(e^{i\theta}\mathbf{I} - A\right)^{-1} b = \gamma\left(e^{i\theta}\right)$. Here $\gamma(z) =$ $c^*(z\mathbf{I} - A)^{-1} b$ is the transfer function of (3.14) that has been mentioned for the continuous time systems in the introductory part of the chapter.

The Plancherel - Parseval equality for DCFT has the form

$$\sum_{-\infty}^{\infty} \alpha_n \beta_n = \frac{1}{2\pi} \operatorname{Re} \int_{-\pi}^{\pi} \tilde{\alpha}(\theta)\,\tilde{\beta}^*(\theta)\,\mathrm{d}\theta$$

hence the first sum in the expression of $\eta(T)$ in (3.24) becomes:

$$\sum_{-\infty}^{\infty} \mu_t^T\left(\mu_t^T/\bar{\varphi} + w_t^T - \delta\mu_t^T\right) = \frac{1}{2\pi} \int_{-\pi}^{\pi}\left[\frac{1}{\bar{\varphi}} + \operatorname{Re}\gamma\left(e^{i\theta}\right) - \delta\right]|\tilde{\mu}_r(\theta)|^2\,\mathrm{d}\theta$$

$$(3.25)$$

Assume now that the following frequency domain inequality holds:

$$\frac{1}{\bar{\varphi}} + \operatorname{Re}\gamma\left(e^{i\theta}\right) > 0\ ,\quad -\pi \le \theta < \pi \qquad (3.26)$$

But $\gamma(z)$ has no singularities on the unit circle because of the assumptions on the eigenvalues of A and $[-\pi, \pi]$ is a compact interval, hence there exists some $\delta_0 > 0$ such that

$$\frac{1}{\bar{\varphi}} + \operatorname{Re}\gamma\left(e^{i\theta}\right) \ge \delta_0 > 0\ ,\quad -\pi \le \theta < \pi \qquad (3.27)$$

Take $\delta = \delta_0$ in (3.20) and because of (3.27) it follows that:

$$\eta(T) \ge \sum_{0}^{T-1} \mu_t^T\left(c^* A^t x_0\right)\ . \qquad (3.28)$$

But if $0 \le t < T - 1$ then $\mu_t^T = -\varphi\left(c^* x_t\right)$ hence (3.15) holds. It follows from (3.20) that:

$$\eta\left(T\right) \leq -\delta_0 \sum_0^{T-1} \left|\varphi\left(c^* x_t\right)\right|^2$$

and combining it with (3.28) the main stability inequality is obtained

$$\delta_0 \sum_0^{T-1} \left|\varphi\left(c^* x_t\right)\right|^2 \leq \sum_0^{T-1} \varphi\left(c^* x_t\right)\left(c^* A^t x_0\right) \leq$$

$$\leq 0.5\delta_0 \sum_0^{T-1} \left|\varphi\left(c^* x_t\right)\right|^2 + \frac{0.5}{\delta_0} \sum_0^{T-1} \left|c^* A^t x_0\right|^2$$

the last inequality following from the elementary fact $\alpha\beta < 0.5\left(|\alpha|^2 + |\beta|^2\right)$. Therefore,

$$\sum_0^{T-1} \left|\varphi\left(c^* x_t\right)\right|^2 \leq \frac{1}{\delta_0^2}\beta_0^2 \left|x_0\right|^2 \sum_0^{T-1} \rho^{2t} \leq \frac{\beta_0^2}{\delta_0^2\left(1 - \delta^2\right)} \left|x_0\right|^2$$

and the sequence $\{\varphi\left(c^* x_t\right)\}_t$ is in ℓ^2. We turn back to (3.17) and have

$$\left|x_t\right| \leq \beta\left|c\right| \left(\rho^t \left|x_0\right| + \left|b\right| \sum_0^{t-1} \rho^{t-1-k} \left|\varphi\left(c^* x_k\right)\right|\right). \qquad (3.29)$$

The use of Cauchy-Schwarz inequality gives

$$\sum_0^{t-1} \rho^{t-1-k} \left|\varphi\left(c^* x_k\right)\right| \leq \frac{1}{\sqrt{1-\rho}} \left(\sum_0^{t-1} \rho^{t-1-k} \left|\varphi\left(c^* x_k\right)\right|^2\right)^{1/2}$$

and this is introduced in (3.29)

$$\left|x_t\right|^2 \leq 2\beta^2 \left|c\right|^2 \left(\rho^{2t} \left|x_0\right|^2 + \frac{\left|b\right|^2}{1-\rho} \sum_0^{t-1} \rho^{t-1-k} \left|\varphi\left(c^* x_k\right)\right|^2\right). \qquad (3.30)$$

From (3.30) two inequalities may be obtained. First

$$\left|x_t\right|^2 \leq 2\beta^2 \left|c\right|^2 \left[1 + \frac{\beta_0^2 \left|b\right|^2}{\delta_0^2 \left(1 - \rho\right)^2 \left(1 + \rho\right)}\right] \left|x_0\right|^2 = K_1 \left|x_0\right|^2 \qquad (3.31)$$

which is nothing else but Liapunov stability of the zero solution and global boundedness of the solutions of (3.12). On the other side we have

$$\sum_0^\infty |x_t|^2 \le 2\beta^2 |c|^2 \left(\frac{|x_0|^2}{1-\rho^2} + \frac{|b|^2}{1-\rho} \sum_0^\infty \sum_0^{t-1} \rho^{t-1-k} |\varphi(c^* x_k)|^2 \right).$$

The double sum is estimated as follows:

$$\sum_0^\infty \sum_0^{t-1} \rho^{t-1-k} |\varphi(c^* x_k)|^2 \;=\; \sum_{k=0}^\infty \left(\sum_{k+1}^\infty \rho^{t-1} \right) \rho^{-k} |\varphi(c^* x_k)|^2 =$$

$$=\; \frac{1}{1-\rho} \sum_0^\infty |\varphi(c^* x_k)|^2 \,.$$

Using the already obtained estimate for the last sum it follows that:

$$\sum_0^\infty |x_t|^2 \le \frac{2\beta^2 |c|^2}{1-\rho^2} \left(1 + \frac{\beta_0^2 |b|^2}{\delta_0^2 (1-\rho) 2} \right) |x_0|^2 \,; \qquad (3.32)$$

hence, $\{x_t\}_t$ is in ℓ^2. The above series being convergent it follows that $\lim_{t\to\infty} |x_t| = 0$. We have thus proved the following result:

Theorem 3.1 *Let system (3.12) satisfy the following assumptions: i) matrix A has its eigenvalues inside the disk $D = \{z \in \mathbb{C} : |z| < 1\}$; ii) the nonlinear function $\varphi(\sigma)$ satisfies (3.13); iii) the frequency domain inequality (3.26) is fulfilled for all $\theta \in [-\pi, \pi)$. Then the zero solution of system (3.12) is globally asymptotically stable that is the frequency domain condition implies absolute stability with respect to the class defined by ii).*

Examples.
   1° Consider first simple second-order system

$$y_{t+2} + \varphi(y_t) = 0, \qquad (3.33)$$

where y is a scalar and $\varphi(\sigma)$ satisfies a sector condition of the form (3.13). If $\varphi(\sigma)$ is linear, i.e., $\varphi(\sigma) = h\sigma$ the above system becomes:

$$y_{t+2} + h y_t = 0$$

with the characteristic equation

$$z^2 + h = 0 \,.$$

Its roots are inside the unit disk provided $|h| < 1$. This means that the stability sector for linear characteristics is $(-1, 1)$. Absolute stability means global asymptotic for **all** functions of some sector; hence, the linear functions are included in the class. Consequently, the largest absolute stability sector could at most be equal to that obtained for linear functions. In fact we have the discrete analogue of **Aizerman's problem** (Aizerman (1948)): when is the absolute stability sector identical to the linear stability sector? The *conjectures* of Aizerman (1948) and Kalman (1957) concerning this subject for continuous time systems were shown to be not always true by counterexamples in the work of Pliss (1958) and Barabanov (1988) but the comparison of the absolute stability sector with the linear stability sector remains a significant test concerning efficiency of various sufficient conditions of absolute stability.

We shall apply now Theorem 3.1 to system (3.33) which may have the form (3.12) provided we denote

$$x_t = \begin{pmatrix} y_t \\ y_{t+1} \end{pmatrix} \text{ then } A = \begin{pmatrix} 0 & 1 \\ 0 & 0 \end{pmatrix}; \ b = \begin{pmatrix} 0 \\ 1 \end{pmatrix}; \ c^* = \begin{pmatrix} 1 & 0 \end{pmatrix}.$$

Clearly the eigenvalues of A, being $\lambda_1 = \lambda_2 = 0$, are inside the unit disk. We have also $\gamma(z) = 1/z^2$ and (3.26) reads

$$\frac{1}{\bar{\varphi}} + \cos 2\theta > 0$$

which gives $\bar{\varphi} < 1$. The sub-sector corresponding to the first and the third quadrants of the plane (σ, φ) is an absolute stability sub-sector that almost coincides with the sub-sector of linear stability.

If $\underline{\varphi}\sigma^2 \le \varphi(\sigma)\sigma \le 0$ then we shall have $-\varphi(\sigma)\sigma \le -\underline{\varphi}\sigma^2$. Denoting $\psi(\sigma) = -\varphi(\sigma)$ and $\bar{\psi} = -\underline{\varphi}$ we have the equation

$$y_{t+2} - \psi(y_t) = 0$$

with $0 \le \psi(\sigma)\sigma \le \bar{\psi}\sigma^2$. We may apply again Theorem 3.1 (Tsypkin criterion) to obtain

$$\frac{1}{\bar{\psi}} - \cos 2\theta > 0$$

which gives again $\bar{\psi} < 1$ that is $\underline{\varphi} > -1$. We have in this way recovered the missing part of the linear stability domain.

Let us remark that the criterion does not allow to consider functions for which

$$-1 < \underline{\varphi} \le \varphi\left(\sigma\right)/\sigma \le \bar{\varphi} < 1$$

although the result is true by a Liapunov function argument.

Indeed, consider the following simple quadratic Liapunov function

$$V\left(y_t, y_{t+1}\right) = y_t^2 + y_{t+1}^2$$

which is strictly positive definite; we shall have, along the solutions of (3.33)

$$V\left(y_{t+1}, y_{t+2}\right) - V\left(y_t, y_{t+1}\right) = y_{t+1}^2 + y_{t+2}^2 - y_t^2 - y_{t+1}^2 = \varphi\left(y_t^2\right) - y_t^2 < 0$$

the last inequality following from the sector condition

$$-1 < \frac{\varphi\left(\sigma\right)}{\sigma} < 1$$

which shows recovering for absolute stability of the entire linear stability sector.

Therefore extension of the criterion of Theorem 3.1 in order to encompass nonlinear functions belonging to sectors that have lower negative (instead of zero) bound appears as useful.

2° Consider now the following second-order equation

$$y_{t+2} + ay_{t+1} + \varphi\left(y_t\right) = 0 \tag{3.34}$$

where $|a| < 1$ and φ satisfies again a sector condition of the form (3.13). If φ is linear the above system becomes

$$y_{t+2} + ay_{t+1} + hy_t = 0$$

with the characteristic equation

$$z^2 + az + h = 0.$$

Its roots are inside the unit disk if the following conditions hold:

$$|h| < 1, \quad |a| < 1 + h.$$

From here we obtain the stability sector for linear characteristics which is $(-1 + |a|, 1)$. Taking into account (3.13) we see that only the sub-sector $(0, 1)$ may be taken into consideration if we want to apply Theorem 3.1. Consider now the nonlinear equation that may be given the form (3.12). The transfer function of the linear part is

$$\gamma(z) = \frac{1}{z(z+a)}$$

whose poles are inside the unit disk because we assumed $|a| < 1$. The frequency domain inequality is

$$\frac{1}{\bar{\varphi}} + \mathrm{Re}\,\gamma\left(e^{i\theta}\right) = \frac{1}{\bar{\varphi}} + \frac{2\cos^2\theta + a\cos\theta - 1}{2a\cos\theta + 1 + a^2} > 0\,, \quad 0 \le \theta < \pi$$

which may be written as

$$\frac{1}{\bar{\varphi}} + \frac{2\lambda^2 + a\lambda - 1}{2a\lambda + 1 + a^2} > 0\,, \quad -1 \le \lambda \le 1\,.$$

The necessary and sufficient condition for the fulfillment of the above inequality is

$$\frac{1}{\bar{\varphi}} + \inf_{\lambda \in [-1,1]} \frac{2\lambda^2 + a\lambda - 1}{2a\lambda + 1 + a^2} > 0\,.$$

If the infimum is nonnegative then $\bar{\varphi}$ may be arbitrarily large. This is not the case: the infimum is the minimum value that corresponds to

$$\lambda = \frac{-\left(1 + a^2\right) + \sqrt{1 - a^2}}{2a}$$

and is negative. The condition on $\bar{\varphi}$ is

$$\bar{\varphi} < -\left[\min \frac{2\lambda^2 + a\lambda - 1}{2a\lambda + 1 + a^2}\right]^{-1} = \frac{2\left(a^2 + 2 + 2\sqrt{1 - a^2}\right)}{a^2 + 8} < 1$$

In this case not only the "negative" stability sector is put aside but even the linear sub-sector $(0, 1)$ is not entirely recovered for absolute stability: larger is $|a|$, more narrow is the absolute stability sector furnished by Theorem 3.1: if $a = 0$ then $\bar{\varphi} < 1$ (see the previous example) and if $|a| = 1$ then $\bar{\varphi} < 2/3$.

Therefore, this example shows that besides considering nonlinear functions in the second and fourth quadrants, it is also necessary to improve the frequency domain inequality in order to obtain larger absolute stability sector.

The Case of the Arbitrary Limits of the Absolute Stability Sector

We shall consider again system (3.12) but with $\varphi(\sigma)$ to the following sector condition:

$$\underline{\varphi}\sigma^2 \leq \varphi(\sigma)\sigma \leq \bar{\varphi}\sigma^2\,, \tag{3.35}$$

where $\underline{\varphi} < 0 < \bar{\varphi}$. The assumption on the eigenvalues of A is the same as previously: they lie inside the unit disk $D = \{z \in \mathbb{C} : |z| < 1\}$; hence, the linear controlled system (3.14) is exponentially stable. The control $\mu = -\varphi(\sigma)$ will satisfy according to (3.35) the following quadratic inequality

$$\left(\mu + \underline{\varphi}\sigma\right)\left(\mu + \bar{\varphi}\sigma\right) \leq 0\,. \tag{3.36}$$

Consider some $T > 0$ and define μ_t^T by (3.16). All estimates on σ_t^T, $c^*A^t x_0$, w_t^T, h_t defined by (3.17)-(3.19) and (3.21)-(3.23) are in power because they are not connected with (3.35) but with the linear system (3.14) and the control function defined by (3.16). We consider now the following index

$$\eta(T) = \sum_0^{T-1} \left[\left(\mu_t + \underline{\varphi}\sigma_t\right)\left(\mu_t/\bar{\varphi} + \sigma_t\right) - \delta\mu_t^2\right]\,, \tag{3.37}$$

where (μ_t, σ_t) might be any input/output pair of (3.14) and $\delta > 0$ will be chosen later. For μ_t^T, σ_t^T defined as above, $\eta(T)$ becomes

$$\eta(T) = \sum_{-\infty}^{\infty} \left[\left(\mu_t^T + \underline{\varphi}w_t^T\right)\left(\mu_t^T/\bar{\varphi} + w_t^T\right) - \delta\left(\mu_t^T\right)^2\right] - \underline{\varphi}\sum_T^{\infty}\left(w_t^T\right)^2 +$$
$$+ \left(1 + \underline{\varphi}/\bar{\varphi}\right)\sum_0^{T-1}\mu_t^T\left(c^*A^t x_0\right) + \underline{\varphi}\sum_0^{T-1}\left(c^*A^t x_0\right)^2$$
$$\tag{3.38}$$

Applying the discrete-to-continuous time Fourier transform DCFT and the corresponding Plancherel-Parseval equality, the first sum in

(3.38) becomes

$$\sum_{-\infty}^{\infty} \left[\left(\mu_t^T + \underline{\varphi} w_t^T \right) \left(\mu_t^T / \bar{\varphi} + w_t^T \right) - \delta \left(\mu_t^T \right)^2 \right] =$$

$$= \frac{1}{2\pi} \int_{-\pi}^{\pi} \left[\frac{1}{\bar{\varphi}} + \left(1 + \underline{\varphi}/\bar{\varphi} \right) \operatorname{Re} \gamma \left(e^{i\theta} \right) + \underline{\varphi} \left| \gamma \left(e^{i\theta} \right) \right|^2 - \delta \right] \left| \tilde{\mu}_T \left(\theta \right) \right|^2 d\theta$$

$$\text{(3.39)}$$

Assume now the following frequency domain inequality holds

$$\frac{1}{\bar{\varphi}} + \left(1 + \underline{\varphi}/\bar{\varphi} \right) \operatorname{Re} \gamma \left(e^{i\theta} \right) + \underline{\varphi} \left| \gamma \left(e^{i\theta} \right) \right|^2 > 0 , \quad 0 \le \theta < \pi \quad \text{(3.40)}$$

With the same arguments as in the previous case we find that there exists some $\delta_0 > 0$ such that

$$\frac{1}{\bar{\varphi}} + \left(1 + \underline{\varphi}/\bar{\varphi} \right) \operatorname{Re} \gamma \left(e^{i\theta} \right) + \underline{\varphi} \left| \gamma \left(e^{i\theta} \right) \right|^2 \ge \delta_0 > 0 , \quad 0 \le \theta < \pi$$

From (3.39) it follows then, by choosing $\delta = \delta_0$,

$$\eta \left(T \right) \ge \left(1 + \underline{\varphi}/\bar{\varphi} \right) \sum_{0}^{T-1} \mu_t^T \left(c^* A^t x_0 \right) + \underline{\varphi} \sum_{0}^{T-1} \left(c^* A^t x_0 \right)^2 - \underline{\varphi} \sum_{T}^{\infty} \left(w_t^T \right)^2$$

$$\text{(3.41)}$$

But if $0 \le t \le T - 1$ then $\mu_t^T = -\varphi \left(c^* x_t \right)$ hence (3.36) holds. Therefore

$$\eta \left(T \right) \le -\delta_0 \sum_{0}^{T-1} \left(\varphi \left(c^* x_t \right) \right)^2$$

and combining it with (3.41) the main stability inequality is obtained:

$$\delta_0 \sum_{0}^{T-1} \left(\varphi \left(c^* x_t \right) \right)^2 \le \underline{\varphi} \sum_{T}^{\infty} \left(w_t^T \right)^2 - \underline{\varphi} \sum_{0}^{T-1} \left(c^* A^t x_0 \right)^2 +$$

$$+ \left(1 + \underline{\varphi}/\bar{\varphi} \right) \sum_{0}^{T-1} \varphi \left(c^* x_t \right) \left(c^* A^t x_0 \right) .$$

Because $\underline{\varphi} < 0$ we shall have

$$\delta_0 \sum_{0}^{T-1} \left| \varphi \left(c^* x_t \right) \right|^2 \le 0.5 \delta_0 \sum_{T}^{\infty} \left| \varphi \left(c^* x_t \right) \right|^2 +$$

$$+ \left(|\underline{\varphi}| + 0.5 \left(1 + |\underline{\varphi}| / \bar{\varphi} \right)^2 / \delta_0 \right) \sum_0^{T-1} \left| c^* A^t x_0 \right|^2 .$$

From now on the proof reproduces the proof of Theorem 3.1 and we shall not repeat it. The result thus proved is the following:

Theorem 3.2   *Let system* (3.12) *satisfy the following assumptions: i) matrix A has its eigenvalues inside the unit disk; ii) the nonlinear function $\varphi(\sigma)$ satisfies* (3.35); *iii) the frequency domain inequality* (3.40) *is fulfilled for all $\theta \in [-\pi, \pi)$. Then the zero solution of system* (3.12) *is globally asymptotically stable that is the frequency domain condition implies absolute stability with respect to the class defined by* (3.35).

Comments and Examples

A. The Tsypkin type criteria proved above belong to the so-called *circle criteria* that were considered for continuous-time systems by Sandberg and Zames. Indeed if we consider the complex plane of $X = \operatorname{Re}\gamma\left(e^{i\theta}\right)$, $Y = \operatorname{Im}\gamma\left(e^{i\theta}\right)$, these two equalities define the hodograph of the transfer function $\gamma(z)$ with respect to the unit circle of the complex plane of the variable z. Taking into account that $\underline{\varphi} < 0$ inequality (3.40) becomes

$$\left[X + \frac{1}{2} \left(\frac{1}{\bar{\varphi}} + \frac{1}{\underline{\varphi}} \right) \right]^2 + Y^2 - \frac{1}{4} \left(\frac{1}{\bar{\varphi}} - \frac{1}{\underline{\varphi}} \right)^2 < 0 .$$

This obviously shows that the hodograph has to lie inside *a disk of the plane* (X, Y), centered at $\left(-\frac{1}{2} \left(\bar{\varphi}^{-1} + \underline{\varphi}^{-1} \right), 0 \right)$ with the radius $0.5 \left(\bar{\varphi}^{-1} - \underline{\varphi}^{-1} \right)$. For $\underline{\varphi} = 0$ the disk degenerates in the half plane $X > -1/\bar{\varphi}$; this half plane is defined in fact by Popov vertical line $X = -1/\bar{\varphi}$ that corresponds to the Popov parameter equal to zero.

B. We shall consider again the example defined by (3.33) and apply the frequency domain inequality (3.40). Because $\gamma(z) = 1/z^2$ if $|z| = 1$ then $|\gamma(z)| = 1$. Therefore the frequency domain inequality becomes

$$\frac{1}{\bar{\varphi}} + \left(1 + \underline{\varphi}/\bar{\varphi} \right) \cos 2\theta + \underline{\varphi} > 0 , \quad 0 \le \theta < \pi$$

which gives the sector $[-1 + \delta_0, 1)$ with $\delta_0 > 0$ arbitrarily small. Remember that the linear stability sector was $(-1, 1)$.

C. Consider again the example defined by (3.34) with $|a| < 1$. Recall the previously established fact that the stability sector for linear functions $\varphi(\sigma) = h\sigma$ is given by $(-1 + |a|, 1)$. The frequency domain inequality (3.40) takes the form

$$\frac{1}{\bar{\varphi}} + \left(1 + \underline{\varphi}/\bar{\varphi}\right) \frac{2\cos^2\theta + a\cos\theta - 1}{2a\cos\theta + 1 + a^2} + \underline{\varphi}\frac{1}{1 + a^2 + 2a\cos\theta} > 0 \,,$$
$$0 \le \theta < \pi$$

or

$$\frac{1}{\bar{\varphi}} + \left(1 + \underline{\varphi}/\bar{\varphi}\right) \frac{2\lambda^2 + a\lambda - 1}{2a\lambda + 1 + a^2} + \underline{\varphi}\frac{1}{1 + a^2 + 2a\lambda} > 0 \,, \ -1 \le \lambda \le 1$$

which takes the form

$$2\left(1 + \frac{\underline{\varphi}}{\bar{\varphi}}\right)\lambda^2 + a\left[\left(1 + \frac{\underline{\varphi}}{\bar{\varphi}}\right) + \frac{2}{\bar{\varphi}}\right]\lambda + \frac{1}{\bar{\varphi}}\left(1 + a^2\right) + \underline{\varphi} - \left(1 + \frac{\underline{\varphi}}{\bar{\varphi}}\right) > 0.$$

An elementary but tedious manipulation leads to the following absolute stability sector

$$|a| - 1 < \frac{\varphi(\sigma)}{\sigma} < \frac{a^2 - 3|a| + 4}{4 - |a|} < 1\,.$$

Again the upper bound of the absolute stability sector is lower than the corresponding one in the linear case while the lower bounds of the sectors coincide. The estimate for the upper bound is slightly improved with respect to the previous case because

$$\frac{a^2 - 3|a| + 4}{4 - |a|} > \frac{2\left(a^2 + 2 + 2\sqrt{1 - a^2}\right)}{a^2 + 8}\,.$$

As previously mentioned, another improvement of the estimates for the largest absolute stability sector might be obtained from the improvement of the frequency domain inequality; this would be possible by incorporating additional information about the nonlinear functions.

3.2. QUADRATIC CONSTRAINTS OF YAKUBOVICH TYPE

The experience of the applications of the Tsypkin condition have indicated the necessity for its improvement. A way of improving the absolute stability conditions that have been successfully applied in the case of continuous-time systems is to make use of additional information about the nonlinear function (see for instance Yakubovich (1967)).This idea is also applicable to discrete systems but it requires one to consider special classes of nonlinearities from the beginning. An explanation is necessary. In the continuous-time case the Tsypkin-type frequency domain inequality corresponds to Popov inequality

$$\frac{1}{\bar{\varphi}} + \mathrm{Re}\left(1 + iwq\right)\gamma\left(iw\right) > 0$$

with the parameter $q = 0$. This means nothing else than not taking into account the fact that we also have the following property of the nonlinear function

$$\int_0^t \varphi\left(\sigma\left(\tau\right)\right)\dot{\sigma}\left(\tau\right)\mathrm{d}\tau = \int_0^{\sigma(t)} \varphi\left(\theta\right)\mathrm{d}\theta - \int_0^{\sigma(0)} \varphi\left(\theta\right)\mathrm{d}\theta = \Phi\left(\sigma\left(t\right)\right) - \Phi\left(\sigma\left(0\right)\right) .$$

$$(3.42)$$

In the discrete-time case this property is no longer valid, generally speaking. Indeed, if the integral is replaced by the sum and the derivative by a forward or a backward difference then the following sums are obtained

$$S_1 = \sum_0^{t-1} \varphi\left(\sigma_k\right)\left(\sigma_{k+1} - \sigma_k\right) , \quad S_2 = \sum_1^{t} \varphi\left(\sigma_k\right)\left(\sigma_k - \sigma_{k-1}\right) .$$

The above integrals were path-independent while the sums S_1 and S_2 are not. This path independence may nevertheless be achieved if some additional information about the nonlinearity is taken into account.

   A. The most useful and mostly used assumption is the *monotonicity assumption*. If $\varphi\left(\sigma\right)$ is monotonically nondecreasing we have the following inequalities

$$\varphi\left(\alpha\right)\left(\beta - \alpha\right) \leq \int_\alpha^\beta \varphi\left(\lambda\right)\mathrm{d}\lambda \leq \varphi\left(\beta\right)\left(\beta - \alpha\right) , \ \forall \alpha, \beta . \qquad (3.43)$$

Therefore,

$$\sum_1^t \varphi(\sigma_k)(\sigma_k - \sigma_{k-1}) \geq \sum_1^t \int_{\sigma_{k-1}}^{\sigma_k} \varphi(\sigma)\,d\sigma =$$

$$= \int_0^{\sigma_t} \varphi(\sigma)\,d\sigma - \int_0^{\sigma_0} \varphi(\sigma)\,d\sigma$$

$$\sum_1^{t-1} \varphi(\sigma_k)(\sigma_k - \sigma_{k+1}) \geq -\sum_0^{t-1} \int_{\sigma_k}^{\sigma_{k+1}} \varphi(\sigma)\,d\sigma =$$

$$= \int_0^{\sigma_0} \varphi(\sigma)\,d\sigma - \int_0^{\sigma_t} \varphi(\sigma)\,d\sigma$$

which are much alike (3.42). As it has been previously shown, the nonlinear part in the system is introduced in the form of the control $\mu_t = -\varphi(\sigma_t)$. The above inequalities take the form

$$\sum_1^t \mu_k (\sigma_k - \sigma_{k-1}) \leq \int_0^{\sigma_0} \varphi(\sigma)\,d\sigma - \int_0^{\sigma_t} \varphi(\sigma)\,d\sigma = \tag{3.44}$$

$$= \Phi(\sigma_0) - \Phi(\sigma_t) \leq \Phi(\sigma_0)$$

$$\sum_0^{t-1} \mu_k (\sigma_k - \sigma_{k+1}) \leq \int_0^{\sigma_t} \varphi(\sigma)\,d\sigma - \int_0^{\sigma_0} \varphi(\sigma)\,d\sigma. \tag{3.45}$$

From (3.13) it follows that $\Phi(\sigma)$ defined in (3.42) is nonnegative which justifies the last inequality in (3.44). On the other side, from (3.13) it follows that $0.5\bar{\varphi}\sigma^2 - \Phi(\sigma)$ is also nonnegative. Therefore, (3.45) may be written as:

$$\sum_0^{t-1} \left(\mu_k + \frac{1}{2}\bar{\varphi}(\sigma_k + \sigma_{k+1}) \right)(\sigma_k - \sigma_{k-1}) \leq 0.5\bar{\varphi}\sigma_0^2 - \Phi(\sigma_0). \tag{3.46}$$

Remark that under the sums we have quadratic forms and the sums are bounded by positive constants that depend on the initial conditions. Previously we obtained a quadratic form in μ_k, σ_k that was negative for any k - see (3.15). Of course, if $W(\mu, \sigma) \le 0$, then obviously

$$\sum_0^t W(\mu_k, \sigma_k) \le 0 . \tag{3.47}$$

B. Consider now the class of nonlinear functions that are no longer monotonic but satisfy the slope constraints

$$-\alpha \le \varphi'(\sigma) \le \beta , \quad \alpha > 0$$

and with the sector condition (3.13). We may write

$$\Phi(\sigma') - \Phi(\sigma'') = (\sigma' - \sigma'')\,\varphi(\sigma'') + \frac{(\sigma' - \sigma'')^2}{2}\varphi'(\sigma_*) , \quad \sigma_* \in (\sigma', \sigma'') ,$$

where

$$\Phi(\sigma) = \int_0^\sigma \varphi(\lambda)\, d\lambda \ge 0$$

Taking $\sigma' = \sigma_{t-1}$, $\sigma'' = \sigma_t$ and making use of the fact that $\varphi'(\sigma) \ge -\alpha$ we have

$$\Phi(\sigma_{t-1}) - \Phi(\sigma_t) \ge (\sigma_{t-1} - \sigma_t)\,\varphi(\sigma_t) - \frac{\alpha}{2}(\sigma_{t-1} - \sigma_t)^2 .$$

Denoting again $\mu_t = -\varphi(\sigma_t)$ we shall have

$$\sum_1^t (\sigma_k - \sigma_{k-1})\left(\mu_k - \frac{\alpha}{2}(\sigma_k - \sigma_{k-1})\right) \le \Phi(\sigma_0) . \tag{3.48}$$

In a similar way, making use of the fact that $\varphi'(\sigma) \le \beta$ and that the sector condition (3.13) will give

$$\frac{1}{2}\bar{\varphi}\sigma^2 - \Phi(\sigma) = \int_0^\sigma (\bar{\varphi}\lambda - \varphi(\lambda))\, d\lambda$$

we shall have

$$\sum_1^t (\sigma_k - \sigma_{k-1})\left[\mu_k + \bar{\varphi}\sigma_k + \frac{1}{2}(\bar{\varphi} - \beta)(\sigma_{k-1} - \sigma_k)\right] \le \frac{1}{2}\bar{\varphi}\sigma_0^2 - \Phi(\sigma_0)$$

$$\tag{3.49}$$

If we consider forward differences taking $\sigma' = \sigma_{t+1}$, $\sigma'' = \sigma_t$ and proceed as above we find

$$\sum_0^{t-1} (\sigma_{k+1} - \sigma_k) \left(\mu_k - \frac{\beta}{2} (\sigma_{k+1} - \sigma_k) \right) \leq \Phi(\sigma_0) \qquad (3.50)$$

$$\sum_0^{t-1} (\sigma_k - \sigma_{k-1}) \left[\mu_k + \bar{\varphi}\sigma_k + \frac{1}{2} (\bar{\varphi} + \alpha)(\sigma_k - \sigma_{k+1}) \right] \leq \frac{1}{2}\bar{\varphi}\sigma_0^2 - \Phi(\sigma_0)$$
$$(3.51)$$

Such (and other) inequalities for quadratic forms were obtained by Yakubovich (1968).

In fact the above considerations show that the conditions on the nonlinear element described by $\mu = -\varphi(\sigma)$ satisfying a sector condition (3.13) or (3.35) and some additional conditions (as monotonicity, slope restrictions) may be written in the form of quadratic constraints

$$W_j(\mu, \sigma) \leq 0, \quad j = 1, ..., q \qquad (3.52)$$

$$\sum_{t_0}^{t-1} W_{q+j}(\mu_k, \sigma_{k-1}, \sigma_k, \sigma_{k+1}) \leq \gamma_j, \quad j = 1, ..., s. \qquad (3.53)$$

Consider now system (3.14) with its input/output pair satisfying (3.52) and (3.53). This is called by Yakubovich a generalized feedback.

We associate with each quadratic form (3.52) a nonnegative parameter $\tau_j \geq 0$ and to each quadratic form (3.53) a nonnegative parameter $v_j \geq 0$, defining the index

$$\eta(t_0, t) = \sum_{t_0}^{t-1} \left[\sum_{j=1}^q W_j(\mu_k, \sigma_k) + \sum_{j=1}^s v_j W_{q+j}(\mu_k, \sigma_{k-1}, \sigma_k, \sigma_{k+1}) \right].$$
$$(3.54)$$

Further we represent the solutions of (3.14) using the variations of constants formula

$$\sigma_t = c^* A^t x_0 + \sum_0^{t-1} (c^* A^{t-1-k} b) \mu_k. \qquad (3.55)$$

The couple "equation plus index" defined by (3.54) and (3.55) represent a mathematical object known as Popov system; the Popov system has many remarkable properties. In the previous section the indices (3.20) defined using the quadratic constraint (3.15) and (3.37) defined using (3.36), coupled with (3.18), had been in fact Popov systems. In the sequel such Popov systems will be constructed quite often.

3.3. AN ABSOLUTE STABILITY CRITERION FOR THE CASE OF NONDECREASING NONLINEAR FUNCTIONS

We shall consider here the system (3.12) with nondecreasing $\varphi(\sigma)$ satisfying (3.13). The problem which we formulate is to find a frequency domain inequality for the absolute stability with respect to this class of nonlinearities.

As previously shown, system (3.12) is replaced by the controlled linear system (3.14) with the control function $\mu_t = -\varphi(\sigma_t)$. The couple (μ, σ) thus defined is subject to the quadratic constraints (3.15) and (3.44). Thus we may associate a Popov system (3.54)-(3.55) as follows:

$$\sigma_t = c^* A^t x_0 + \sum_0^{t-1} \left(c^* A^{t-1-k} b \right) \mu_k \qquad (3.56)$$

$$\eta(T) = \tau \sum_{t_0}^{T-1} \mu_t \left(\frac{\mu_t}{\bar{\varphi}} + \sigma_t \right) + \upsilon \sum_0^{T-1} \mu_t (\sigma_t - \sigma_{t-1}) - \delta \sum_0^{T-1} \mu_t^2, \ (3.57)$$

where $\tau \geq 0$, $\upsilon \geq 0$, $\delta > 0$ are arbitrary parameters. Here $T > 0$ is arbitrary. Define the truncated control μ_t^T as previously, i.e., by (3.16) and the corresponding output σ_t^T by (3.18). Assuming again that the eigenvalues of A are located inside the unit disk of the complex plane, the estimates for $c^* A^t x_0$, for the weighting patterns h_t and for σ_t^T obtained in Section 3.1 are valid. Remember that $\mu_t^T = 0$ for $t < 0$ and for $t \geq T$; we take $\sigma_t^T = 0$, $t < 0$, as previously and if we write $\eta(T)$ for the pair $\left(\mu_t^T, \sigma_t^T \right)$ thus defined we have

$$\eta(T) = \sum_{-\infty}^{\infty} \mu_t^T \left[\tau \left(\mu_t^T / \bar{\varphi} + w_t^T \right) + \upsilon \left(w_t^T - w_{t-1}^T \right) - \delta \mu_t^T \right] +$$

$$+ \sum_{1}^{T-1} \mu_t^T \left(\tau c^* A^t x_0 + \upsilon c^* A^t x_0 - \upsilon c^* A^t x_0 \right) + \mu_0^T \left(\tau + \upsilon \right) c^* x_0 \quad (3.58)$$

As previously we are in position to apply DCFT and corresponding Plancherel-Parseval equality, all formulae of Section 3.1 are valid and also

$$\sum_{-\infty}^{\infty} w_{t-1}^T e^{-i\theta t} = \sum_{-\infty}^{\infty} w_k^T e^{-i\theta(k+1)} = e^{-i\theta} \tilde{w}_T \left(\theta \right) .$$

Therefore, the first sum in the expression of $\eta \left(T \right)$ in (3.58) becomes

$$\sum_{-\infty}^{\infty} \mu_t^T \left[\tau \left(\mu_t^T / \bar{\varphi} + w_t^T \right) + \upsilon \left(w_t^T - w_{t-1}^T \right) - \delta \mu_t^T \right] =$$

$$= \frac{1}{2\pi} \int_{-\pi}^{\pi} \left[\tau / \bar{\varphi} + \mathrm{Re} \left(\tau + \upsilon \left(1 - e^{-i\theta} \right) \right) \gamma \left(e^{i\theta} \right) - \delta \right] \left| \tilde{\mu}_T \left(\theta \right) \right|^2 d\theta .$$

$$(3.59)$$

Assume now that there exist parameters $\tau \geq 0$, $\upsilon \geq 0$ such that the following frequency domain inequality holds

$$\frac{\tau}{\bar{\varphi}} + \mathrm{Re} \left(\tau + \upsilon \left(1 - z^{-1} \right) \right) \gamma \left(z \right) > 0 , \quad |z| = 1 , \quad (3.60)$$

where again $\gamma \left(z \right) = c^* \left(z \mathbf{I} - A \right)^{-1} b$ is the transfer function of the linear controlled system (3.14). Because $\gamma \left(z \right)$ has no singularities on the unit circle $|z| = 1$ due to the assumptions on the eigenvalues of A and the unit circle is a compact set there exists some $\delta_0 > 0$ such that

$$\frac{\tau}{\bar{\varphi}} + \mathrm{Re} \left(\tau + \upsilon \left(1 - z^{-1} \right) \right) \gamma \left(z \right) \geq \delta_0 > 0 , \quad |z| = 1 , \quad (3.61)$$

Taking $\delta = \delta_0$ in (3.57) the frequency domain inequality (3.61), equality (3.59) and (3.58) will give

$$\eta \left(T \right) \geq \sum_{0}^{T-1} \left(\tau + \theta \right) \mu_t^T \left(c^* A^t x_0 \right) - \theta \sum_{1}^{T-1} \mu_t^T c^* A^{t-1} x_0 . \quad (3.62)$$

But if $0 \leq t \leq T - 1$ then $\mu_t^T = -\varphi \left(c^* x_t \right)$ hence (3.15) and (3.45) hold. It follows that

$$
\eta(T) \leq -\delta_0 \sum_{0}^{T-1} |\varphi(c^*x_t)|^2 + \upsilon[\Phi(c^*x_0) - \Phi(c^*x_{T-1})] -
$$
$$
-\upsilon\varphi(c^*x_0)(c^*x_0)
$$

and combining it with (3.62) the main stability inequality is obtained

$$
\delta_0 \sum_{0}^{T-1} |\varphi(c^*x_t)|^2 + \upsilon\Phi(c^*x_{T-1}) \leq
$$
$$
\leq \sum_{0}^{T-1} \varphi(c^*x_t)\left[(\tau + \upsilon)c^*A^tx_0 - \upsilon c^*A^{t-1}x_0\right] +
$$
$$
+\upsilon[\Phi(c^*x_0) - \varphi(c^*x_0)(c^*x_0)] \leq
$$
$$
\leq 0.5\delta_0 \sum_{0}^{T-1} |\varphi(c^*x_t)|^2 + \frac{0.5}{\delta_0} \sum_{0}^{T-1} \left|(\tau + \upsilon)c^*A^tx_0 - \upsilon c^*A^{t-1}x_0\right|^2 +
$$
$$
+\upsilon[\Phi(c^*x_0) - \varphi(c^*x_0)(c^*x_0)] .
$$

$$(3.63)$$

In the above inequalities $\Phi(\sigma)$ is the one defined by (3.42) and the term $c^*A^{t-1}x_0$ is missing for $t = 0$ - see (3.62). From the main stability inequality (3.63) we have

$$
0.5\delta_0 \sum_{0}^{T-1} |\varphi(c^*x_t)|^2 \leq \frac{0.5}{\delta_0} \sum_{1}^{T-1} \left|(\tau + \upsilon)c^*A^tx_0 - \upsilon c^*A^{t-1}x_0\right|^2 +
$$
$$
+\frac{0.5}{\delta_0}(\tau + \theta)|c^*x_0|^2 + \upsilon\Phi(c^*x_0) .
$$

$$(3.64)$$

From now on the proof is identical to the proof of Theorem 3.1 and, as in the case of Theorem 3.2, we shall not repeat it; it is fully based on the estimate

$$
\left|c^*A^tx_0\right| \leq \beta|c||x_0|\rho^t , \quad 0 < \rho < 1
$$

and consists of some straightforward estimates.

We have, in fact, obtained the following result:

Theorem 3.3 *Let system* (3.12) *satisfy the following assumptions: i) matrix A has its eigenvalues inside the unit disk; ii) the nonlinear function $\varphi(\sigma)$ satisfies* (3.13) *and is nondecreasing; iii) there exist $\tau \geq 0$, $\upsilon \geq 0$ such that the frequency domain inequality* (3.60) *holds* *for all $|z| = 1$. Then the zero solution of system (3.12) is globally asymptotically stable that is the frequency domain condition implies absolute stability with respect to the class defined by ii).*

Comments

Only the backward difference constraint (3.45) has been used in obtaining the absolute stability criterion. Handling the forward difference is not an easy job in the framework of the classical technique of obtaining frequency domain inequalities for absolute stability. It can be done however by Liapunov techniques or by more recent technique of *"saturability"* (Popov (1977 a)). Consider now again the example described by (3.34) where, as shown previously, the linear stability sector was $(-1 + |a|, 1)$. Our aim is to obtain the largest $\bar\varphi \le 1$ for the absolute stability using the frequency domain inequality of Theorem 3.3. Because the assumption of nondecreasing function is not compatible with the sectors of the type $(\underline\varphi, 0)$ where $\underline\varphi < 0$, the result of Theorem 3.3 is significant only for the positive part of the sector.

The frequency domain condition in Theorem 3.3 reduces to the one in Theorem 3.1 if $v = 0$. Recall that when applying Theorem 3.1 to get the maximal sector we had

$$\frac{1}{\bar\varphi} + \min_{|z|=1} \operatorname{Re} \gamma(z) > 0$$

that is

$$\bar\varphi < \left[- \inf_{|z|=1} \operatorname{Re} \gamma(z) \right]^{-1}.$$

From the frequency condition in Theorem 3.3 we obtain the constraint

$$\bar\varphi < \left[\sup_{v>0} \inf_{|z|=1} \operatorname{Re} \left(1 + v\left(1 - z^{-1}\right)\right) \gamma(z) \right]^{-1}$$

which is obviously larger than the previous one.

3.4. THE BROCKETT-WILLEMS TYPE CRITERION FOR SYSTEMS WITH NONDECREASING NONLINEARITY

In a joint cycle of two papers, Brockett and Willems (1965) were able, by using the additional information about monotonicity, to improve the usual Popov frequency domain inequality in the continuous time case. Instead of the Popov multiplier $1 + iwq$ a general multiplier $Z(iw)$ was introduced This multiplier depends on several free parameters; consequently, the stability inequality might improve the stability sector. The price to be paid is the difficulty in handling a larger number of parameters. A corresponding criterion will be obtained here for discrete-time systems.

Consider the system described by (3.12) with nondecreasing $\varphi(\sigma)$ satisfying (3.13). Replacing (3.12) by the controlled linear system (3.14) with the control function $\mu_t = -\varphi(\sigma_t)$, the couple (μ, σ) is subject to the quadratic constraints (3.15) and (3.44):

$$W_1(\mu, \sigma) \equiv \mu(\mu/\bar{\varphi} + \sigma) \leq 0 \qquad (3.65)$$

$$\sum_1^t W_2(\mu_k, \sigma_k, \sigma_{k-1}) \equiv \sum_1^t \mu_k(\sigma_k - \sigma_{k-1}) \leq \Phi(\sigma_0), \qquad (3.66)$$

where $\Phi(\sigma) = \int_0^\sigma \varphi(\lambda)\mathrm{d}\lambda$.

Following Brockett and Willems (1965) we consider the system with an augmented dimension of the state space

$$x_{t+1} = Ax_t - b\varphi(c^* x_t)$$
$$\xi_{t+1}^j = \frac{1}{\gamma_j + \rho_j}\left[\alpha_j\xi_t^j + c^* Ax_t - c^* b\varphi(c^* x_t)\right], \quad j = \overline{1, p} \qquad (3.67)$$

Obviously the added part is controlled by the basic subsystem. We shall assume $\rho_j > 0$, $\gamma_j > \alpha_j \geq 0$ to be arbitrary parameters; the above assumptions ensure in any case the inherent stability of the added linear subsystem.

Using the added part and the monotonicity of $\varphi(\sigma)$ we may obtain some new quadratic constraints

$$\sum_{k=1}^t W_{2+j}\left(\mu_k, \sigma_k, \xi_k^j\right) \equiv \sum_{k=1}^t \mu_k\left(\sigma_k - \rho_j\xi_k^j\right) =$$
$$= -\sum_{k=0}^{t-1} \varphi(\sigma_{k+1})\left(\sigma_{k+1} - \rho_j\xi_{k+1}^j\right), \ j = \overline{1, p}$$

It is easily seen that in (3.67) the variable ξ^j satisfies the equation

$$\xi_{t+1}^j = \frac{1}{\gamma_j + \rho_j} \left(\alpha_j \xi_t^j + \sigma_{t+1} \right);$$

hence,

$$\sigma_{t+1} - \rho_j \xi_{t+1}^j = \gamma_j \xi_{t+1}^j - \alpha_j \xi_t^j.$$

Therefore,

$$-\sum_{k=0}^{t-1} \varphi\left(\sigma_{k+1}\right)\left(\sigma_{k+1} - \rho_j \xi_{k+1}^j\right) =$$

$$= -\sum_{k=0}^{t-1} \left[\varphi\left(\sigma_{k+1}\right) - \varphi\left(\rho_j \xi_{k+1}^j\right)\right]\left(\sigma_{k+1} - \rho_j \xi_{k+1}^j\right) -$$

$$-\sum_{k=0}^{t-1} \varphi\left(\rho_j \xi_{k+1}^j\right)\left(\sigma_{k+1} - \rho_j \xi_{k+1}^j\right) \le$$

$$\le -\sum_{k=0}^{t-1} \varphi\left(\rho_j \xi_{k+1}^j\right)\left(\gamma_j \xi_{k+1}^j - \alpha_j \xi_k^j\right) =$$

$$= -\left(\gamma_j - \alpha_j\right)\frac{1}{\rho_j}\sum_{k=0}^{t-1} \varphi\left(\rho_j \xi_{k+1}^j\right)\rho_j \xi_{k+1}^j -$$

$$-\frac{\alpha_j}{\rho_j}\sum_{k=0}^{t-1} \varphi\left(\rho_j \xi_{k+1}^j\right)\left(\rho_j \xi_{k+1}^j - \rho_j \xi_k^j\right) \le \frac{\alpha_j}{\rho_j}\Phi\left(\rho_j \xi_0^j\right).$$

We obtained in fact the following quadratic constraints

$$\sum_{k=1}^{t} W_{2+j}\left(\mu_k, \sigma_k, \xi_k^j\right) \equiv \sum_{k=1}^{t} \mu_k\left(\sigma_k - \rho_j \xi_k^j\right) \le \frac{\alpha_j}{\rho_j}\Phi\left(\rho_j \xi_0^j\right), \quad j = \overline{1,p}$$

$$(3.68)$$

where again $\Phi\left(\sigma\right) = \int\limits_{0}^{\sigma} \varphi\left(\lambda\right)\mathrm{d}\lambda$.

In obtaining (3.68) use was made of the monotonicity, of the sector condition (3.13) that implies $\varphi\left(\sigma\right)\sigma \ge 0$ and also of the assumption on the parameters $\rho_j, \alpha_j, \gamma_j$.

We are now in position to associate a Popov system of the form

$$\sigma_t = c^* A^t x_0 + \sum_0^{t-1} \left(c^* A^{t-1-k} b \right) \mu_k \qquad (3.69)$$

$$\xi_t^j = \left(\frac{\alpha_j}{\gamma_j + \rho_j} \right)^t \xi_0^j + \sum_0^{t-1} \left(\frac{\alpha_j}{\gamma_j + \rho_j} \right)^{t-1-k} \frac{1}{\gamma_j + \rho_j} \sigma_{k+1} , \; j = \overline{1,p} \qquad (3.70)$$

$$\eta\left(T\right) = \tau \sum_1^{T-1} \mu_t \left(\mu_t / \bar{\varphi} + \sigma_t \right) + \upsilon \sum_1^{T-1} \mu_t \left(\sigma_t - \sigma_{t-1} \right) +$$
$$+ \sum_{t=1}^{T-1} \sum_{j=1}^p \delta_j \mu_t \left(\sigma_t - \rho_j \xi_t^j \right) - \delta \sum_1^{T-1} \mu_t^2 . \qquad (3.71)$$

In the above expressions the following parameters are arbitrary: $p \geq 1$, $\tau \geq 0$, $\upsilon \geq 0$, $\delta_j \geq 0$, $\alpha_j \geq 0$, $\gamma_j \geq \alpha_j$, $\rho_j > 0$, $\delta > 0$.

We define the truncated control μ_t^T as previously, i.e., by (3.16) and the corresponding outputs as follows

$$\sigma_t^T = c^* A^t x_0 + \sum_0^{t-1} \left(c^* A^{t-1-k} b \right) \mu_k^T , \quad t \geq 1 \qquad (3.72)$$

$$\xi_t^{jT} = \left(\frac{\alpha_j}{\gamma_j + \rho_j} \right)^t \xi_0^j + \sum_0^{t-1} \left(\frac{\alpha_j}{\gamma_j + \rho_j} \right)^{t-1-k} \frac{1}{\gamma_j + \rho_j} \sigma_{k+1}^T ,$$
$$t \geq 1, \; j = \overline{1,p}. \qquad (3.73)$$

We assume again that the eigenvalues of A lie inside the unit disk of the complex plane. Therefore, the estimates for $c^* A^t x_0$, for the weighting patterns h_t and for σ_t^T obtained in Section 3.1 are valid. We recall that $\mu_t^T = 0$ for $t < 0$ and $t \geq T$ and take $\sigma_t^T = 0$ for $t < 0$.

Concerning the variables ξ_t^{jT}, they play an auxiliary role and are arbitrarily defined. Without any loss of generality for the stability problem we may take $\xi_0^j = 0$, $j = \overline{1,p}$. Therefore, ξ_t^{jT} is a discrete convolution. Clearly $0 \leq \alpha_j \left(\gamma_j + \rho_j \right)^{-1} < 1$. Denoting as previously

$$w_t^T = \sum_0^{t-1} \left(c^* A^{t-1-k} b \right) \mu_k^T = \sum_{-\infty}^{\infty} h_{t-k} \mu_k^T$$

with $h_t = 0$, $t \le 0$ we may write

$$\sigma_t^T = c^* A^t x_0 + w_t^T .$$

Consequently we have

$$
\begin{aligned}
\xi_t^{jT} &= \sum_0^{t-1} \left(\frac{\alpha_j}{\gamma_j + \rho_j} \right)^{t-1-k} \frac{1}{\gamma_j + \rho_j} w_{k+1}^T + \\
&\quad + \sum_0^{t-1} \left(\frac{\alpha_j}{\gamma_j + \rho_j} \right)^{t-1-k} \frac{1}{\gamma_j + \rho_j} c^* A^t x_0 \\
&= \left(\xi_t^{jT} \right)' + \left(\xi_t^{jT} \right)''
\end{aligned}
$$

Using a standard technique based on the Cauchy-Schwarz inequality (see also Section 1) we obtain that both $\left(\xi_t^{jT} \right)'$ and $\left(\xi_t^{jT} \right)''$ are in $\ell^1 \cap \ell^2$ for all j. Indeed

$$
\begin{aligned}
\sum_0^\infty \left| \left(\xi_t^{jT} \right)' \right| &\le \frac{1}{\gamma_j + \rho_j} \sum_{t=0}^\infty \sum_{k=0}^{t-1} \left(\frac{\alpha_j}{\gamma_j + \rho_j} \right)^{t-1-k} \left| w_{k+1}^T \right| = \\
&= \frac{1}{\gamma_j + \rho_j} \sum_{k=0}^\infty \left(\sum_{k+1}^\infty \left(\frac{\alpha_j}{\gamma_j + \rho_j} \right)^{t-1} \right) \left(\frac{\alpha_j}{\gamma_j + \rho_j} \right)^{-k} \left| w_{k+1}^T \right| = \\
&= \frac{1}{\gamma_j + \rho_j - \alpha_j} \sum_0^\infty \left| w_{k+1}^T \right|
\end{aligned}
$$

The convergence of the last series follows from the definition of w_t^T as a discrete convolution, also from the facts that μ_t^T is a finite sequence and A has the eigenvalues inside the unit disk. Analogously

$$
\sum_0^\infty \left| \left(\xi_t^{jT} \right)'' \right| \le \frac{1}{\gamma_j + \rho_j - \alpha_j} \sum_0^\infty |c| \, \beta \, |x_0| \, \rho^{k+1} , \quad 0 < \rho < 1
$$

Similarly we obtain

$$
\sum_0^\infty \left| \left(\xi_t^{jT} \right)' \right|^2 \le \frac{\gamma_j + \rho_j}{(\gamma_j + \rho_j + \alpha_j)(\gamma_j + \rho_j - \alpha_j)} \sum_0^\infty \left| w_{k+1}^T \right|^2
$$

$$
\sum_0^\infty \left| \left(\xi_t^{jT} \right)'' \right|^2 \le \frac{(\gamma_j + \rho_j) \, \beta_1 \, |c|^2 \, |x_0|^2}{(\gamma_j + \rho_j + \alpha_j)(\gamma_j + \rho_j - \alpha_j)} \sum_0^\infty \rho^{2(k+1)}
$$

The index $\eta(T)$ may be written as follows

$$\eta(T) = \sum_{-\infty}^{\infty} \mu_t^T \left[\tau \left(\frac{\mu_t^T}{\bar{\varphi}} + w_t^T \right) + v \left(w_t^T - w_{t-1}^T \right) + \right.$$

$$+ \sum_{j=1}^{p} \delta_j \left(w_t^T - \rho_j \left(\xi_t^{jT} \right)' \right) - \delta \mu_t^T \right] + \tau \varphi(\sigma_0) \left(\sigma_0 - \frac{\varphi(\sigma_0)}{\bar{\varphi}} \right) +$$

$$+ \left(v + \sum_{1}^{p} \delta_j \right) \varphi(\sigma_0) \sigma_0 + \delta \varphi^2(\sigma_0) +$$

$$+ \sum_{1}^{T-1} \mu_t^T \left[\left(\tau + v + \sum_{1}^{p} \delta_j \right) c^* A^t x_0 - v c^* A^{t-1} x_0 - \sum_{j=1}^{p} \delta_j \rho_j \left(\xi_t^{jT} \right)'' \right]$$

$$\geq \sum_{-\infty}^{\infty} \mu_t^T \left[\tau \left(\frac{\mu_t^T}{\bar{\varphi}} + w_t^T \right) + v \left(w_t^T - w_{t-1}^T \right) + \right.$$

$$+ \sum_{j=1}^{p} \delta_j \left(w_t^T - \rho_j \left(\xi_t^{jT} \right)' \right) - \delta \mu_t^T \right] +$$

$$+ \sum_{1}^{T-1} \mu_t^T \left[\left(\tau + v + \sum_{1}^{p} \delta_j \right) c^* A^t x_0 - v c^* A^{t-1} x_0 - \right.$$

$$- \sum_{j=1}^{p} \frac{\delta_j \rho_j}{\gamma_j + \rho_j} \sum_{k=0}^{t-1} \left(\frac{\alpha_j}{\gamma_j + \rho_j} \right)^{t-1-k} c^* A^{k+1} x_0 \right] + \delta \varphi^2(\sigma_0)$$

The last inequality has been deduced by making use of the sector condition (3.13). Denote by $\eta_1(T)$ the first sum in the RHS of the above inequality; it will be expressed using DCFT and Plancherel-Parseval equality. We shall have

$$\tilde{w}_T(\theta) = \sum_{-\infty}^{\infty} w_t^T e^{-i\theta t} = \tilde{h}(\theta) \tilde{\mu}_T(\theta) ,$$

$$\tilde{h}(\theta) = \sum_{-\infty}^{\infty} h_t e^{-i\theta t} = \sum_{1}^{\infty} c^* A^{t-1} b e^{-i\theta t} =$$

$$= c^* \left(e^{i\theta} I - A \right)^{-1} b = \gamma(e^{i\theta})$$

$$\sum_{-\infty}^{\infty} w_{t-1}^T e^{-i\theta t} = e^{-i\theta} \tilde{w}_T(\theta) = e^{-i\theta} \gamma\left(e^{i\theta}\right) \tilde{\mu}_T(\theta)$$

$$\sum_{-\infty}^{\infty} \left(\xi_t^{jT}\right)'' e^{-i\theta t} = \sum_{t=0}^{\infty} \left(\sum_{k=0}^{t-1} \left(\frac{\alpha_j}{\gamma_j + \rho_j}\right)^{t-1-k} \frac{1}{\gamma_j + \rho_j} w_{k+1}^T \right) e^{-i\theta t} =$$

$$= \frac{1}{\gamma_j + \rho_j} \sum_{k=0}^{\infty} \left(\sum_{t=k+1}^{\infty} \left(\frac{\alpha_j}{\gamma_j + \rho_j}\right)^{t-1-k} e^{-i\theta t} \right) w_{k+1}^T =$$

$$= \frac{1}{\gamma_j + \rho_j - \alpha_j e^{-i\theta}} \sum_{1}^{\infty} w_k^T e^{-i\theta k} = \frac{1}{\gamma_j + \rho_j - \alpha_j e^{-i\theta}} \gamma\left(e^{i\theta}\right) \tilde{\mu}_T(\theta)$$

$$\eta_1(T) = \frac{1}{2\pi} \int_{-\pi}^{\pi} \left\{ \frac{\tau}{\bar{\varphi}} + \mathrm{Re}\left[\tau + \upsilon\left(1 - e^{-i\theta}\right) + \right.\right.$$
$$\left.\left. + \sum_{j=1}^{p} \delta_j \frac{\gamma_j - \alpha_j e^{-i\theta}}{\gamma_j + \rho_j - \alpha_j e^{-i\theta}} \right] \gamma\left(e^{i\theta}\right) - \delta \right\} |\tilde{\mu}_T(\theta)|^2 \, d\theta$$

$$(3.74)$$

Assume now there exist the integer $p \geq 1$ and the parameters $\tau \geq 0$, $\upsilon \geq 0$, $\delta_j \geq 0$, $\gamma_j \geq \alpha_j \geq 0$, $\rho_j > 0$, $j = \overline{1,p}$ such that the following frequency domain inequality holds

$$\frac{\tau}{\bar{\varphi}} + \mathrm{Re}\left[\tau + \upsilon\left(1 - z^{-1}\right) + \sum_{j=1}^{p} \delta_j \frac{\gamma_j - \alpha_j z^{-1}}{\gamma_j + \rho_j - \alpha_j z^{-1}}\right] \gamma(z) > 0 \ , \ |z| = 1.$$

$$(3.75)$$

Therefore, there exists some $\delta_0 > 0$ such that

$$\frac{\tau}{\bar{\varphi}} + \mathrm{Re}\left[\tau + \upsilon\left(1 - z^{-1}\right) + \sum_{j=1}^{p} \delta_j \frac{\gamma_j - \alpha_j z^{-1}}{\gamma_j + \rho_j - \alpha_j z^{-1}}\right] \gamma(z) \geq \delta_0 > 0 \ ,$$

$$|z| = 1$$
$$(3.76)$$

Taking $\delta = \delta_0$ in (3.71), the frequency domain inequality (3.76) will give $\eta_1(T) \geq 0$; hence

$$
\eta(T) \geq \delta\varphi^2(\sigma_0) + \sum_1^{t-1} \mu_t^T \left[\left(\tau + \upsilon + \sum_1^p \delta_j \right) c^* A^t x_0 - \upsilon c^* A^{t-1} x_0 - \right.
$$
$$
\left. - \sum_{j=1}^p \frac{\delta_j \rho_j}{\gamma_j + \rho_j} \sum_{k=0}^{t-1} \left(\frac{\alpha_j}{\gamma_j + \rho_j} \right)^{t-1-k} c^* A^{k+1} x_0 \right]
$$
$$(3.77)$$

On the other side if $1 \leq t \leq T-1$ then $\mu_t^T = -\varphi(\sigma_t)$ and (3.65), (3.66), (3.67) hold. It follows that:

$$
\eta(T) \leq -\delta_0 \sum_1^{T-1} |\varphi(\sigma_t)|^2 + \upsilon\Phi(\sigma_0) + \sum_1^p (\delta_j \alpha_j/\rho_j) \Phi\left(\rho_j \xi_0^j\right) \quad (3.78)
$$

Combining (3.77) and (3.78) together with the choice $\xi_0^j = 0$, $j = \overline{1,p}$ we obtain the main stability inequality

$$
\delta_0 \sum_0^{T-1} |\varphi(\sigma_t)|^2 \leq \upsilon\Phi(\sigma_0) + \sum_1^{T-1} \varphi(\sigma_t) \left[\left(\tau + \upsilon + \sum_1^p \delta_j \right) c^* A^t x_0 - \right.
$$
$$
\left. -\upsilon c^* A^{t-1} x_0 - \sum_{j=1}^p \frac{\delta_j \rho_j}{\gamma_j + \rho_j} \sum_{k=0}^{t-1} \left(\frac{\alpha_j}{\gamma_j + \rho_j} \right)^{t-1-k} c^* A^{k+1} x_0 \right]
$$

From here we obtain as previously

$$
0.5\delta_0 \sum_0^{T-1} |\varphi(\sigma_t)|^2 \leq \upsilon\Phi(\sigma_0) + \frac{0.5}{\delta_0} \sum_1^{T-1} \left| \left(\tau + \upsilon + \sum_1^p \delta_j \right) c^* A^t x_0 - \right.
$$
$$
\left. -\upsilon c^* A^{t-1} x_0 - \sum_{j=1}^p \frac{\delta_j \rho_j}{\gamma_j + \rho_j} \sum_{k=0}^{t-1} \left(\frac{\alpha_j}{\gamma_j + \rho_j} \right)^{t-1-k} c^* A^{k+1} x_0 \right|^2
$$
$$(3.79)$$

This inequality is of the type previously obtained. From now on the proof is identical to the proofs of the previous theorems and it will not be repeated. In fact the estimate

$$
|c^* A^t x_0| \leq \beta |c| |x_0| \rho^t, \quad 0 < \rho < 1
$$

allows one to obtain from (3.79) that

$$\sum_{0}^{T-1} |\varphi(\sigma_t)|^2 \leq K(|x_0|)$$

with $\lim_{\lambda \to 0} K(\lambda) = 0$, hence $\{\varphi(\sigma_t)\}_t \in \ell^2$; this will give $|x_t| \leq K_1(|x_0|)$ with $K_1(0) = 0$ and this is Liapunov stability. Further we obtain $\{x_t\}_t \in \ell^2$ and from here $x_t \to 0$, i.e., asymptotic stability.

Therefore, the following result has been obtained:

Theorem 3.4 *Let system* (3.12) *satisfy the following assumptions: i) matrix A has its eigenvalues inside the unit disk; ii) the nonlinear function $\varphi(\sigma)$ satisfies* (3.13) *and is nondecreasing; iii) there exist the integer $p \geq 1$ and the real parameters $\tau \geq 0$, $\upsilon \geq 0$, $\gamma_j \geq \alpha_j \geq 0$, $\rho_j > 0$, $j = \overline{1,p}$ such that the frequency domain inequality* (3.76) *holds for all $|z| = 1$. Then the zero solution of system* (3.12) *is globally asymptotically stable that is the frequency domain condition implies absolute stability with respect to the class defined by ii).*

Comments

The problem of multiplier significance and choice in absolute stability theory by frequency domain inequalities has received some attention in the past especially in the framework of graphical representations. After the papers of Brockett and Willems (1965) it became clear that the multiplier $Z(s)$ in the frequency domain inequality

$$\frac{1}{k} + \text{Re}\, Z(i\omega)\, \gamma(i\omega) > 0$$

plays the role of a correcting device that makes the linear part of the system become passive (having the modified transfer function positive real). In particular, the classical Popov multiplier $1 + qs$ is called sometimes PD - proportional derivative - multiplier because it corresponds to the so-called ideal PD controller described by the input/output relation

$$u(t) = k_p \left[e(t) + T_d \frac{de}{dt}(t) \right],$$

where $u(t)$ is the output signal of the controller and $e(t)$, called the control error, is the input signal of the controller. Obviously the transfer function of this controller is

$$\frac{\tilde{u}(s)}{\tilde{e}(s)} = k_p (1 + T_d s) .$$

It is an improper transfer function corresponding to a so-called non-causal system. In fact this is an idealized version of the PD controller, it realistic version being described for instance by

$$\frac{\tilde{u}(s)}{\tilde{e}(s)} = k_p \frac{1 + T_d s}{1 + \varepsilon s} ,$$

where $\varepsilon > 0$ is a small parameter. In the Brockett-Willems criterion $Z(s)$ is a proper transfer function corresponding to a causal multiplier but there exist absolute stability criteria based on noncausal multipliers (e.g., Venkatesh (1986)). These interpretations of the stability multipliers are valid for the discrete-time systems also. Indeed, let us consider the stability inequality (3.60) for $\tau = 1$:

$$\frac{1}{\bar{\varphi}} + \mathrm{Re}\left(1 + v\frac{z-1}{z}\right)\gamma(z) > 0 , \quad |z| = 1$$

and also the input/output relation of the PD controller written at $t = k\delta$ using backward differences for the derivative. Denoting $u(k\delta) = u_k$, $e(k\delta) = e_k$ we obtain the discrete input-output relation

$$u_k = k_p \left[e_k + T_d \frac{e_k - e_{k-1}}{\delta}\right] .$$

Applying the Z-transform we find

$$\frac{\tilde{u}(z)}{\tilde{e}(z)} = k_p \left[1 + \frac{T_p}{\delta} \cdot \frac{z-1}{z}\right]$$

which corresponds to the multiplier of the stability inequality (3.60). With respect to this we may consider the extended Tsypkin criterion (3.60) as a discrete version of the Popov criterion. Remark that the multiplier of PD-type is proper and causal in the discrete case; this is due to the fact that we used backward differences.

The multiplier in the discrete Brockett-Willems type criterion (3.75) is also causal and proper due to the backward differences used in writing down the augmented system (3.67).

Forward differences are likely to lead to non-causal multipliers. From here are the difficulties in handling such cases; they may possibly be overcome by the "saturability technique" of Popov (1977).

3.5. LIAPUNOV FUNCTIONS AND FREQUENCY DOMAIN INEQUALITIES IN ABSOLUTE STABILITY. THE KSPY LEMMA (SINGLE-INPUT CASE)

The problem of absolute stability was studied about 50 years ago in the continuous time case by Lurie and others using the Liapunov functions of a special form (quadratic form in the state variables and, possibly, an integral of the nonlinearity). This has been one of the first cases of applying the "second method of Liapunov" in modern engineering. The approach was quite successful and for some cases it gave very good information about the stability domain in the parameter space of the system. In 1958-1959 V.M. Popov in Romania (Popov, 1959) discovered a new (at that time) method for studying this problem and gave frequency domain criteria for absolute stability. His method, based on integral representations for the solutions, was extensively developed by Halanay, Corduneanu, Sandberg, Zames, and many others. At that time Popov was able to prove that if there exists a Liapunov function of the form used by Lurie and others then his frequency domain condition is satisfied. In this way it became clear that frequency domain criteria gave results no worse than the method of Liapunov functions and after the results of Popov became known it was the end (for a while) of the Liapunov method in the problem of absolute stability. This might explain the fact that for discrete time systems the study of the absolute stability started directly in the framework of the frequency domain inequalities.

3.5.1. For and Against Liapunov Functions in the Study of Nonlinear Systems

As it was pointed out previously the early results of Popov put an end to the use of Liapunov functions in absolute stability. Nevertheless, Yakubovich (1962 a), then Kalman (1963) and finally Popov himself (1964) proved an important algebraic lemma from which it followed that if the frequency domain condition of Popov was satisfied then there existed a Liapunov function of the form considered. This lemma signified not only the perfect equivalence between the two methods (for finite dimensional systems) but also that the frequency domain condition corresponded to the best choice of the Liapunov function.

The Kalman - Yakubovich - Popov lemma (sometimes called "- Positive Real Lemma") had a considerable impact on further development in stability and other problems of the theory of dynamical and control systems (optimal control, disturbance attenuation, H_∞ -control, nonlinear vibrations). This explains its extensions to other classes of systems as discrete time or infinite dimensional.

A most significant application of the lemma was the result of Yakubovich (1964) giving frequency domain conditions for the existence of periodic or almost periodic solutions of nonlinear systems. In fact there was a rather common idea that Liapunov functions may be used to prove the existence of periodic and sometimes even almost periodic solutions. This idea is very transparent, for instance, in the two books of Yoshizawa (1966, 1975). In the framework of the "-Positive Real Lemma" a frequency domain inequality which gives the existence of a suitable Liapunov function - even a quadratic form - will give a condition for the existence of periodic or almost periodic solutions. Such conditions work very well even in specific nonlinear vibrations problems; the reader is sent to the joint papers of Barbălat and Halanay (1971, 1974). All these results speak at the same time for and against Liapunov functions: the theoretical results, use Liapunov functions in the proofs; the existence is ensured by the frequency domain condition via Yakubovich - Kalman - Popov lemma but the result itself does not contain any Liapunov function and works better than the direct use of Liapunov function.

At the middle of the 1970s it was considered as a well established fact that the use of Liapunov functions as a theoretical tool supported by the Lemma gave much more possibilities in establishing sufficient conditions for various qualitative behavior: stability and instability, forced and self-sustained oscillations (Yakubovich (1977 a), Gelig, Leonov and Yakubovich (1978)). On the other hand, it was also considered as a well established fact that the method of V.M. Popov is more suitable for applications to time-delay systems and systems described by integral, integro-differential or partial differential equations for which the Lemma did not work. In this direction we have to cite the basic papers of Kurzweil (1967) and Halanay (1967) on invariant manifolds for flows on Banach spaces that allowed the applications of Popov methods to the problem of forced oscillations in systems with time lag (Halanay (1969), Halanay and Răsvan (1977, 1979)). Also Popov (1974) in a seminal paper showed the way of obtaining dichotomy criteria without using Liapunov functions.

In the same period the extensions of the Lemma to infinite dimensional systems were obtained (Yakubovich (1974, 1975), Likhtarnikov (1976), Likhtarnikov and Yakubovich (1977)) including their discrete

counter part (Antonov, Likhtarnikov and Yakubovich (1975), Likhtarnikov, Ponomarenko and Yakubovich (1976)) which gave a new impulse to the research on qualitative behavior for such systems. Worthy
of mention is that the Yakubovich - Kalman - Popov Lemma is the
basis of hyperstability theory of Popov (1973), strongly connected,
as we have shown in the previous chapters, with dissipativity and
passivity (Răsvan (1978)).

3.5.2. The Lemma for Discrete Systems

Many of the problems that may be formulated for continuous-time
systems are the same for discrete-time systems. For instance, the
simplest frequency-domain inequality of Tsypkin - see Theorem 3.1 -
is a necessary condition for the existence of a Liapunov function in
the class of the quadratic forms. Indeed, if we consider again (3.12)
with $\varphi(\sigma)$ subject to sector conditions (3.13) and with A having its
eigenvalues inside the unit disk, we may try to obtain the absolute
stability condition with a quadratic Liapunov function of the form

$$V(x) = x^* P x \qquad (3.80)$$

assuming that $P > 0$. Along the solutions of (3.12) we shall have

$$V(x_{t+1}) - V(x_t) = (Ax_t - b\varphi(c^* x_t))^* P x_t + \\ + x_t^* P (Ax_t - b\varphi(c^* x_t)) - x_t^* P x_t = \\ = x_t^* (A^* PA - P) x_t - \varphi(c^* x_t) b^* P x_t - x_t^* P b \varphi(c^* x_t)$$

A sufficient condition for global asymptotic stability is the choice
of such a $P > 0$ that the following quadratic form

$$W(x, \mu) = (x^* \ \mu^*) \begin{pmatrix} A^* PA - P & -Pb \\ -b^* P & 0 \end{pmatrix} \begin{pmatrix} x \\ \mu \end{pmatrix}$$

is negative definite on the set where the quadratic constraint (3.15)
is fulfilled. Studying the sign of a quadratic form on some set defined
by quadratic inequalities is not easy. Usually this problem is replaced
by another applying the so-called $S - procedure$; by writing

$$W(x, \mu) = W(x, \mu) - \tau\mu(\mu + \bar{\varphi}c^* x) + \tau\mu(\mu + \bar{\varphi}c^* x) \ ,$$

where $\tau > 0$, it is required that the quadratic form

$$S(x, \mu) = W(x, \mu) - \tau\mu(\mu + \bar{\varphi}c^* x) \qquad (3.81)$$

is negative definite for all (x, μ). One may ask whether this procedure does not provide more restrictive conditions for absolute stability but a rather general result of Yakubovich (1971) shows that this is not the case.

Assume that $P > 0$ is such that $S(x, \mu) < 0$ for any x, μ. In particular this is true for complex $\tilde{x}, \tilde{\mu}$ satisfying

$$z\tilde{x} = A\tilde{x} + b\tilde{\mu}, \quad |z| = 1.$$

Therefore,

$$
\begin{aligned}
S(\tilde{x}, \tilde{\mu}) &= (\tilde{x}^* \; \tilde{\mu}^*) \left(
\begin{array}{cc}
A^* PA - P & -Pb - \tfrac{1}{2}\tau\bar{\varphi}c \\
-(Pb + \tfrac{1}{2}\tau\bar{\varphi}c)^* & \tau
\end{array}
\right) \left(
\begin{array}{c}
\tilde{x} \\
\tilde{\mu}
\end{array}
\right) = \\
&= (\tilde{x}^* A^* + \tilde{\mu}^* b^*) P (A\tilde{x} + b\tilde{\mu}) - \tilde{x}^* P\tilde{x} - \tau |\tilde{\mu}|^2 - \tfrac{1}{2}\tau\bar{\varphi}\tilde{\mu}^* c^* \tilde{x} - \\
&\quad - \tfrac{1}{2}\tau\bar{\varphi}\tilde{x}^* c\tilde{\mu} = \tilde{x}^* P\tilde{x} |z|^2 - \tilde{x}^* P\tilde{x} - \tau |\tilde{\mu}|^2 - \mathrm{Re}\,[\tau\bar{\varphi}\tilde{\mu}^* \gamma(z)\tilde{\mu}] = \\
&= \tilde{x}^* P\tilde{x} \left(|z|^2 - 1\right) - [\tau + \mathrm{Re}\,\tau\bar{\varphi}\tilde{\mu}^* \gamma(z)] |\tilde{\mu}|^2 < 0
\end{aligned}
$$

Since $|z| = 1$ the first term vanishes; the choice $\tau = 1/\bar{\varphi}$ shows that for the fulfillment of the above inequality it is necessary that

$$\frac{1}{\bar{\varphi}} + \mathrm{Re}\,\gamma(z) > 0, \quad |z| = 1$$

which is nothing else than the frequency domain inequality of Theorem 3.1.

The converse assertion obviously leads to the discrete counterpart of the Lemma. The Lemma for discrete-time systems was proved in the papers of Szegö and Kalman (1963), and Szegö (1963 a) , using the same method applied by Kalman (1963) for continuous-time systems. This method is based on the factorization of polynomials and polynomial matrices and was applied and developed in the work of Popov (1964, 1973) and Yakubovich (1973); in particular both Popov (1973) and Yakubovich are considering the continuous-time and discrete-time cases in a unifying framework.

Besides the method based on factorization there exist several other methods of proof as mentioned in the book of Gelig, Leonov and Yakubovich (1978); algebraic methods, due to Yakubovich (1970) and Kalman (1970); based on Hahn-Banach theorem, due to Nudelman and Svartsman (1975) and especially the method based on solving an auxiliary optimization problem via dynamic programming (Yakubovich (1974, 1977), Likhtarnikov (1977), Likhtarnikov and Yakubovich (1976, 1977), Brusin (1976 a, b)). Among these methods the last one showed itself to be the most suitable for infinite dimensional systems; within its framework factorization is avoided and in fact it follows as a consequence of the main result of the Lemma. As already mentioned, for discrete systems this approach occurs in the papers of Antonov, Likhtarnikov and Yakubovich (1975) and Likhtarnikov, Ponomarenko and Yakubovich (1976).

3.5.3. Problem Statement

In the following we shall give the Lemma for discrete-time systems as it occurs in Popov (1973). The statement as well as the proof will follow §8 of the cited book of Popov.

We shall first define the framework. The main object of study will be the set (A, b, κ, ℓ, M) where A, M are $n \times n$ matrices, b and ℓ are n-vectors and κ a scalar. All the items of the set are complex; it is assumed that M is Hermitian and κ is real. The set is associated to the following Popov system

$$x_{t+1} = Ax_t + b\mu_t \tag{3.82}$$

$$\eta(T) = \sum_0^{T-1} \mathcal{F}(x_t, \mu_t) \tag{3.83}$$

$$\mathcal{F}(x, \mu) = x^* M x + x^* \ell \mu + \mu^* \ell^* x + \mu^* \kappa \mu \tag{3.84}$$

which is defined on the pair (μ_t, x_t). We associate also *the character- istic function* $\chi(\lambda, \sigma)$ of the system defined by

$$\begin{aligned}\chi(\lambda, \sigma) &= \kappa + \ell^* (\sigma \mathbf{I} - A)^{-1} b + b^* (\lambda \mathbf{I} - A^*)^{-1} \ell + \\ &\quad + b^* (\lambda \mathbf{I} - A^*)^{-1} M (\sigma \mathbf{I} - A)^{-1} b\end{aligned} \tag{3.85}$$

and also the characteristic polynomial

$$\pi(\lambda, \sigma) = \frac{1}{\nu} \det(\lambda \mathbf{I} - A^*) \det(\sigma \mathbf{I} - A) \chi(\lambda, \sigma) \tag{3.86}$$

where ν is a scaling factor. Remark that

$$\chi(\lambda, \sigma) = \overline{\chi(\bar{\sigma}, \bar{\lambda})}, \quad \pi(\lambda, \sigma) = \overline{\pi(\bar{\sigma}, \bar{\lambda})}. \tag{3.87}$$

From this it follows that if $\pi\left(\lambda,\sigma\right)$ is written as

$$\pi\left(\lambda,\sigma\right) = \frac{1}{\nu}\sum_{0}^{n}\sum_{0}^{n}\kappa_{jk}\lambda^{j}\sigma^{k}\,, \qquad (3.88)$$

then $\kappa_{jk} = \bar{\kappa}_{kj}$. By choosing $\nu = \max|\kappa_{jk}|$ if there exists some non-zero κ_{jk} and $\nu = 1$ if $\pi\left(\lambda,\sigma\right) \equiv 0$ all coefficients of $\pi\left(\lambda,\sigma\right)$ are less than 1. Remark that in most cases it is not necessary to define $\nu > 0$ because it is simplified during the operations with $\pi\left(\lambda,\sigma\right)$.

We shall use the following

Definition 3.1 System (3.82)-(3.84) is called *positive in the sense of Popov* if $\eta\left(T\right)$ can be given the form

$$\eta\left(T\right) = \alpha\left(x_{T}\right) - \alpha\left(x_{0}\right) + \sum_{0}^{T-1}\beta\left(x_{t},\mu_{t}\right)\,, \qquad (3.89)$$

where $\beta\left(x,\mu\right) \geq 0$.

We are now in a position to state

Lemma 3.5 (Positivity Lemma for single-input discrete systems) *If $\left(A,b\right)$ is a controllable pair then the following properties are equivalent:*
1^{0} *System (3.82)-(3.84) is positive in the sense of Definition 3.1.*
2^{0} *The inequality*

$$\chi\left(\frac{1}{z},z\right) \geq 0 \qquad (3.90)$$

holds for any complex z such that $|z| = 1$ and $\det\left(z\mathbf{I} - A\right) \neq 0$.
3^{0} *There exists at least a polynomial $\psi\left(\sigma\right)$ such that*

$$\pi\left(\frac{1}{\sigma},\sigma\right) = \bar{\psi}\left(\frac{1}{\sigma}\right)\psi\left(\sigma\right) \qquad (3.91)$$

and for any polynomial $\psi\left(\sigma\right)$ verifying the above factorization there exist a complex number γ and a complex vector w such that $\chi\left(\frac{1}{\sigma},\sigma\right)$ may be factorized as

$$\chi\left(\frac{1}{\sigma},\sigma\right) = \bar{\varphi}\left(\frac{1}{\sigma}\right)\varphi\left(\sigma\right) \qquad (3.92)$$

with $\varphi(\sigma) = \gamma + w^*(\sigma I - A)^{-1}b$, γ *and* w *verifying also*

$$\gamma + w^*(\sigma I - A)^{-1}b = \sqrt{\nu}\frac{\psi(\sigma)}{\det(\sigma I - A)} . \tag{3.93}$$

There exists also a Hermitian matrix H *such that the following hold*

$$\begin{array}{c} \kappa + b^*Hb = \gamma^*\gamma \\ \ell + A^*Hb = w\gamma \\ M + A^*HA - H = ww^* \end{array} \tag{3.94}$$

If the items of (A, b, κ, ℓ, M) *are real and if* $\psi(\sigma)$ *in the factorization is chosen such that its coefficients are real then* γ, w, H *can be chosen to be real.*

4^0 *There exist a Hermitian matrix* H, *a number* γ *and a vector* w *such that*

$$\eta(T) = x_0^*Hx_0 - x_T^*Hx_T + \sum_0^{T-1}|\gamma\mu_t + w^*x_t|^2 . \tag{3.95}$$

5^0 *There exists a Hermitian matrix* H *such that the matrix*

$$\begin{pmatrix} \kappa + b^*Hb & \ell^* + b^*HA \\ \ell + A^*Hb & M + A^*HA - H \end{pmatrix}$$

is nonnegative definite.

Remark that the equivalence of the frequency domain inequality (3.90) and of the matrix equalities (3.94) - the discrete counterpart of the so-called Lurie quadratic equations - represents the early (Kalman-Szegö) version of the Lemma.

3.5.4. Proof of the Lemma - the Stable Case

We shall perform the proof according to the following scheme:

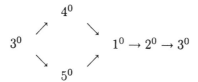

beginning the proof with the simplest implication in the above chain.

Property 4^0 represents a direct consequence of (3.94). Indeed

$$x_0^* H x_0 - x_T^* H x_T = - \sum_0^{T-1} \left(x_{t+1}^* H x_{t+1} - x_t^* H x_t \right)$$

and because (x_t, μ_t) is a solution of (3.82) it follows that

$$\eta(T) = x_0^* H x_0 - x_T^* H x_T + \sum_0^{T-1} [x_t^* (M + A^* H A - H) x_t +$$
$$+ x_t^* (\ell + A^* H b) \mu_t + \mu_t^* (\ell^* + b^* H A) x_t + \mu_t^* (\kappa + b^* H b) \mu_t] =$$
$$= x_0^* H x_0 - x_T^* H x_T + \sum_0^{T-1} [x_t^* w w^* x_t + x_t^* w \gamma \mu_t + \mu_t^* \gamma^* w^* x_t +$$
$$+ \mu_t^* \gamma^* \gamma \mu_t] = x_0^* H x_0 - x_T^* H x_T + \sum_0^{T-1} |\gamma \mu_t + w^* x_t|^2 .$$

Property 1^0 obviously follows from (3.95).

From *Property* 3^0, more precisely from (3.94), it follows that the matrix written in the text of Property 5^0 has the form

$$\begin{pmatrix} \gamma^* \gamma & \gamma^* w^* \\ w \gamma & w w^* \end{pmatrix} = \begin{pmatrix} \gamma^* \\ w \end{pmatrix} \begin{pmatrix} \gamma & w^* \end{pmatrix} ,$$

being nonnegative definite. *Property* 5^0 *is thus proved.*

Assume now *Property* 5^0 *holds* and let H be the matrix whose existence is ensured by it. Therefore,

$$\eta(T) = \sum_0^{T-1} [x_t^* H x_t - (A x_t + b \mu_t)^* H (A x_t + b \mu_t)] +$$
$$+ \sum_0^{T-1} [\mu_t^* (\kappa + b^* H b) \mu_t + \mu_t^* (\ell + A^* H b)^* x_t +$$
$$+ x_t^* (\ell + A^* H b) \mu_t + x_t^* (M + A^* H A - H) x_t] .$$

But (x_t, μ_t) satisfies (3.82); we may write

$$\eta(T) = - \left. x_t^* H x_t \right|_0^T + \sum_0^{T-1} \beta(x_t, \mu_t) ,$$

where the quadratic form $\beta(x, \mu)$ is given by

$$\beta(x, \mu) = \begin{pmatrix} \mu^* & x^* \end{pmatrix} \begin{pmatrix} \kappa + b^* H b & \ell^* + b^* H A \\ \ell + A^* H b & M + A^* H A - H \end{pmatrix} \begin{pmatrix} \mu \\ x \end{pmatrix}$$

and $\beta(x, \mu) \geq 0$ because of *Property* 5^0. But this is nothing else than *Property* 1^0 which is thus proved.

Assume now that *Property* 1^0 *holds* and consider the following pair of solutions

$$\mu_t = z^t, \qquad x_t = (z\mathbf{I} - A)^{-1} b z^t,$$

where $|z| = 1$ and $\det(z\mathbf{I} - A) \neq 0$. Substituting in (3.83) and taking into account (3.85) we find

$$\eta(T) = \sum_0^{T-1} \mathcal{F}(x_t, \mu_t) = \sum_0^{T-1} \chi\left(\frac{1}{z}, z\right) = \chi\left(\frac{1}{z}, z\right) T.$$

Since the system is positive we have on the other hand

$$\eta(T) = \alpha(x_T) - \alpha(x_0) + \sum_0^{T-1} \beta(x_t, \mu_t).$$

Choose now T in order that $x_T = x_0$ $(T = 2\pi(\arg z)^{-1}$ if $z \neq 1$ and arbitrary if $z = 1)$ to get

$$\sum_0^{T-1} \beta(x_t, \mu_t) = T\chi\left(z^{-1}, z\right).$$

Since $\beta(x, \mu) \geq 0$ we deduce (3.90) i.e. *Property* 2^0 follows.

Assume now that *Property* 2^0 holds. We shall prove that this implies *Property* 3^0 in several steps.

A. We have for z such that $|z| = 1$

$$\pi\left(\frac{1}{z}, z\right) = \frac{1}{\nu} |\det(z\mathbf{I} - A)|^2 \chi\left(\frac{1}{z}, z\right) \geq 0.$$

If $\pi\left(\frac{1}{\sigma}, \sigma\right)$ is constant then this constant is nonnegative and we can write

$$\pi\left(\frac{1}{\sigma}, \sigma\right) = \bar{\alpha}_0 \alpha_0.$$

Let us note that for all w we have $w^* (\sigma \mathbf{I} - A)^{-1} b = \dfrac{\pi (\sigma)}{\det (\sigma \mathbf{I} - A)}$

where $\pi (\sigma)$ is a polynomial of degree at most $n - 1$. Because (A, b) is controllable, w is uniquely determined by $\pi (\sigma)$ (Popov (1973), Appendix A). It follows that there exists w such that

$$w^* (\sigma \mathbf{I} - A)^{-1} b = \frac{\alpha_0}{\det (\sigma \mathbf{I} - A)}$$

and using this w we have

$$\chi \left(\frac{1}{\sigma}, \sigma \right) = \bar{\varphi} \left(\frac{1}{\sigma} \right) \varphi (\sigma) , \quad \varphi (\sigma) = w^* (\sigma \mathbf{I} - A)^{-1} b .$$

This simple case shows how one should proceed in the general one. Suppose we obtained

$$\pi \left(\frac{1}{\sigma}, \sigma \right) = \bar{\psi} \left(\frac{1}{\sigma} \right) \psi (\sigma)$$

and take

$$\varphi (\sigma) = \sqrt{\nu} \frac{\psi (\sigma)}{\det (\sigma \mathbf{I} - A)} .$$

Then (3.92) holds. But $\sqrt{\nu} \psi (\sigma)$ is a polynomial of degree at most n; define:

$$\gamma = \lim_{\sigma \to \infty} \sqrt{\nu} \frac{\psi (\sigma)}{\det (\sigma \mathbf{I} - A)} .$$

Then $\tilde{\psi} (\sigma) = \sqrt{\nu} \psi (\sigma) - \gamma \det (\sigma \mathbf{I} - A)$ is of degree less than n and from the controllability we deduce that there exists w such that

$$w^* (\sigma \mathbf{I} - A)^{-1} b = \frac{\tilde{\psi} (\sigma)}{\det (\sigma \mathbf{I} - A)} = \varphi (\sigma) - \gamma$$

and (3.93) is thus obtained.

B. We have still to derive the factorization formula (3.91). We can assume that $\pi \left(\dfrac{1}{\sigma}, \sigma \right)$ is not constant since we have already considered that case. From (3.88) it follows that

$$\pi \left(\frac{1}{\sigma}, \sigma \right) = \frac{1}{\nu} \sum_{k=0}^{n} \sum_{j=0}^{n} \kappa_{jk} \sigma^{k-j} = \sum_{-n}^{n} \alpha_i \sigma^i$$

with

$$
\alpha_i = \begin{cases} \displaystyle\sum_{k=i}^{n} \kappa_{k-i,k} & , \quad i \geq 0 \\[2ex] \displaystyle\sum_{k=0}^{n+i} \kappa_{k-i,k} & , \quad i < 0 \end{cases}
$$

We may now follow the line of Popov (1973, §37, Proposition 1). From (3.87) it follows that $\pi\left(\dfrac{1}{\sigma},\sigma\right) = \overline{\pi\left(\bar\sigma,\dfrac{1}{\bar\sigma}\right)}$. Assume now that $\sigma_1 \neq 0$ is a root of the polynomial $\sigma^n \pi\left(\dfrac{1}{\sigma},\sigma\right)$. Therefore $\pi\left(\dfrac{1}{\sigma_1},\sigma_1\right) = 0$ and the above equality obtained from (3.87) gives $\overline{\pi\left(\bar\sigma_1,\dfrac{1}{\bar\sigma_1}\right)} = 0$; this shows that $1/\bar\sigma_1$ is also a root of $\sigma^n \pi\left(\dfrac{1}{\sigma},\sigma\right)$. If $|\sigma_1| \neq 1$ the two roots are different and the expression

$$
q(\sigma) = \frac{\sigma^n \pi\left(\dfrac{1}{\sigma},\sigma\right)}{(\sigma - \sigma_1)\left(\sigma - \dfrac{1}{\bar\sigma_1}\right)}
$$

is a polynomial. This is true also when $|\sigma_1| = 1$ but in this case the root is multiple; this can be proved by showing that $q_1(\sigma_1) = 0$ where $q_1(\sigma) = \dfrac{\sigma^n \pi\left(\dfrac{1}{\sigma},\sigma\right)}{\sigma - \sigma_1}$. Let σ be on the unit circle: since $|\sigma_1| = 1$, $\sigma = \sigma_1 e^{i\varphi}$ where φ is real. Therefore

$$
\pi\left(\frac{1}{\sigma_1}e^{-i\varphi}, \sigma_1 e^{i\varphi}\right) = \frac{e^{-in\varphi}\left(e^{i\varphi}-1\right)}{\sigma_1^{n-1}} q_1\left(\sigma_1, e^{i\varphi}\right)
$$

Taking the Taylor expansion around $\varphi = 0$ we shall have

$$
\pi\left(\frac{1}{\sigma_1}e^{-i\varphi}, \sigma_1 e^{i\varphi}\right) = \frac{q_1(\sigma_1)}{\sigma_1^{n-1}} i\varphi + o(\varphi)
$$

but on the unit circle $\pi\left(\dfrac{1}{\sigma},\sigma\right)$ must be real and nonnegative what gives $q_1(\sigma_1) = 0$.

We have thus proved that for a nonzero root σ_1 of $\pi\left(\frac{1}{\sigma},\sigma\right)$ the expression

$$\pi_1(\sigma) = \frac{\pi\left(\frac{1}{\sigma},\sigma\right)}{(\sigma - \sigma_1)\left(\frac{1}{\sigma} - \bar{\sigma}_1\right)}$$

is such that $\sigma^{n-1}\pi_1(\sigma)$ is a polynomial; also $\pi_1(\sigma)$ is of the form of $\pi\left(\frac{1}{\sigma},\sigma\right)$ but of degree $n-1$. On the unit circle $\pi_1(\sigma) \geq 0$. Therefore, we are in position to repeat the above argument. Repeating it as many times as possible we obtain that the nonzero roots occur in pairs of the form $(\sigma_1, \sigma_{-1}), (\sigma_2, \sigma_{-2}), \cdots, (\sigma_{n_0}, \sigma_{-n_0})$ where $\sigma_{-k} = 1/\bar{\sigma}_k$; here $n_0 \leq n$.

We may now rearrange these roots in the form

$$\sigma_{i_1}, \quad \sigma_{i_2}, \quad \cdots \quad , \sigma_{i_{n_0}}$$
$$\sigma_{-i_1}, \quad \sigma_{-i_2}, \quad \cdots \quad , \sigma_{-i_{n_0}}$$

by choosing them arbitrarily, one from each pair; i.e., the index i_j may take one of the values $+j$ and $-j$. Assuming that the polynomial $\sigma^n\pi\left(\frac{1}{\sigma},\sigma\right)$ has some zero roots we have

$$\sigma^n\pi\left(\frac{1}{\sigma},\sigma\right) = \mu_1\sigma^p\prod_{j=1}^{n_0}(\sigma - \sigma_{i_j})(\sigma - \sigma_{-i_j}),$$

where $\mu_1 \neq 0$ is a complex constant. From the property of the nonzero roots of $\sigma^n\pi\left(\frac{1}{\sigma},\sigma\right)$ it follows that

$$(\sigma - \sigma_{-i_j}) = -\frac{\sigma}{\bar{\sigma}_{i_j}}\left(\frac{1}{\sigma} - \bar{\sigma}_{i_j}\right)$$

and this will give the following factorization

$$\pi\left(\frac{1}{\sigma},\sigma\right) = \mu(\sigma)\rho_0(\sigma)\bar{\rho}_0\left(\frac{1}{\sigma}\right),$$

where

$$\rho_0\left(\sigma\right) = \prod_{j=1}^{n_0}\left(\sigma - \sigma_{i_j}\right) , \quad \mu\left(\sigma\right) = (-1)^{n_0}\,\sigma^{p+n_0-n}\,\frac{\mu_1}{\bar{\sigma}_{i_1}\bar{\sigma}_{i_2}\cdots\bar{\sigma}_{i_{n_0}}}$$

and we denoted $\bar{\rho}\left(\lambda\right) = \overline{\rho\left(\bar{\lambda}\right)}$.

Now on the unit circle we have $\bar{\sigma} = 1/\sigma$ hence the product $\rho_0\left(\sigma\right)\overline{\rho_0\left(\sigma^{-1}\right)}$ is nonnegative. But $\pi\left(\sigma^{-1},\sigma\right) \geq 0$ on the same unit circle and this is true only if $\mu\left(\sigma\right) \geq 0$ on the unit circle; this last condition is satisfied if $\mu\left(\sigma\right)$ equals a real nonnegative constant what happens if $p = n - n_0$ i.e. $\pi\left(\sigma^{-1},\sigma\right)$ has a zero root of even multiplicity and

$$\mu\left(\sigma\right) \equiv \mu_0 = \frac{(-1)^{n_0}\,\mu_1}{\bar{\sigma}_{i_1}\bar{\sigma}_{i_2}\cdots\bar{\sigma}_{i_{n_0}}} \geq 0 , \quad \mathrm{Im}\,\mu_0 = 0$$

By taking

$$\psi\left(\sigma\right) = \kappa\rho_0\left(\sigma\right) , \quad \kappa = \sqrt{|\mu_0|}e^{i\varphi}$$

with φ an arbitrary real constant, we have obtained the factorization formula (3.91).

Remark that the choice of $\rho_0\left(\sigma\right)$, hence of $\psi\left(\sigma\right)$, is somehow arbitrary depending on the choice of the roots σ_{i_j}. In particular $\psi\left(\sigma\right)$ may be chosen with all its roots inside the unit disk (a Schur polynomial) or outside this disk. Also if $\psi\left(\sigma\right)$ satisfies the factorization, any polynomial of the form $\sigma^r\psi\left(\sigma\right)$ with arbitrary integer $r > 0$ satisfies the factorization.

Also, if $\pi\left(\sigma^{-1},\sigma\right)$ has real coefficients, $\sigma^n\pi\left(\sigma^{-1},\sigma\right)$ has real coefficients and its non-real roots occur in pairs of complex conjugate roots. However, it can be easily seen that no pair of roots of the form $(\sigma_j,\sigma_{-j}) = (\sigma_j,1/\bar{\sigma}_j)$ may contain two non-real, complex conjugate roots. Therefore, all the pairs (σ_j,σ_{-j}) which contain non-real roots, can be arranged in couples of pairs

$$\left(\sigma_k,\sigma_{-k}\right),\left(\sigma_{\bar{k}},\sigma_{-\bar{k}}\right) , \quad \sigma_k = \bar{\sigma}_{\bar{k}} .$$

Using this property the roots of $\psi\left(\sigma\right)$ can be chosen such that for every non-real root the complex conjugate root occurs also. The constant κ may be chosen real. *Then the polynomial $\psi\left(\sigma\right)$ has real coefficients if $\pi\left(\sigma^{-1},\sigma\right)$ is as such.*

C. We shall assume now that *the eigenvalues of matrix A are inside the unit disk,* i.e., system $x_{t+1} = Ax_t$ is stable. Therefore, the discrete Liapunov equation

$$A^* H A - H = w w^* - M \,,$$

where w is the one in the factorization (3.92) has a unique Hermitian solution. We shall prove that this H together with w and γ determined from the factorization satisfy the discrete Lurie equations (3.94). We have the two expressions of $\chi\left(\sigma^{-1}, \sigma\right)$ - the one from its definition (3.85), and the one from the factorization (3.92); they have to be identical. Therefore,

$$\chi\left(\frac{1}{\sigma}, \sigma\right) = \kappa + \ell^*\left(\sigma I - A\right)^{-1} b + b^*\left(\frac{1}{\sigma} I - A^*\right)^{-1} \ell +$$
$$+ b^*\left(\frac{1}{\sigma} I - A^*\right)^{-1} M\left(\sigma I - A\right)^{-1} b \equiv \bar{\varphi}\left(\frac{1}{\sigma}\right) \varphi(\sigma) \equiv$$
$$\equiv \left[\gamma^* + b^*\left(\frac{1}{\sigma} I - A^*\right)^{-1} w\right]\left[\gamma + w^*\left(\sigma I - A\right)^{-1} b\right]$$

and from here

$$\kappa - \gamma^* \gamma + (\ell - w\gamma)^*\left(\sigma I - A\right)^{-1} b + b^*\left(\frac{1}{\sigma} I - A^*\right)^{-1}(\ell - w\gamma) +$$
$$+ b^*\left(\frac{1}{\sigma} I - A^*\right)^{-1}(M - ww^*)\left(\sigma I - A\right)^{-1} b \equiv 0 \,.$$

Taking into account the Liapunov equation it follows that:

$$\kappa - \gamma^* \gamma + (\ell - w\gamma)^*\left(\sigma I - A\right)^{-1} b + b^*\left(\frac{1}{\sigma} I - A^*\right)^{-1}(\ell - w\gamma) -$$
$$- b^*\left(\frac{1}{\sigma} I - A^*\right)^{-1}(A^* H A - H)\left(\sigma I - A\right)^{-1} b \equiv 0 \,.$$

We now make use of the obvious identities:

$$-A\left(\sigma I - A\right)^{-1} \equiv I - \sigma\left(\sigma I - A\right)^{-1}$$
$$-\left(\frac{1}{\sigma} I - A^*\right)^{-1} A^* \equiv I - \frac{1}{\sigma}\left(\frac{1}{\sigma} I - A^*\right)^{-1}$$

and obtain

$$\kappa + b^* H b - \gamma^* \gamma + (\ell - w\gamma + A^* H b)^*\left(\sigma I - A\right)^{-1} b +$$
$$+ b^*\left(\frac{1}{\sigma} I - A^*\right)^{-1}(\ell + A^* H b - w\gamma) \equiv 0$$

From here we may write

$$\gamma^*\gamma - \kappa - b^*Hb \equiv (\ell - w\gamma + A^*Hb)^* (\sigma I - A)^{-1} b +$$
$$+ b^* \left(\frac{1}{\sigma}I - A^*\right)^{-1} (\ell + A^*Hb - w\gamma) .$$

Since the eigenvalues of A are located inside the unit disk, $(\sigma I - A)^{-1}$ has its poles there while $\left(\frac{1}{\sigma}I - A^*\right)^{-1}$ has its poles outside the unit disk. We deduce that if $(\ell - w\gamma + A^*Hb)^* (\sigma I - A)^{-1} b \not\equiv 0$ the LHS of the above identity has poles inside the unit disk and outside it. Therefore, it cannot be a constant what would not be true. We deduce that $(\ell - w\gamma + A^*Hb)^* (\sigma I - A)^{-1} b \equiv 0$ and from the controllability of (A, b) we have finally that

$$\ell + A^*Hb = w\gamma$$

and

$$\kappa + b^*Hb = \gamma^*\gamma ;$$

hence (3.94) are obtained.

To end the proof we still have to show that γ, w, H can be chosen real if the items of A, b, κ, ℓ, M are real and $\psi(\sigma)$ in the factorization is chosen with real coefficients.

If $\psi(\sigma)$ has real coefficients and A is a real matrix then γ is real and $\tilde{\psi}(\sigma) = \psi(\sigma) - \gamma \det(\sigma I - A)$ is also a polynomial with real coefficients. Further, by writing down the equality

$$w^* (\sigma I - A)^{-1} b = \frac{\tilde{\psi}(\sigma)}{\det(\sigma I - A)}$$

for some real σ_j, $j = \overline{1, n}$ we shall have

$$\frac{1}{2}(w - \bar{w})^* (\sigma_j I - A)^{-1} b = 0 , \quad j = \overline{1, n} ,$$

where $*$ denotes transpose. Since (A, b) is controllable the linear vectors $(\sigma_j I - A)^{-1} b$ are linearly independent. Therefore, $w - \bar{w} = 0$ hence w is a real vector.

From the representation of the solution of the Liapunov equation

$$H = \sum_{0}^{\infty} (A^*)^k (M - ww^*) A^k$$

it follows that H is real what ends the proof in the stable case.

3.5.5. Proof of the Lemma - the General Case

The only point of the proof where the assumption about the eigenvalues of A is used is the existence of matrix A satisfying the Liapunov equation

$$A^* H A - H = w w^* - M \,.$$

Therefore, we have to reformulate the proof from this point.

Since (A, b) is controllable there exists q such that $A + b q^*$ has the eigenvalues inside the unit disk. Therefore, we shall consider the set $\left(\hat{A}, b, \kappa, \hat{\ell}, \hat{M} \right)$, where

$$\hat{A} = A + b q^* \,, \; \hat{\ell} = \ell + \kappa q \,, \; \hat{M} = M + \kappa q q^* + \ell q^* + q \ell^*$$

defined using vector q and the initial set. A straightforward computation shows the following relation between the characteristic functions of the two systems

$$\hat{\chi}(\lambda, \sigma) = \kappa + \hat{\ell}^* \left(\sigma \mathbf{I} - \hat{A} \right)^{-1} b + b^* \left(\lambda \mathbf{I} - \hat{A}^* \right)^{-1} \hat{\ell} +$$

$$+ b^* \left(\lambda \mathbf{I} - \hat{A}^* \right)^{-1} \hat{M} \left(\sigma \mathbf{I} - \hat{A} \right)^{-1} b =$$

$$= \frac{\chi(\lambda, \sigma)}{\left[1 - b^* (\lambda \mathbf{I} - A^*)^{-1} q \right] \left[1 - q^* (\sigma \mathbf{I} - A)^{-1} b \right]}$$

A quite well known identity based on Schur theorem on determinants for block matrices is the following

$$\det \left(\sigma \mathbf{I} - A - b q^* \right) \equiv \det \left(\sigma \mathbf{I} - A \right) \left(1 - q^* \left(\sigma \mathbf{I} - A \right)^{-1} b \right) \,.$$

This identity shows that on the unit circle, for those z, $|z| = 1$, such that $\det (z \mathbf{I} - A) \neq 0$ we cannot have $1 - q^* (z \mathbf{I} - A)^{-1} b = 0$ because the eigenvalues of $A + b q^*$ are inside the unit disk. Therefore, for such z we may write

$$\hat{\chi} \left(z^{-1}, z \right) = \frac{\chi \left(z^{-1}, z \right)}{\left| 1 - q^* (z \mathbf{I} - A)^{-1} b \right|^2} \geq 0$$

due to the fact that (3.90) holds for those z considered above. From the controllability of (A, b) it follows the controllability of $(A + bq^*, b)$, i.e., of $\left(\hat{A}, b\right)$ and since $\hat{A}$ has the eigenvalues inside the unit disk, we can apply to the set $\left(\hat{A}, b, \kappa, \hat{\ell}, \hat{M}\right)$ the just proved Lemma: there exist $\hat{\gamma}, \hat{w}, \hat{M}$ (with $\hat{M}$ a Hermitian matrix) such that

$$
\begin{aligned}
\kappa + b^* \hat{H} b &= \hat{\gamma}^* \hat{\gamma} \\
\hat{\ell} + \hat{A}^* \hat{H} b &= \hat{w} \hat{\gamma} \\
\hat{M} + \hat{A}^* \hat{H} \hat{A} - \hat{H} &= \hat{w} \hat{w}^* .
\end{aligned}
$$

Using the expressions of $\hat{\ell}, \hat{A}, \hat{M}$ the last two equalities become

$$
\ell + \left(\kappa + b^* \hat{H} b\right) q + A^* \hat{H} b = \hat{w} \hat{\gamma}
$$

$$
M + A^* \hat{H} A - \hat{H} + \left(\kappa + b^* \hat{H} b\right) q q^* +
$$

$$
+ \left(\ell + A^* \hat{H} b\right) q^* + q \left(\ell + A^* \hat{H} b\right) = \hat{w} \hat{w}^* .
$$

Taking into account the first equality and introducing it into the second one, then introducing the second one into the third we obtain the Kalman - Szegö - Popov - Yakubovich equations as follows

$$
\begin{aligned}
\kappa + b^* \hat{H} b &= \hat{\gamma}^* \hat{\gamma} \\
\ell + A^* \hat{H} b &= (\hat{w} - q \hat{\gamma}^*) \hat{\gamma} \\
M + A^* \hat{H} A - \hat{H} &= (\hat{w} - q \hat{\gamma}^*) (\hat{w} - q \hat{\gamma}^*)^*
\end{aligned}
$$

and it is obvious that (3.94) are verified by

$$
\gamma = \hat{\gamma} \, , \quad w = \hat{w} - q \hat{\gamma}^* \, , \quad H = \hat{H}
$$

which ends the proof of the Lemma.

3.6. A NEW CONDITION OF ABSOLUTE STABILITY FOR SYSTEMS WITH SLOPE RESTRICTED NONLINEARITY

Usually when discussing frequency domain conditions for absolute stability one keeps in mind the classical Popov condition for continuous-time systems that may be found in most textbooks and monographs

$$
\frac{1}{k} + \mathrm{Re} \, (1 + iwq) \, \gamma \, (iw) \geq 0 \, , \quad w \in \mathbb{R}_+
$$

with its PD-multiplier $1 + qs$ (see the comments ending Section 3.4).

The restriction of the class of nonlinearities (e.g., monotonic, odd-monotonic, slope-restricted etc.) leads to other frequency domain criteria containing several free parameters (Yakubovich (1962), (1965a, b), (1967), Brockett and Willems (1965)). In the early papers by Yakubovich (1962 b, 1965 a,b) as well as in a paper on oscillations (Yakubovich (1977 b)) a specific combination of parameters involved leads to a condition of the form

$$\text{Re}\left\{\left(1 - \frac{q}{i\omega}\right)\left(\frac{1}{k} + \gamma(i\omega)\right)\right\} \geq \delta\,|\gamma(i\omega)|^2 \qquad (3.96)$$

explicitly stated by Barabanov and Yakubovich (1979). A similar condition is considered by Lellouche (1970) for feedback systems described by the integral equation

$$\sigma(t) = \rho(t) - \int\limits_0^t h(t - \tau)\,\varphi(\sigma(\tau))\,d\tau \qquad (3.97)$$

to obtain oscillatory behavior of the solutions. Later Popov himself (Popov (1977 b)) considered the same integral equation and using a new technique called "saturability" was able to prove under slightly different assumptions than those of Lellouche that the frequency domain inequality (3.96) ensures absolute stability - all solutions of (3.97) are bounded and tend asymptotically to zero. We would like to remark the form of the multiplier, $1 - q\,(i\omega)^{-1}$ which, in the framework described in the comments ending Section 3.4 may be called a PI (proportional integral) multiplier because it corresponds to the so-called ideal PI controller described by the input-output relation

$$u(t) = k_p\left[e(t) + \frac{1}{T_i}\int\limits_0^t e(\tau)\,d\tau\right].$$

Obviously the transfer function of this controller is

$$\frac{\tilde{u}(s)}{\tilde{e}(s)} = k_p\left[1 + \frac{1}{sT_i}\right]$$

and is identical with the above mentioned multiplier.

The problem of using this PI type multiplier got renewed attention with the paper of Singh (1984). Its purpose was to obtain a new frequency domain inequality for absolute stability incorporating only slope information about the nonlinear element. Singh's arguments were far from convincing so Halanay and Răsvan (1991) extended his idea to the case of systems with several slope restricted nonlinearities on a rigorous base. This was done using the properties of the first integral that occurs during the proofs and associating a Liapunov function with the Yakubovich - Kalman - Popov lemma, provided a frequency domain inequality with a PI multiplier was fulfilled. More recently, Haddad and Kapila (1995) returned to the Singh's paper and showed that the problem could be solved by the so-called *kinetic Liapunov function*. The proof of their result needs to be improved (for instance, some necessary conditions are just omitted) but finally it turns to be exactly the condition of Yakubovich and Barabanov (1979) extended to the case of several nonlinear elements.

In the case of discrete time-systems there exist the frequency domain inequality of Tsypkin where no free parameter occurs (Section 3.1), the modified Tsypkin criterion for nondecreasing nonlinearities with a discrete PD type multiplier including a free parameter (Section 3.2) and many other criteria with free parameters (see the paper of Yakubovich (1968)). For nondecreasing nonlinearities the Brockett-Willems type frequency domain inequality introduces a rather general multiplier with several free parameters. In this context it is natural to ask whether the introduction of a PI type multiplier is possible also in the discrete case. The answer is positive: using the same approach as in their paper on the continuous-time case, Halanay and Răsvan (1990) obtained a frequency domain inequality with a discrete PI type multiplier incorporating only slope information about the nonlinear element.

3.6.1. System Description. The Class of Nonlinearities. The Associated System and the First Integral

We shall consider again system (3.12), i.e., the system

$$x_{t+1} = Ax_t - b\varphi\left(c^*x_t\right) \tag{3.98}$$

with the usual notations under the following basic assumptions: *i)* $\det\left(A - \mathbf{I}\right) \neq 0$; *ii)* (A, b) is a controllable pair and (c^*, A) is an observable pair; *iii)* $c^*\left(\mathbf{I} - A\right)^{-1}b \neq 0$.

These assumptions have a well-defined significance in System Theory: the second assumption allows to specify the linear part only by its rational strictly proper transfer function $\gamma\left(\sigma\right) = c^*\left(\sigma\mathbf{I} - A\right)^{-1}b$;

given $\gamma(\sigma)$, the triple (A, b, c) might be any minimal state realization of $\gamma(\sigma)$; the third assumption (which implies the first one) means $\gamma(1) \neq 0$, i.e., a static gain can be defined for the linear part.

The nonlinear function $\varphi(\sigma)$ will be considered to be such that $\varphi(\sigma) = 0$ iff $\sigma = 0$ and satisfy

$$0 \leq \underline{\varphi} < \frac{\varphi(\sigma_1) - \varphi(\sigma_2)}{\sigma_1 - \sigma_2} < \bar{\varphi} \leq +\infty, \quad \sigma_1 \neq \sigma_2. \tag{3.99}$$

Obviously $\varphi(\sigma)$ is nondecreasing.

Let x_t be a solution of (3.98). We shall have

$$x_{t+1} - x_t = (A - I) x_t - b\varphi(c^* x_t)$$

and denoting $z_t = x_{t+1} - x_t$ this new variable will satisfy

$$\begin{aligned} z_{t+1} &= (A - I) x_{t+1} - b\varphi(c^* x_{t+1}) = \\ &= (A - I)(z_t + x_t) - b\varphi(c^* z_t + c^* x_t) = \\ &= Az_t - x_{t+1} + x_t + Ax_t - x_t - b\varphi(c^* z_t + c^* x_t) = \\ &= Az_t - b[\varphi(c^* z_t + c^* x_t) - \varphi(c^* x_t)] . \end{aligned}$$

Denoting also $\xi_t = c^* x_t$ the pair (z_t, ξ_t) defined starting from the solution x_t of (3.98) will satisfy the system

$$\begin{aligned} z_{t+1} &= Az_t - b[\varphi(c^* z_t + \xi_t) - \varphi(\xi_t)] \\ \xi_{t+1} &= c^* z_t + \xi_t . \end{aligned} \tag{3.100}$$

In this way we associated with system (3.98) a system with an augmented dimension. This new system also has the origin as equilibrium and if the equilibrium at zero is globally asymptotically stable for system (3.98), then obviously this property is true for system (3.100). This simple fact follows from the definition formulae

$$\begin{aligned} z_t &= (A - I) x_t - b\varphi(c^* x_t) \\ \xi_t &= c^* z_t . \end{aligned} \tag{3.101}$$

Suppose now that we want to study absolute stability for system (3.98) using the properties of (3.100); in this case we need a way of defining the solutions of (3.98) starting from the solutions of (3.100). Since (3.101) is of higher dimension than (3.98) the solutions of (3.98) have to be located on some invariant set of (3.100). Indeed we have

Proposition 3.6 *If x_t is a solution of (3.98) then (z_t, ξ_t) defined*

by (3.101) *is a solution of* (3.100) *satisfying also*

$$F\left(z_t, \xi_t\right) \equiv \xi_t - c^* \left(A - I\right)^{-1} \left[z_t + b\varphi\left(\xi_t\right)\right] \equiv 0 \qquad (3.102)$$

Conversely, if $\left(z_t, \xi_t\right)$ *is a solution of* (3.100) *satisfying* (3.102) *then* x_t *defined by*

$$x_t = \left(A - I\right)^{-1} \left[z_t + b\varphi\left(\xi_t\right)\right] \qquad (3.103)$$

is a solution of (3.98).

Proof. We have already proved when introducing system (3.100) that if x_t is a solution of (3.98) and $\left(z_t, \xi_t\right)$ is defined by (3.101) then $\left(z_t, \xi_t\right)$ is a solution of (3.100). We have still to show that z_t and ξ_t thus defined satisfy (3.102). By direct computation we shall have

$$\xi_t - c^* \left(A - I\right)^{-1} \left[z_t + b\varphi\left(\xi_t\right)\right] = c^* x_t - c^* \left(A - I\right)^{-1} \left[\left(A - I\right) x_t - \right.$$
$$\left. -b\varphi\left(c^* x_t\right) + b\varphi\left(c^* x_t\right)\right] \equiv 0 .$$

Conversely, let $\left(z_t, \xi_t\right)$ be a solution of (3.100) satisfying also (3.102); define x_t by (3.103). By direct computation we find

$$x_{t+1} = \left(A - I\right)^{-1} \left[z_{t+1} + b\varphi\left(\xi_{t+1}\right)\right] =$$
$$= \left(A - I\right)^{-1} \left[Az_t - b\varphi\left(c^* z_t + \xi_t\right) + b\varphi\left(\xi_t\right) + b\varphi\left(c^* z_t + \xi_t\right)\right] =$$
$$= \left(A - I\right)^{-1} \left[A\left(A - I\right) x_t - Ab\varphi\left(\xi_t\right) + b\varphi\left(\xi_t\right)\right] = Ax_t - b\varphi\left(\xi_t\right)$$

$\xi_t \equiv c^* \left(A - I\right)^{-1} \left[z_t + b\varphi\left(\xi_t\right)\right] \equiv c^* x_t$, from (3.102). Therefore,

$$x_{t+1} = Ax_t - b\varphi\left(c^* x_t\right)$$

which ends the proof.

3.6.2. The Stability Criterion

We are now in position to state the main result; i.e., the stability criterion.

Theorem 3.7 *Consider system* (3.98) *under the following assumptions: i)* $\det\left(A - I\right) \neq 0$; *ii)* (A, b) *is a controllable pair and* (c^*, A) *is an observable pair; iii)* $c^* \left(I - A\right)^{-1} b \neq 0$; *iv) there exist* $\tau \geq 0$, $q > 0$

such that the following frequency domain inequality holds:

$$\frac{\tau}{\bar{\varphi}} + \text{Re}\left(\tau\left(1 + \underline{\varphi}/\bar{\varphi}\right) + \frac{z}{z-1}q\right)\gamma(z) + \left(\tau\underline{\varphi}\right)|\gamma(z)|^2 \geq 0, \ |z| = 1$$

$$(3.104)$$

where $\underline{\varphi}$ and $\bar{\varphi}$ are those of (3.99) and $\gamma(z) = c^*(z\mathbf{I} - \mathbf{A})^{-1}b$ is the transfer function of the linear part of (3.98); v) there exists $\tilde{\varphi} \in (\underline{\varphi}, \bar{\varphi})$ such that $\mathbf{A} - \tilde{\varphi}bc^*$ has the eigenvalues inside the unit disk.

Then the zero solution of (3.98) is asymptotically stable for all functions $\varphi(\sigma)$ satisfying (3.99) with $\varphi(\sigma) = 0$ iff $\sigma = 0$ and

$$\Omega(\sigma) = \frac{1}{\gamma(1)}\left[\frac{\psi^2(\sigma)}{2(1 + \gamma(1)\tilde{\varphi})} - \int_0^\sigma \psi(\lambda)\,d\lambda\right] > 0, \quad \sigma \neq 0 \quad (3.105)$$

where

$$\psi(\sigma) = \sigma + \gamma(1)\varphi(\sigma) \tag{3.106}$$

If additionally the functions φ are such that

$$\lim_{\sigma \to \pm\infty} \Omega(\sigma) = \infty \tag{3.107}$$

then the asymptotic stability of the zero solution is global - system (3.98) is absolutely stable for the class of nonlinearities defined by (3.99) and satisfying (3.105) - (3.107).

Remark. Among the functions satisfying (3.99) and (3.105) one can mention:

1) For $\gamma(1) > 0$ the differentiable functions satisfying the slope restrictions

$$\tilde{\varphi} < \varphi'(\sigma) < \bar{\varphi};$$

2) For $\gamma(1) < 0$ and $\tilde{\varphi} > -1/\gamma(1)$ the differentiable functions satisfying the slope restrictions

$$\min\left(\underline{\varphi}, -1/\gamma(1)\right) < \varphi'(\sigma) < -1/\gamma(1)$$

or

$$\tilde{\varphi} < \varphi'(\sigma) < \bar{\varphi};$$

3) For $\gamma(1) < 0$ and $\tilde{\varphi} < -1/\gamma(1)$ the differentiable functions satis-

fying the slope restrictions

$$\tilde{\varphi} < \varphi'(\sigma) < \min\left(\bar{\varphi}, -1/\gamma(1)\right).$$

Proof. We shall perform the proof in several steps.

A. We associate with system(3.100) the following Popov system "difference equation plus index":

$$z_{t+1} = Az_t + b\mu_t$$
$$\xi_{t+1} = c^* z_t + \xi_t$$
$$\eta(T) = \sum_{0}^{T-1} \left[\tau\left(\mu_t + \underline{\varphi} c^* z_t\right)\left(\mu_t/\bar{\varphi} + c^* z_t\right) + q\mu_t\left(c^* z_t + \xi_t\right)\right],$$

$$(3.108)$$

where τ, q are those of the frequency domain inequality (3.104).

We make the following notations:

$$y = \begin{pmatrix} z \\ \xi \end{pmatrix}, \quad \kappa = \frac{\tau}{\bar{\varphi}}, \quad \ell = \frac{1}{2}\begin{pmatrix} \left(\tau\left(1+\underline{\varphi}/\bar{\varphi}\right)+q\right)c \\ q \end{pmatrix}$$

$$\mathcal{M} = \begin{pmatrix} \tau\underline{\varphi}cc^* & 0 \\ 0 & 0 \end{pmatrix}, \quad \mathcal{A} = \begin{pmatrix} A & 0 \\ c^* & 1 \end{pmatrix}, \quad \mathbf{b} = \begin{pmatrix} b \\ 0 \end{pmatrix}$$

and system (3.108) reads as follows

$$y_{t+1} = \mathcal{A}y_t + \mathbf{b}\mu_t$$
$$\eta(T) = \sum_{0}^{T-1} \left[\kappa\mu_t^2 + \mu_t^*\ell^* y_t + y_t^*\ell\mu_t + y_t^*\mathcal{M}y_t\right].$$

$$(3.109)$$

This system is of the form (3.82) - (3.84) considered in Lemma 3.5. In order to apply it we shall check if the assumptions of the Lemma are fulfilled. Because (A, b) is a controllable pair, a straightforward result (see Halany and Răsvan (1990), the Appendix) shows that $(\mathcal{A}, \mathbf{b})$ defined above is a controllable pair. Also $\kappa \geq 0$ and $\mathcal{M} \geq 0$. The characteristic function

$$\chi(\lambda, \sigma) = \kappa + \ell^*\left(\sigma\mathbf{I} - \mathcal{A}\right)^{-1}\mathbf{b} + \mathbf{b}^*\left(\lambda\mathbf{I} - \mathcal{A}\right)^{-1}\ell +$$
$$+\mathbf{b}^*\left(\lambda\mathbf{I} - \mathcal{A}\right)^{-1}\mathcal{M}\left(\sigma\mathbf{I} - \mathcal{A}\right)^{-1}\mathbf{b}$$

$$(3.110)$$

has the following form on the unit circle

$$\chi\left(z^{-1}, z\right) = \frac{\tau}{\bar{\varphi}} + \mathrm{Re}\left(\tau\left(1+\underline{\varphi}/\bar{\varphi}\right) + \frac{z}{z-1}q\right)\gamma(z) + \left(\tau\underline{\varphi}\right)|\gamma(z)|^2,$$
$$|z| = 1.$$

From the frequency domain inequality (3.104) we have that $\chi\left(z^{-1}, z\right) \geq 0$ for all z such that $|z| = 1$. From the Lemma we deduce existence of the scalar γ_0, of the $(n+1)$-dimensional vector w and of the Hermitian $(n+1) \times (n+1)$ matrix $\mathcal{P}$ such that

$$\kappa - \mathbf{b}^* \mathcal{P} \mathbf{b} = |\gamma_0|^2$$
$$\ell - \mathcal{A}^* \mathcal{P} \mathbf{b} = w\gamma_0 \qquad (3.111)$$
$$-\mathcal{A}^* \mathcal{P} \mathcal{A} + \mathcal{P} + \mathcal{M} = ww^*$$

(with respect to (3.94) we have here $-\mathcal{P}$ instead of H). By splitting w and $\mathcal{P}$ as follows

$$w = \begin{pmatrix} w_1 \\ \gamma_2 \end{pmatrix}, \quad \mathcal{P} = \begin{pmatrix} P & p_{12} \\ p_{12}^* & \pi_{22} \end{pmatrix},$$

where w_1 and p_{12} are n-vectors, γ_2 and π_{22} are scalars and P is a $n \times n$ matrix, equalities (3.111) will give

$$\gamma_2 = 0, \ \pi_{22} = (1/2) q \left[c^* (\mathbf{I} - A)^{-1} b \right]^{-1} = (1/2) q (\gamma(1))^{-1},$$
$$p_{12} = (1/2) q (\gamma(1))^{-1} (\mathbf{I} - A^*)^{-1} c.$$

In fact the assumptions of the Theorem ensure via the Positivity Lemma the existence of the symmetric $n \times n$ matrix P, of the n-vector w_1 and the scalar γ_0 such that

$$\tau / \bar{\varphi} - b^* P b = |\gamma_0|^2$$
$$\tfrac{1}{2} \tau \left(1 + \varphi / \bar{\varphi}\right) c - A^* P b = w_1 \gamma_0$$
$$\tau \underline{\varphi} c c^* - A^* P A + P - \tfrac{1}{2} q (\gamma(1))^{-1} \left[c c^* (\mathbf{I} - A)^{-1} A + \right.$$
$$\left. + A^* (\mathbf{I} - A^*)^{-1} c c^* + c c^* \right] = w_1 w_1^*.$$

The above equations allow the following form of $\eta(T)$:

$$\eta(T) = z_T^* P z_T - z_0^* P z_0 + \sum_0^{T-1} |\gamma_0 \mu_t + w_1^* z_t|^2 + q \sum_0^{T-1} \mu_t (c^* z_t + \xi_t) +$$
$$+ \tfrac{1}{2} q (\gamma(1))^{-1} \sum_0^{T-1} z_t^* \left[c c^* (\mathbf{I} - A)^{-1} A + A^* (\mathbf{I} - A^*)^{-1} c c^* + c c^* \right] z_t$$

$$(3.112)$$

We use now (3.108) to obtain

$$c^* (\mathbf{I} - A)^{-1} z_{t+1} = c^* (\mathbf{I} - A)^{-1} z_t + \gamma(1) \mu_t.$$

Therefore,

$$
q\mu_t \left(c^* z_t + \xi_t\right) + \tfrac{1}{2} q \left(\gamma\left(1\right)\right)^{-1} z_t^* \left[cc^* \left(\mathbf{I} - A\right)^{-1} A + \right.
$$
$$
\left. + A^* \left(\mathbf{I} - A^*\right)^{-1} cc^* + cc^*\right] z_t =
$$
$$
= q\mu_t \xi_t + \tfrac{1}{2} q \left(\gamma\left(1\right)\right)^{-1} \left[z_t^* cc^* \left(\mathbf{I} - A\right)^{-1} z_{t+1} + \right.
$$
$$
\left. + z_{t+1}^* \left(\mathbf{I} - A^*\right)^{-1} cc^* z_t + z_t^* cc^* z_t\right] =
$$

$$
= q \left(\gamma\left(1\right)\right)^{-1} \left[\xi_{t+1} c^* \left(\mathbf{I} - A\right)^{-1} z_{t+1} - \xi_t c^* \left(\mathbf{I} - A^*\right)^{-1} z_t - \right.
$$
$$
\left. - \xi_t c^* \left(\mathbf{I} - A\right)^{-1} \left(z_{t+1} - z_t\right) + \tfrac{1}{2} \left(\xi_{t+1} - \xi_t\right)^2 + \gamma\left(1\right) \mu_t \xi_t\right] =
$$

$$
= q \left(\gamma\left(1\right)\right)^{-1} \left[\xi_{t+1} c^* \left(\mathbf{I} - A\right)^{-1} z_{t+1} - \xi_t c^* \left(\mathbf{I} - A^*\right)^{-1} z_t + \right.
$$
$$
\left. + \xi_t c^* z_t + \tfrac{1}{2} \left(\xi_{t+1} - \xi_t\right)^2\right] =
$$
$$
= q \left(\gamma\left(1\right)\right)^{-1} \left[\xi_{t+1} \left(c^* \left(\mathbf{I} - A\right)^{-1} z_{t+1} + \tfrac{1}{2}\xi_{t+1}\right) - \right.
$$
$$
\left. - \xi_t \left(c^* \left(\mathbf{I} - A^*\right)^{-1} z_t + \tfrac{1}{2}\xi_t\right)\right]
$$

Introducing this in (3.112) we have

$$
\eta\left(T\right) = z_T^* P z_T + q \left(\gamma\left(1\right)\right)^{-1} \xi_T \left(c^* \left(\mathbf{I} - A\right)^{-1} z_T + \tfrac{1}{2}\xi_T\right) - z_0^* P z_0 -
$$
$$
- q \left(\gamma\left(1\right)\right)^{-1} \xi_0 \left(c^* \left(\mathbf{I} - A\right)^{-1} z_0 + \tfrac{1}{2}\xi_0\right) + \sum_0^{T-1} \left|\gamma_0 \mu_t + w_1^* z_t\right|^2
$$

$$
(3.113)
$$

B. Consider now a solution of (3.98), associate the corresponding solution (z_t, ξ_t) of (3.100) satisfying also (3.102) and take the input function μ_t in (3.108) as

$$
\mu_t = -\varphi \left(c^* z_t + \xi_t\right) + \varphi\left(\xi_t\right) \qquad (3.114)
$$

with $\varphi\left(\sigma\right)$ subject to (3.99). With this choice the corresponding solution of the Popov system (3.108) coincides with the solution of (3.100) and satisfies also (3.102). Therefore, $\eta(T)$ can be written as follows

$$
\eta(T) = \tau \sum_{0}^{T-1} \left[\underline{\varphi} - \left(\varphi\left(\xi_{t+1}\right) - \varphi\left(\xi_t\right) \right)\left(\xi_{t+1} - \xi_t\right)^{-1} \right] \times
$$
$$
\times \left[1 - \left(\varphi\left(\xi_{t+1}\right) - \varphi\left(\xi_t\right)\right)\left(\bar{\varphi}\left(\xi_{t+1} - \xi_t\right)\right)^{-1} \right](c^* z_t)^2 - \qquad (3.115)
$$
$$
- q \sum_{0}^{T-1} \left(\varphi\left(\xi_{t+1}\right) - \varphi\left(\xi_t\right)\right)\xi_{t+1}
$$

A straightforward manipulation shows that

$$
\sum_{0}^{T-1} \xi_{t+1}\left(\varphi\left(\xi_{t+1}\right) - \varphi\left(\xi_t\right)\right) = \xi_T \varphi\left(\xi_T\right) - \xi_0 \varphi\left(\xi_0\right) - \int_{\xi_0}^{\xi_T} \varphi\left(s\right)\,ds +
$$
$$
+ \sum_{0}^{T-1} \left[\int_{\xi_t}^{\xi_{t+1}} \varphi\left(s\right)\,ds - \left(\xi_{t+1} - \xi_t\right)\varphi\left(\xi_t\right) \right] \geq
$$
$$
\geq \xi_T \varphi\left(\xi_T\right) - \xi_0 \varphi\left(\xi_0\right) - \int_{\xi_0}^{\xi_T} \varphi\left(s\right)\,ds
$$

The last inequality followed from the monotonicity of $\varphi(\sigma)$ (remember that $\underline{\varphi} \geq 0$).

We equate (3.113) and (3.114) and take into account (3.102) and the above inequality, obtaining

$$
V\left(z_T,\xi_T\right) \leq V\left(z_0,\xi_0\right) - \sum_{0}^{T-1} \left|\gamma_0\varphi\left(\xi_t\right) - \gamma_0\varphi\left(c^* z_t + \xi_t\right) + w_1^* z_t\right|^2 -
$$
$$
- \tau \sum_{0}^{T-1} \left[\left(\varphi\left(\xi_{t+1}\right) - \varphi\left(\xi_t\right)\right)\left(\xi_{t+1} - \xi_t\right)^{-1} - \underline{\varphi} \right] \times
$$
$$
\times \left[1 - \left(\varphi\left(\xi_{t+1}\right) - \varphi\left(\xi_t\right)\right)\left(\bar{\varphi}\left(\xi_{t+1} - \xi_t\right)\right)^{-1} \right](c^* z_t)^2 \ ,
$$
$$
\qquad (3.116)
$$

where

$$
V\left(z,\xi\right) = z^* P z - q \left(\frac{1}{2\gamma(1)}\xi^2 + \int_{0}^{\xi} \varphi\left(s\right)\,ds \right)
$$

is a *candidate Liapunov function*. From (3.99) and $\tau \geq 0$ we deduce that $V\left(z,\xi\right)$ *is nonincreasing* along the solutions of (3.100) that satisfy also (3.102), i.e., *along those solutions of (3.100) that are located in the invariant manifold defined by the first integral (3.102).*

C. For a given $\tilde{x}_0$ define

$$\tilde{x}_t = (A - \tilde{\varphi}bc^*)^t \tilde{x}_0 \ , \ \tilde{\xi}_t = c^*\tilde{x}_t \ , \ \tilde{z}_t = (A - \mathbf{I} - \tilde{\varphi}bc^*) \tilde{x}_t \quad (3.117)$$

Obviously the pair $\left(\tilde{z}_t, \tilde{\xi}_t \right)$ thus defined is a solution of (3.100) satisfying (3.102) and corresponding to $\varphi(\sigma) = \tilde{\varphi}\sigma$. Because $\underline{\varphi} < \tilde{\varphi} < \bar{\varphi}$ we obtain from (3.116) that

$$V\left(\tilde{z}_t, \tilde{\xi}_t \right) \le V\left(\tilde{z}_0, \tilde{\xi}_0 \right) .$$

But $A - \tilde{\varphi}bc^*$ has the eigenvalues inside the unit disk. By letting $t \to \infty$ it follows that

$$V\left(\tilde{z}_0, \tilde{\xi}_0 \right) = \tilde{z}_0^* P \tilde{z}_0 - q \frac{1 + \tilde{\varphi}\gamma(1)}{2\gamma(1)} \tilde{\xi}_0^2 \ge 0$$

Replacing $\tilde{z}_0$ and $\tilde{\xi}_0$ from (3.117) and taking into account that $\tilde{x}_0$ is arbitrary it follows, after some manipulation that

$$P - \frac{q}{2\gamma(1)(1 + \tilde{\varphi}\gamma(1))} (A^* - \mathbf{I})^{-1} cc^* (\mathbf{I} - A)^{-1} \ge 0. \quad (3.118)$$

Using the expression (3.106) of $\psi(\sigma)$ we have

$$V(z, \xi) = z^* P z - \frac{q}{\gamma(1)} \int_0^{\xi} \psi(\sigma) \, d\sigma , \quad (3.119)$$

i.e., $V(z, \xi)$ is a quadratic form plus an integral of some suitably chosen nonlinear function. Such Liapunov functions are basic in absolute stability. On the invariant set (3.102) we have

$$\psi(\xi_t) = \xi_t + \gamma(1)\varphi(\xi_t) \equiv c^* (A - \mathbf{I})^{-1} z_t$$

and taking into account (3.118) it follows that

$$z_t^* P z_t \ge \frac{q}{2\gamma(1)(1 + \tilde{\varphi}\gamma(1))} \psi^2(\xi_t) .$$

If this inequality is combined with (3.119) and with the fact that $V(z, \xi)$ is nonincreasing along solutions of (3.100) satisfying (3.102), we shall have

$$q\Omega(\xi_t) \le V(z_0, \xi_0) , \quad (3.120)$$

where $\Omega(\sigma)$ is defined by (3.105). By assumption only those functions $\Omega(\sigma)$ are allowed for which $\Omega(\sigma) > 0$, $\sigma \neq 0$; also $\Omega(0) = 0$. Therefore, $\Omega(\sigma) \geq \alpha(|\sigma|)$ where $\alpha(\rho) = \max\{\Omega(\rho), \Omega(-\rho)\}$. One can see that $\alpha(\rho)$ is increasing for $\rho > 0$ and $\alpha(0) = 0$ i.e. is a Massera function.

Therefore,

$$|\xi_t| \leq \alpha^{-1}(V(z_0, \xi_0)/q).$$

Replace in (3.119) z_0 and ξ_0 from (3.101) obtaining

$$V(z_0, \xi_0) \leq \beta(|x_0|), \quad \lim_{\rho \to \infty} \beta(\rho) = 0$$

which gives

$$|\xi_t| \leq \alpha^{-1}(\beta(|x_0|)/q).$$

D. We rewrite (3.98) in the form

$$x_{t+1} = (A - \tilde{\varphi}bc^*)x_t + b(\tilde{\varphi}c^*x_t - \varphi(c^*x_t)).$$

But $c^*x_t \equiv \xi_t$; hence

$$x_t = (A - \tilde{\varphi}bc^*)^t x_0 + \sum_0^{t-1}(A - \tilde{\varphi}bc^*)^{t-1-k}b(\tilde{\varphi}\xi_t - \varphi(\xi_t))$$

and because $A - \tilde{\varphi}bc^*$ has eigenvalues inside the unit disk it follows that

$$|x_t| \quad \leq \quad \beta_0\rho^t|x_0| + \beta_0\left|b(\bar{\varphi} - \tilde{\varphi})\alpha^{-1}(\beta(|x_0|)/q)\right| \sum_0^{t-1}\rho^k \leq$$

$$\leq \quad \beta_0\left[|x_0| + |b|(\bar{\varphi} - \tilde{\varphi})\alpha^{-1}(\beta(|x_0|)/q)(1-\rho)^{-1}\right] = \delta(|x_0|)$$

where $\lim_{\rho \to \infty} \delta(\rho) = 0$. In this way Liapunov stability of the zero solution for (3.98) has been obtained.

E. Consider again a solution of (3.98) with x_0 such that $|x_t| \leq \delta(|x_0|)$ holds. The solution is thus bounded and the corresponding ω-limit set is nonempty and consists only of trajectories. Associate to x_t the solution of (3.100) satisfying (3.102). Consider also $\bar{x}_t$ a solution from the ω-limit set of x_t and $(\bar{z}_t, \bar{\xi}_t)$ associated to $\bar{x}_t$ by the same procedure. The sequence $V_t^* = V(z_t, \xi_t)$ is nonincreasing and bounded from below (it is positive since $V(z, \xi) \geq q\Omega(\xi) > q\alpha(|\xi|) > 0$); hence there exists $\lim_{t \to \infty} V_t^* = V_\infty$. Consider an arbitrary integer $m > 0$ and a sequence of integers $\{t_k\}$, $t_k \to \infty$ such that $\lim\limits_{k\to\infty} x_{t_k} = \bar{x}_m$ ($\bar{x}_t$ is a trajectory from the ω -limit set hence $\bar{x}_m$ is a ω -limit point). Therefore, $\lim\limits_{k\to\infty} z_{t_k} = \bar{z}_m$, $\lim\limits_{k\to\infty} \xi_{t_k} = \bar{\xi}_m$ and from the continuity of V with respect to its arguments it follows that $\lim\limits_{k\to\infty} V_{t_k}^* = V\left(\bar{z}_m, \bar{\xi}_m\right) = V_\infty$; m being arbitrary, $V\left(\bar{z}_t, \bar{\xi}_t\right)$ is constant. From (3.116) and from the fact that inequalities (3.99) are strict we deduce that $c^* \bar{z}_t \equiv 0$; hence $\bar{\xi}_t \equiv const.$ and $\bar{z}_t$ is a solution of

$$z_{t+1} = A z_t$$

with $c^* \bar{z}_t \equiv 0$. But (c^*, A) is observable what gives $\bar{z}_t \equiv 0$. Therefore $c^* \left(\mathbf{I} - A\right)^{-1} \bar{z}_t \equiv 0$ hence $\psi\left(\bar{\xi}_t\right) \equiv 0$. This implies $\bar{\xi}_t \equiv 0$ hence $\bar{x}_t \equiv 0$.

Actually it has been shown, applying in the given case the arguments of the Barbashin - Krasovskii - La Salle invariance principle for discrete systems (La Salle (1976)), that the ω -limit set of a bounded solution x_t coincides with the zero solution thus obtaining asymptotic stability.

F. The asymptotic stability is global if (3.107) holds. Indeed in this case $\alpha\left(\rho\right)$ defined above has the property $\lim\limits_{\rho\to\infty} \alpha\left(\rho\right) = +\infty$ and the reasoning above is true for arbitrary initial conditions which ends the proof.

3.6.3. An Example

Consider the discrete dynamical system described by

$$y_{t+2} - 2y_{t+1} \cos\delta + y_t + \varphi\left(2\left(y_{t+1}\cos 2\delta - y_t \cos\delta\right)\right) = 0$$

with $0 < \delta < \pi/3$.

For linear functions $\varphi\left(\sigma\right) = \mu\sigma$ the stability sector can be found by Schur-Cohn conditions (2.11) applied to the linear equation

$$y_{t+2} + 2\left(\mu\cos 2\delta - \cos\delta\right) y_{t+1} + \left(1 - 2\mu\cos\delta\right) y_t = 0 .$$

We have $1 > 0$, $|1 - 2\mu\cos\delta| < 1$ which gives $0 < \mu < \left(\cos\delta\right)^{-1}$ and $2\left|\mu\cos 2\delta - \cos\delta\right| < 2\left(1 - \mu\cos\delta\right)$. It is easily found that the linear system is stable provided

$$0 < \mu < \left(1 + 2\cos\delta\right)^{-1} .$$

This is the Schur-Cohn sector. In the nonlinear case, the transfer function of the linear part

$$y_{t+2} - 2y_{t+1} \cos \delta + y_t = \mu_t$$
$$\sigma_t = 2 \left(y_{t+1} \cos 2\delta - y_t \cos \delta \right)$$

takes the form

$$\gamma(z) = 2 \frac{z \cos 2\delta - \cos \delta}{z^2 - 2z \cos \delta + 1}$$

and its poles are clearly on the unit circle, i.e., the *nonlinear system is in a critical case*.

The frequency domain stability inequalities proved in the previous sections have been obtained with the assumptions that the system was in the *stable case* i.e. $\gamma(z)$ had its poles inside the unit disk. This assumption was essential in order to apply Fourier techniques and Plancherel - Parseval equality. Nevertheless, we shall assume them extended to critical cases for the sake of comparison with the criterion obtained in this section.

a. The criterion of Tsypkin given in Section 3.1 reads

$$\frac{1}{\bar{\varphi}} + \frac{\cos 2\delta - (\cos \delta) \cos \theta}{\cos \theta - \cos \delta} = \frac{1}{\bar{\varphi}} - \cos \delta - \frac{1 - \cos \delta}{\cos \theta - \cos \delta} > 0 \,.$$

Remark that the ratio $(1 - \cos \delta)(\cos \theta - \cos \delta)^{-1}$ will change sign in the neighborhood of $\theta = \delta$ by taking arbitrarily large negative and positive values. Therefore, this inequality is never fulfilled for the considered system.

b. The modified Tsypkin criterion of Section 3.1 which in fact acts as a discrete counterpart of the classical Popov condition (due to its PD multiplier) and has been obtained also by Jury and Lee (1964) reads

$$\frac{1}{\bar{\varphi}} + \mathrm{Re} \left(1 + \left(1 - e^{-i\theta} \right) q \right) \gamma \left(e^{i\theta} \right) =$$
$$= \frac{1}{\bar{\varphi}} + (\cos \theta - \cos \delta)^{-1} \left[(1 + q \left(1 - \cos \theta \right)) \left(\cos 2\delta - \cos \delta \cos \theta \right) - \right.$$
$$\left. -q \cos \delta \sin^2 \theta \right] > 0 \,.$$

The unique choice for q in order that the above inequality holds for all $\theta \in [0, \pi)$ is that allowing to simplify $(\cos \theta - \cos \delta)$ that is $q = -1$.

With this choice the above inequality becomes

$$\frac{1}{\bar{\varphi}} - (1 + 2 \cos \delta \cos \theta) > 0$$

which gives $\bar{\varphi} < (1 + 2 \cos \delta)^{-1}$, i.e., the absolute stability sector coincides with the linear stability one.

Unfortunately the frequency condition applied here is proved only for $q > 0$; hence it cannot be applied to this case. As is mentioned in the paper of Popov (1977) the proof for $q < 0$ would require the use of forward differences which are more difficult to manipulate. In the cited paper it is suggested that the difficulties could be overcome by the so-called "saturability technique". The question is still open.

c. Apply now the condition of Theorem 3.7 with $\tau = 1$, $\underline{\varphi} = 0$; we have

$$\frac{1}{\bar{\varphi}} + \mathrm{Re}\left(1 + q\left(1 - e^{-i\theta}\right)^{-1}\right)\gamma\left(e^{i\theta}\right) =$$

$$= \frac{1}{\bar{\varphi}} + \frac{1}{2(1-\cos\theta)(\cos\theta - \cos\delta)}\left[(q+2)(1 - \cos\theta)(\cos 2\delta - \cos\delta\cos\theta) + \right.$$
$$\left. + q\sin^2\theta\cos\delta\right] =$$

$$= \frac{1}{\bar{\varphi}} + \frac{1}{2(\cos\theta - \cos\delta)}\left[q(\cos\delta + 1)(2\cos\delta - 1) + 4\cos^2\delta - \right.$$
$$\left. -2 - 2\cos\delta\cos\theta\right].$$

In order to simplify $(\cos\theta - \cos\delta)$ which introduces sign changes when $\theta \to \delta$ we shall choose q from the equality

$$q(\cos\delta + 1)(2\cos\delta - 1) + 4\cos^2\delta - 2 - 2\cos^2\delta = 0$$

which gives

$$q = (1 - \cos\delta)(2\cos\delta - 1)^{-1} > 0$$

and the frequency domain inequality holds provided $\bar{\varphi} \leq (\cos\delta)^{-1}$. But $\gamma(1) = -(1 + 2\cos\delta) < 0$ and because the Schur - Cohn sector is bounded from below by 0 we shall have $\check{\varphi} > 0$ arbitrarily small hence lower than $-1/\gamma(1)$. Applying Theorem 3.7 we find that our system is absolutely stable, e.g., for all differentiable functions satisfying

$$0 < \varphi'(\sigma) < -(\gamma(1))^{-1} = (1 + 2\cos\delta)^{-1}$$

which is the entire Schur - Cohn linear stability sector.

This example shows that the proposed stability condition of Theorem 3.7 can give an absolute stability sector as large as that of the linear case while rather standard stability criteria cannot be applied at all.

3.6.4. Concluding Remarks

A. Usefulness and efficiency in applications of the proposed frequency domain inequality have been illustrated by an example. From the theoretical point of view the application of the Kalman - Szegö - Popov - Yakubovich lemma is allowed to combine refinements of the Liapunov function method (used here as a mathematical proof instrument) with simplicity of the frequency domain method. In fact the main result does not mention any Liapunov function but only a frequency inequality and some additional assumptions on the linear part and on the class of nonlinearities. In particular the assumption on the location of the eigenvalues of A inside the unit disk is considerably weakened here; matrix A may have its eigenvalues any part of the complex plane but satisfy the assumption $v)$ - an output stability assumption and at the same time some kind of "minimal stability". For the absolute stability we ask the system to be stable for *one* linear function in the sector.
      B. It is worth mentioning the invariant manifold (3.102) once more. The two systems (basic and associated) are equivalent on the manifold. Also the good properties of the Liapunov function are valid on this manifold only. The application of the kinetic Liapunov function may avoid the consideration of this manifold but the proofs (Haddad and Kapila (1995)) need improvement, even in the continuous - time case). In the discrete - time case the same authors published a paper (Kapila and Haddad (1996)) where the same kinetic Liapunov function was used but the criterion is the one of Popov (1977) containing a usual PD multiplier. Therefore, the problem considered here is still of interest although it already has a history.

3.7. THE KSPY LEMMA (MULTI-INPUT CASE)

The main object of study will be here the set of matrices (A, B, K, L, M) where A, M are $n \times n$ matrices, K is a $m \times m$ matrix and B, L are $n \times m$ matrices. All the items of the set are complex; it is assumed that M and K are Hermitian: $M = M^*$, $K = K^*$. The set is associated to the following multi-input Popov system

$$x_{t+1} = Ax_t + Bu_t \tag{3.121}$$

$$\eta_T = \sum_0^{T-1} \mathcal{F}(x_t, u_t) \tag{3.122}$$

$$\mathcal{F}(x, u) = x^* M x + x^* L u + u^* L^* x + u^* K u \tag{3.123}$$

η_T is defined for pairs satisfying (3.121). We associate also the characteristic matrix $T(\lambda, \sigma)$ of the system, defined by

$$\begin{aligned} T(\lambda, \sigma) = K &+ L^* (\sigma \mathbf{I} - A)^{-1} B + B^* (\lambda \mathbf{I} - A^*)^{-1} L + \\ &+ B^* (\lambda \mathbf{I} - A^*)^{-1} M (\sigma \mathbf{I} - A)^{-1} B. \end{aligned} \tag{3.124}$$

Remark that $T(\lambda, \sigma) = \overline{T(\bar{\lambda}, \bar{\sigma})}$.

As in Section 3.5 we shall call Kalman-Szegö-Popov-Yakubovich lemma the connection between the following properties:

(F) The frequency condition $T(\sigma^{-1}, \sigma) \geq 0$ holds for all $|\sigma| = 1$ for which $\det(\sigma \mathbf{I} - A) \neq 0$;

(L) There exist matrices V, W and $H = H^*$ such that the following matrix equalities hold

$$\begin{aligned} K + B^* H B &= V^* V \\ L + A^* H B &= W V \\ M + A^* H A - H &= W W^* \end{aligned} \tag{3.125}$$

In order to obtain such connection we need some preliminaries.

3.7.1. Weak and Strong Frequency Domain Conditions

Consider the following real numbers

$$\alpha' = \inf_{\substack{x \in \mathbb{C}^n \\ u \in \mathbb{C}^m}} \left\{ \frac{\mathcal{F}(x, u)}{|x|^2 + |u|^2} \; ; \; \sigma x = Ax + Bu, \; \sigma \in \mathbb{C}, \; |\sigma| = 1 \right\} \tag{3.126}$$

$$\alpha'' = \inf_{u \in \mathbb{C}^m} \left\{ \frac{\mathcal{F}\left((\sigma \mathbf{I} - A)^{-1} Bu, u \right)}{|u|^2} \; , \; \sigma \in \mathbb{C}, \; |\sigma| = 1 \right\}. \tag{3.127}$$

where α'' is defined if $\det(\sigma \mathbf{I} - A) \neq 0$ for $|\sigma| = 1$.

Remark that if α'' is well defined then $\mathcal{F}\left((\sigma I - A)^{-1} Bu, u\right) = u^* T\left(\sigma^{-1}, \sigma\right) u$ hence $\alpha'' \geq 0$ means fulfillment of the frequency domain condition (F). We have

Proposition 3.8   If $\det (\sigma I - A) \neq 0$ for $|\sigma| = 1$ then $\alpha'' > 0$ implies $\alpha' > 0$.

Proof. Define $x = (\sigma I - A)^{-1} Bu$ for $u \in \mathbb{C}^m$, $\sigma \in \mathbb{C}$, $|\sigma| = 1$; we shall have

$$|x|^2 + |u|^2 \quad \leq \quad \left[1 + \left|(\sigma I - A)^{-1} B\right|^2\right] |u|^2 \leq \beta |u|^2 \ ,$$

$$\beta \quad = \quad 1 + \sup_{|\sigma|=1} \left|(\sigma I - A)^{-1} B\right|^2 .$$

For these u and x we shall have

$$\frac{\mathcal{F}(x, u)}{|x|^2 + |u|^2} \geq \frac{\mathcal{F}(x, u)}{\beta |u|^2} \ ; \quad \text{hence} \quad \alpha' \geq \frac{\alpha''}{\beta} > 0 .$$

If $\alpha' > 0$ (or if α'' is well defined, $\alpha'' > 0$) we shall say that *strong* frequency domain inequality holds; if $\alpha' \geq 0$ ($\alpha'' \geq 0$) we shall say that *weak* frequency domain inequality holds.

3.7.2. An Abstract Minimization Problem

We shall consider here a complex Hilbert space $\mathcal{H}$, $\mathcal{M}_0 \subset \mathcal{H}$ a closed linear subspace of $\mathcal{H}$ (i.e., a linear closed subset of $\mathcal{H}$ with the same scalar product and norm) and let $\mathcal{M} = f + \mathcal{M}_0$ where $f \in \mathcal{H}$. We denote by $(\cdot, \cdot)$ the scalar product on $\mathcal{H}$ and by $|\cdot| = (\cdot, \cdot)^{1/2}$ the norm on $\mathcal{H}$. Let $\mathcal{J}(h) = (Gh, h) + (g, h) + (h, g)$, where $g \in \mathcal{H}$ and G is a bounded self-adjoint operator, be a functional on $\mathcal{H}$. Let
$$\alpha = \inf_{h \in \mathcal{M}_0} \frac{(Gh, h)}{|h|^2} .$$
We may state now

Lemma 3.9   *The necessary and sufficient condition for the existence of an element $\tilde{h} \in \mathcal{M}$ such that $\mathcal{J}(h) \geq \mathcal{J}\left(\tilde{h}\right)$ for any $h \in \mathcal{M}$ is that $\alpha \geq 0$ and $G\tilde{h} + g$ be orthogonal to $\mathcal{M}_0$ $\left(G\tilde{h} + g \perp \mathcal{M}_0\right)$; any $\tilde{h}$ satisfying this orthogonality condition is an optimal element. If*

$\alpha > 0$ *then the optimal element exists and is unique. If* $\alpha < 0$ *then*
$\inf\limits_{h \in \mathcal{M}} \mathcal{J}(h) = -\infty.$

Proof. A. Let $\alpha < 0$ hence there exists $h_0 \in \mathcal{M}_0$ such that $(Gh_0, h_0) < 0$. Let $h_\lambda = \lambda h_0 + f \in \mathcal{M}$ where λ is a real number. We shall have

$$
\begin{aligned}
\mathcal{J}(h_\lambda) &= (Gh_\lambda, h_\lambda) + (g, h_\lambda) + (h_\lambda, g) = \\
&= \lambda^2 (Gh_0, h_0) + \lambda \left[(Gh_0, f) + (Gf, h_0) + (g, h_0) + (h_0, g) \right] + \\
&\quad + (Gf, f) + (g, f) + (f, g)
\end{aligned}
$$

and obviously $\lim\limits_{\lambda \to \infty} \mathcal{J}(h_\lambda) = -\infty$. This proves the last sentence of the lemma and the necessity of $\alpha \ge 0$.

B. Let now $\tilde{h} \in \mathcal{M}$ be an optimal element and $h \in \mathcal{H}$. We shall have

$$
\mathcal{J}\left(\tilde{h} + h\right) - \mathcal{J}\left(\tilde{h}\right) = \left(G\tilde{h} + g, h\right) + \left(h, G\tilde{h} + g\right) + (Gh, h)
$$

since G is self-adjoint. Let $h = \varepsilon h_0$, $h_0 \in \mathcal{M}_0$, ε be a real number. Since $\tilde{h}$ is optimal $\mathcal{J}\left(\tilde{h} + h\right) - \mathcal{J}\left(\tilde{h}\right) \ge 0$. It follows that

$$
2\varepsilon \operatorname{Re}\left(G\tilde{h} + g, h_0\right) + \varepsilon^2 (Gh_0, h_0) \ge 0
$$

and since ε is arbitrary we deduce that $\operatorname{Re}\left(G\tilde{h} + g, h_0\right) = 0$ for $h_0 \in \mathcal{M}_0$. Taking $h = i\varepsilon h_0$ we deduce $\operatorname{Im}\left(G\tilde{h} + g, h_0\right) = 0$. Therefore, $G\tilde{h} + g \perp \mathcal{M}_0$ is a necessary condition of optimality.

Conversely, let $\tilde{h} \in \mathcal{M}$ be such that $G\tilde{h} + g \perp \mathcal{M}_0$ and let $\alpha \ge 0$. Let $h \in \mathcal{M}_0$. We shall have

$$
\mathcal{J}\left(\tilde{h} + h\right) - \mathcal{J}\left(\tilde{h}\right) = (Gh, h) \ge 0 \,, \quad h \in \mathcal{M}_0.
$$

Let $h \in \mathcal{M}$ hence there exists $h_0 \in \mathcal{M}_0$ such that $h = f + h_0$. Also since $\tilde{h} \in \mathcal{M}$ there exists $\tilde{h}_0 \in \mathcal{M}_0$ such that $\tilde{h} = f + \tilde{h}_0$. We shall have

$$
\begin{aligned}
\mathcal{J}(h) - \mathcal{J}\left(\tilde{h}\right) &= \mathcal{J}(f + h_0) - \mathcal{J}\left(\tilde{h}\right) = \\
&= \mathcal{J}\left(f + \tilde{h}_0 + h_0 - \tilde{h}_0\right) - \mathcal{J}\left(\tilde{h}\right) = \\
&= \mathcal{J}\left(\tilde{h} + h_0 - \tilde{h}_0\right) - \mathcal{J}\left(\tilde{h}\right) =
\end{aligned}
$$

$$= \left(G \left(h_0 - \tilde{h}_0 \right), h_0 - \tilde{h}_0 \right) \geq 0$$

the last equality following from the previous one since $\left(h_0 - \tilde{h}_0 \right) \in \mathcal{M}_0$. We proved that $\mathcal{J}(h) - \mathcal{J}\left(\tilde{h} \right) \geq 0$ for any $h \in \mathcal{M}$.

   C. Assume now that $\alpha > 0$. Let $\tilde{h} \in \mathcal{M}$ such that $G\tilde{h}+g \perp \mathcal{M}_0$ and let $\mathcal{P}$ be the orthogonal projector on $\mathcal{M}_0$. We shall have $\mathcal{P}\left(G\tilde{h} + g \right) = 0$ and since $\tilde{h} = f + \tilde{h}_0$, $\tilde{h}_0 \in \mathcal{M}_0$ it follows that $\mathcal{P}G\tilde{h}_0 + \mathcal{P}(Gf + g) = 0$; but $\mathcal{P}\tilde{h}_0 = \tilde{h}_0$; hence $\mathcal{P}G\mathcal{P}\tilde{h}_0 + \mathcal{P}(Gf + g) = 0$. But $\alpha > 0$ implies that the restriction of the operator $\mathcal{P}G\mathcal{P}$ to $\mathcal{M}_0$ has a bounded inverse. We deduce $\tilde{h}_0 = -(\mathcal{P}G\mathcal{P})^{-1}_{\mathcal{M}_0} \mathcal{P}(Gf + g)$ and $\tilde{h} = \tilde{h}_0 + f$ is the optimal element. Indeed we have $G\tilde{h} + g = Gf + g + G\tilde{h}_0 = Gf + g + G\mathcal{P}\tilde{h}_0$, $\mathcal{P}\left(G\tilde{h} + g \right) = \mathcal{P}(Gf + g) + \mathcal{P}G\mathcal{P}\tilde{h}_0 = \mathcal{P}(Gf + g) - \mathcal{P}(Gf + g) = 0$ hence $G\tilde{h} + g \perp \mathcal{M}_0$.

   The uniqueness of the optimal element follows by contradiction. Assume existence of some $\hat{h} \in \mathcal{M}$ that is also optimal. A straightforward manipulation will give

$$\mathcal{J}\left(\tilde{h} \right) - \mathcal{J}\left(\hat{h} \right) = \left(G\left(\tilde{h} - \hat{h} \right), \tilde{h} - \hat{h} \right) + \left(G\hat{h} + g, \tilde{h} - \hat{h} \right) + $$
$$ + \left(\tilde{h} - \hat{h}, G\hat{h} + g \right).$$

But $\tilde{h} - \hat{h} \in \mathcal{M}_0$ and since $\hat{h}$ is optimal $\left(G\hat{h} + g, \tilde{h} - \hat{h} \right) = 0$. It follows that $\mathcal{J}\left(\tilde{h} \right) - \mathcal{J}\left(\hat{h} \right) = \left(G\left(\tilde{h} - \hat{h} \right), \tilde{h} - \hat{h} \right) = 0$ since both $\tilde{h}$ and $\hat{h}$ are optimal. But $\alpha > 0$ and the above equality requires $\hat{h} = \tilde{h}$. The proof is complete.

   Consider now that in the above lemma the element $f \in \mathcal{H}$ which defines $\mathcal{M}$ is given by Sx where $x \in X$, X being a Hilbert space and $S : X \rightarrow \mathcal{H}$ is a linear bounded operator. We denote $\mathcal{M}_x = \mathcal{M}_0 + Sx$. We may state

Lemma 3.10 *Consider the functional $\mathcal{J}(h)$ defined as above and assume that $\alpha = \inf\limits_{h \in \mathcal{M}_0} \dfrac{(Gh, h)}{|h|^2} > 0$. Let $\tilde{h}(x)$ be the unique optimal element in $\mathcal{M}_x = \mathcal{M}_0 + Sx$ whose existence is ensured from Lemma 3.9. Then there exist bounded linear operators $Q : X \rightarrow \mathcal{H}$ and $R : \mathcal{H} \rightarrow \mathcal{H}$ such that $\tilde{h}(x) = Qx + Rg$.*

Proof. We have from Lemma 3.9 that if $\alpha > 0$ then $\tilde{h} = f - (\mathcal{P}G\mathcal{P})_{\mathcal{M}_0}^{-1}\mathcal{P}(Gf + g)$; here $f = Sx$ hence

$$\tilde{h} = Sx - (\mathcal{P}G\mathcal{P})_{\mathcal{M}_0}^{-1}\mathcal{P}GSx - (\mathcal{P}G\mathcal{P})_{\mathcal{M}_0}^{-1}\mathcal{P}g$$

and we deduce $Q = S - (\mathcal{P}G\mathcal{P})_{\mathcal{M}_0}^{-1}\mathcal{P}GS$, $R = -(\mathcal{P}G\mathcal{P})_{\mathcal{M}_0}^{-1}\mathcal{P}$. Boundedness of these operators follows from the boundedness of $(\mathcal{P}G\mathcal{P})_{\mathcal{M}_0}^{-1}$, S and G.

3.7.3. A Linear Quadratic Minimization Problem

We shall consider here the problem of minimizing the index

$$\mathcal{J}(x,u) = \eta_\infty = \sum_{t=0}^\infty \mathcal{F}(x_t, u_t) \qquad (3.128)$$

on the set of all pairs $(x_t, u_t) \in \ell_+^2(\mathbb{C}^n) \times \ell_+^2(\mathbb{C}^m)$ satisfying the dynamics (3.121) with $x_0 = x$. We denote by $\ell_+^2(\mathbb{C}^n)$ the space of sequences $\{x_t\}_t$, $t = 0, 1, ...$, with values in $\mathbb{C}^n$ satisfying $\sum_0^\infty |x_t|^2 < \infty$, with the usual norm; $\ell_+^2(\mathbb{C}^m)$ has the same meaning for $\mathbb{C}^m$ - valued sequences on $(0, \infty)$. Our aim is to study this minimization problem as an application of previous abstract results. We take $\mathcal{H} = \ell_+^2(\mathbb{C}^n) \times \ell_+^2(\mathbb{C}^m)$

$$\mathcal{M}_x = \{(x_t, u_t) \in \mathcal{H} : x_{t+1} = Ax_t + Bu_t , t \geq 0 , x_0 = x\} \quad (3.129)$$

$$\mathcal{M}_0 = \{(x_t, u_t) \in \mathcal{H} : x_{t+1} = Ax_t + Bu_t , t \geq 0 , x_0 = 0\} . \quad (3.130)$$

The two sets are nonempty since they contain at least the pairs that are zero for $t > T$, T finite. The dynamics equation is linear hence $\mathcal{M}_x$ and $\mathcal{M}_0$ are linear manifolds in $\ell_+^2(\mathbb{C}^n) \times \ell_+^2(\mathbb{C}^m)$.

Proposition 3.11   *The sets $\mathcal{M}_x$ and $\mathcal{M}_0$ are closed in $\ell_+^2(\mathbb{C}^n) \times \ell_+^2(\mathbb{C}^m)$.*

Proof. Let $\{x_t^k, u_t^k\}_k$ be a sequence of elements of $\mathcal{M}_x$ that converges for $k \to \infty$ to (x_t, u_t), the convergence being taken in the sense of $\ell_+^2(\mathbb{C}^n) \times \ell_+^2(\mathbb{C}^m)$

$$\lim_{k \to \infty} \left(\sum_{t=0}^\infty |x_t^k - x_t|^2 + \sum_{t=0}^\infty |u_t^k - u_t|^2 \right) = 0 . \qquad (3.131)$$

We compute

$$x_t - A^t x - \sum_{j=0}^{t-1} A^{t-1-j} B u_j = x_t - x_t^k - \sum_{j=0}^{t-1} A^{t-1-j} B \left(u_j - u_j^k \right)$$

and therefore

$$\left| x_t - A^t x - \sum_{j=0}^{t-1} A^{t-1-j} B u_j \right|^2 \leq$$

$$\leq 2 \left[\left| x_t - x_t^k \right|^2 + \left(\sum_{j=0}^{t-1} \left| A^{t-1-j} B \right|^2 \right) \left(\sum_{j=0}^{t-1} \left| u_j - u_j^k \right|^2 \right) \right]$$

Letting $k \to \infty$ and taking into account (3.131) we obtain

$$x_t - A^t x - \sum_{j=0}^{t-1} A^{t-1-j} B u_j = 0 \, ;$$

hence $(x_t, u_t) \in \mathcal{M}_x$. If $x = 0$ we have $\left(x_t^k, u_t^k \right) \in \mathcal{M}_0$ and $(x_t, u_t) \in \mathcal{M}_0$.

Proposition 3.12   *Assume that (A, B) is stabilizable i.e. there exists a matrix F of appropriate dimensions such that the eigenvalues of $A + BF$ are inside the unit disk. Then $\mathcal{M}_x = \mathcal{M}_0 + Sx$ where $S : \mathbb{C}^n \to \ell_+^2 (\mathbb{C}^n) \times \ell_+^2 (\mathbb{C}^m)$ is a bounded linear operator.*

Proof. Let $(x_t, u_t) \in \mathcal{M}_x$, i.e., $(x_t, u_t) \in \ell_+^2 (\mathbb{C}^n) \times \ell_+^2 (\mathbb{C}^m)$, $x_{t+1} = Ax_t + Bu_t$, $x_0 = x$. We define $x_t^F = (A + BF)^t x$, $u_t^F = F (A + BF)^t x$ and since $A + BF$ defines an exponentially stable evolution we have $(x_F, u_F) \in \ell_+^2 (\mathbb{C}^n) \times \ell_+^2 (\mathbb{C}^m)$. Consider the pair $x_t^0 = x_t - x_t^F$, $u_t^0 = u_t - u_t^F$; we have obviously $\left(x_t^0, u_t^0 \right) \in \ell_+^2 (\mathbb{C}^n) \times \ell_+^2 (\mathbb{C}^m)$ and $x_0^0 = 0$. Also

$$
\begin{aligned}
x_{t+1}^0 &= x_{t+1} - x_{t+1}^F = Ax_t + Bu_t - (A + BF)(A + BF)^t x = \\
&= Ax_t + Bu_t - A(A + BF)^t x - BF(A + BF)^t x = \\
&= A \left(x_t - x_t^F \right) + B \left(u_t - u_t^F \right) = Ax_t^0 + Bu_t^0 \, .
\end{aligned}
$$

We proved that $\left(x_t^0, u_t^0 \right) \in \mathcal{M}_0$. Since (x_t, u_t) was arbitrary we obtained $\mathcal{M}_x = \left(x^F, u^F \right) + \mathcal{M}_0$. The definition equalities for $\left(x_t^F, u_t^F \right)$ show that we defined a bounded linear operator from $\mathbb{C}^n$ to $\ell_+^2 (\mathbb{C}^n) \times$

$\ell_+^2(\mathbb{C}^m)$. Linearity is obvious and boundedness follows from the fact that $A + BF$ has its eigenvalues inside the unit disk.

We may now state the existence result for the problem of minimizing $J(x, u)$ defined by (3.128) on the subspace $\mathcal{M}_x$ defined by (3.129).

Proposition 3.13 *Assume that (A, B) is stabilizable and $\alpha' > 0$ where α' is defined by (3.126) (or $\alpha'' > 0$ when α'' defined by (3.127) is well defined). Then for any $x_0 \in \mathbb{C}^n$ there exists a unique optimal pair $(\tilde{x}_t, \tilde{u}_t)$ satisfying the dynamics (3.121) with $\tilde{x}_0 = x_0$. Moreover, this unique optimal pair is defined by a bounded linear operator that applies $\mathbb{C}^n$ into $\ell_+^2(\mathbb{C}^n) \times \ell_+^2(\mathbb{C}^m)$ and the optimal index is a quadratic form on $\mathbb{C}^n$.*

Proof.   Existence and uniqueness of the optimal pair will follow from Lemma 3.9. We have to obtain from the assumption $\alpha' > 0$ (or $\alpha'' > 0$) that $\alpha > 0$ for $J(x, u)$ on $\mathcal{M}_0$ i.e. on those pairs in $\ell_+^2(\mathbb{C}^n) \times \ell_+^2(\mathbb{C}^m)$ that are solutions of (3.121) with zero initial conditions. We have

$$\alpha = \inf_{\mathcal{M}_0} \frac{1}{\|x\|^2 + \|u\|^2} \sum_0^\infty \mathcal{F}(x_t, u_t) ,$$

where $\|x\|^2 = \sum_0^\infty |x_t|^2$, $\|u\|^2 = \sum_0^\infty |u_t|^2$. Let (x_t, u_t) be some pair from $\mathcal{M}_0$ (i.e., $x_0 = 0$, $x_{t+1} = Ax_t + Bu_t$). Extending x_t and u_t by 0 for $t < 0$ we define

$$\hat{u}(\omega) = \sum_{-\infty}^\infty u_t e^{-i\omega t} , \quad \hat{x}(\omega) = \sum_{-\infty}^\infty x_t e^{-i\omega t} ,$$

where the convergence is in the ℓ^2 sense. We shall have

$$\mathcal{F}(\hat{x}(\omega), \hat{u}(\omega)) =$$
$$= \sum_{k=-\infty}^\infty \sum_{l=-\infty}^\infty e^{-i\omega(k-l)} [u_l^* K u_k + u_l^* L^* x_k + x_l^* L u_k + x_l^* M x_k]$$

Integrating from $-\pi$ to π and taking into account orthogonality of

$\left\{ e^{-i\omega k} \right\}_k$ we find

$$\frac{1}{2\pi} \int_{-\pi}^{\pi} \mathcal{F}\left(\hat{x}\left(\omega\right), \hat{u}\left(\omega\right)\right) d\omega = \sum_{0}^{\infty} \mathcal{F}\left(x_t, u_t\right) \qquad (3.132)$$

equality (3.132) representing Plancherel equality for DCFT - Discrete to Continuous Fourier Transform - (see also Section 3.1 and the following ones).

We have also

$$\begin{aligned}
e^{i\omega} \hat{x}\left(\omega\right) &= \sum_{t=0}^{\infty} e^{-i\omega(t-1)} x_t = \sum_{1}^{\infty} e^{-i\omega(t-1)} x_t = \sum_{0}^{\infty} e^{-i\omega t} x_{t+1} = \\
&= \sum_{0}^{\infty} e^{-i\omega t} \left(A x_t + B u_t \right) = A\hat{x}\left(\omega\right) + B\hat{u}\left(\omega\right).
\end{aligned}$$

It follows that $\hat{x}\left(\omega\right), \hat{u}\left(\omega\right)$ thus defined verify $e^{i\omega} \hat{x}\left(\omega\right) = A\hat{x}\left(\omega\right) + B\hat{u}\left(\omega\right)$. Since $\alpha' > 0$ we deduce that

$$\mathcal{F}\left(\hat{x}\left(\omega\right), \hat{u}\left(\omega\right)\right) \geq \alpha' \left[\left|\hat{x}\left(\omega\right)\right|^2 + \left|\hat{u}\left(\omega\right)\right|^2 \right] > 0$$

because (x_t, u_t) is a nonzero pair. Therefore,

$$\sum_{t=0}^{\infty} \mathcal{F}\left(x_t, u_t\right) \geq \alpha' \frac{1}{2\pi} \int_{-\pi}^{\pi} \left[\left|\hat{x}\left(\omega\right)\right|^2 + \left|\hat{u}\left(\omega\right)\right|^2 \right] d\omega = \alpha' \left[\|x\|^2 + \|u\|^2 \right].$$

It follows that $\alpha \geq \alpha' > 0$ and Lemma 3.9 gives existence and uniqueness of the optimal pair.

Stabilizability of (A, B) implies fulfillment of the assumptions of Proposition 3.12. We have $\mathcal{M}_{x_0} = \mathcal{M}_0 + S x_0$ where S is bounded and linear. Since $\alpha > 0$ we may apply abstract Lemma 3.10 with $g = 0$ (the functional is quadratic in $x \times u$). We deduce that the optimal pair is linear in x_0 being defined by

$$\tilde{x}\left(x_0\right) = Q^1 x_0, \quad \tilde{u}\left(x_0\right) = Q^2 x_0,$$

where $Q^1 : \mathbb{C}^n \to \ell_+^2\left(\mathbb{C}^n\right)$ and $Q^2 : \mathbb{C}^n \to \ell_+^2\left(\mathbb{C}^m\right)$ are bounded linear operators. Since $\tilde{x}_t$ and $\tilde{u}_t$ are finite dimensional we shall have $\tilde{x}_t\left(x_0\right) = \left(Q^1 x_0\right)_t = Q_t^1 x_0$, $\tilde{u}_t\left(x_0\right) = Q_t^2 x_0$, the matrix sequences $\left\{Q_t^1\right\}_{t \geq 0}$ and $\left\{Q_t^2\right\}_{t \geq 0}$ being in $\ell_+^2\left(\mathbb{C}^{n \times n}\right)$ and in $\ell_+^2\left(\mathbb{C}^{m \times n}\right)$ respectively. We deduce, denoting by $V\left(x_0\right)$ the optimal (minimal) performance index

$$V(x_0) = \sum_{t=0}^{\infty} \mathcal{F}(\tilde{x}_t(x_0), \tilde{u}_t(x_0)) =$$

$$= x_0^* \left[\sum_{t=0}^{\infty} \left((Q_t^1)^* M Q_t^1 + (Q_t^1)^* L Q_t^2 + (Q_t^2)^* L^* Q_t^1 + \right. \right.$$

$$\left. \left. + (Q_t^2)^* K Q_t^2 \right) \right] x_0 = x_0^* H x_0 \qquad (H = H^*) .$$

The proposition is completely proved.

We shall need now the following result:

Proposition 3.14 *The optimal pair has the semi-group property, i.e.,* $\tilde{x}_{t+s}(x_0) = \tilde{x}_t(\tilde{x}_s(x_0)),$ $\tilde{u}_{t+s}(x_0) = \tilde{u}_t(\tilde{x}_s(x_0))$ *for all* $t \geq 0,\ s \geq 0.$

Proof. Consider the fundamental inequality

$$\sum_{t=0}^{\infty} \mathcal{F}(x_t, u_t) \geq \sum_{t=0}^{\infty} \mathcal{F}(\tilde{x}_t(x_0), \tilde{u}_t(x_0))$$

that holds for any pair $(x_t, u_t) \in \mathcal{M}_{x_0}.$ Take u_t such that $u_t = \tilde{u}_t(x_0),$ $0 \leq t < s.$ The above inequality becomes

$$\sum_{s}^{\infty} \mathcal{F}(x_t, u_t) \geq \sum_{s}^{\infty} \mathcal{F}(\tilde{x}_t(x_0), \tilde{u}_t(x_0)) \qquad (3.133)$$

for all admissible pairs $(x_t, u_t) \in \mathcal{M}_{x_0},$ $t \geq s.$ Remark that $\tilde{x}_t(x_0)$ and $\tilde{u}_t(x_0)$ satisfy $x_{t+1} = A x_t + B u_t,$ $T \geq s$ with the initial condition $\tilde{x}_s(x_0).$ The above inequality shows that $(\tilde{x}_t(x_0), \tilde{u}_t(x_0))$ is the optimal pair for the functional starting from $t = s.$ On the other hand, if we take $t = k+s,$ $x_t = \hat{x}_{t-s},$ $u_t = \hat{u}_{t-s}$ we reduce the minimization problem on $[s, \infty)$ to one on $[0, \infty)$ with the same initial condition. Its solution will be the optimal pair $\tilde{x}_{t-s}(\tilde{x}_s(x_0)),$ $\tilde{u}_{t-s}(\tilde{x}_s(x_0)).$ But the optimal pair is unique; hence

$$\tilde{x}_{t-s}(\tilde{x}_s(x_0)) = \tilde{x}_t(x_0) , \quad \tilde{u}_{t-s}(\tilde{x}_s(x_0)) = \tilde{u}_t(x_0)$$

and if we replace $t - s$ by k we shall have

$$\tilde{x}_k(\tilde{x}_s(x_0)) = \tilde{x}_{k+s}(x_0) , \quad \tilde{u}_k(\tilde{x}_s(x_0)) = \tilde{u}_{k+s}(x_0)$$

which ends the proof.

Remark. Inequality (3.133) represents what is usually called *Bell-man optimality principle: the final segment of an optimal trajectory is an optimal trajectory.*

3.7.4. The KSPY Lemma under the Strong Frequency Domain Inequality

Consider some integer $s > 0$ and some $v \in \mathbb{C}^m$. Suppose that (A, B) is stabilizable and $\alpha' > 0$; i.e., the assumptions of Proposition 3.13 hold and let $(\tilde{x}(x_0), \tilde{u}(x_0))$ be the optimal pair associated to the initial condition x_0. Let $\hat{x}_0 = \tilde{x}_s(x_0)$, $\hat{x} = A\hat{x}_0 + Bv$. Define the input sequence

$$u_t = \begin{cases} \tilde{u}_t(x_0) & , \quad 0 \le t < s \\ v & , \quad t = s \\ \tilde{u}_{t-s-1}(\hat{x}) & , \quad t \ge s+1 \, , \end{cases} \qquad (3.134)$$

where $\tilde{u}_t(x)$, $t \ge 0$ is the optimal input sequence corresponding to the initial condition $\hat{x}$.

The solution x_t corresponding to the input u_t defined by (3.134) will be defined by

$$x_t = \begin{cases} \tilde{x}_t(x_0) & , \quad 0 \le t \le s \\ \hat{x} & , \quad t = s+1 \\ \tilde{x}_{t-s-1}(\hat{x}) & , \quad t \ge s+2 \, . \end{cases} \qquad (3.135)$$

We have to check only the last equality

$$x_{s+2} = Ax_{s+1} + Bu_{s+1} = A\hat{x} + B\tilde{u}_0(\hat{x}) = \tilde{x}_1(\hat{x})$$

and for the next indices the procedure is the same. We compute the index $J(x, u)$ with the pair (x, u) constructed above

$$J(x, u) = \sum_0^\infty \mathcal{F}(x_t, u_t) = \sum_{t=0}^{s-1} \mathcal{F}(\tilde{x}_t(x_0), \tilde{u}_t(x_0)) + \mathcal{F}(\tilde{x}_s(x_0), v) +$$

$$+ \mathcal{F}(\hat{x}, \tilde{u}(\hat{x})) + \sum_{t=s+2}^\infty \mathcal{F}(\tilde{x}_{t-s-1}(\hat{x}), \tilde{u}_{t-s-1}(\hat{x})) =$$

$$= \sum_{t=0}^{s-1} \mathcal{F}(\tilde{x}_t(x_0), \tilde{u}_t(x_0)) + \mathcal{F}(\tilde{x}_s(x_0), v) + \sum_{k=0}^\infty \mathcal{F}(\tilde{x}_k(\hat{x}), \tilde{u}_k(\hat{x})) \, .$$

On the other hand we have

$$
\begin{aligned}
V(x_0) \;&=\; J(\tilde{x}(x_0), \tilde{u}(x_0)) = \\
&=\; \sum_{t=0}^{s-1} \mathcal{F}(\tilde{x}_t(x_0), \tilde{u}_t(x_0)) + \sum_{t=s}^{\infty} \mathcal{F}(\tilde{x}_t(x_0), \tilde{u}_t(x_0)) = \\
&=\; \sum_{t=0}^{s-1} \mathcal{F}(\tilde{x}_t(x_0), \tilde{u}_t(x_0)) + \sum_{k=0}^{\infty} \mathcal{F}(\tilde{x}_{k+s}(x_0), \tilde{u}_{k+s}(x_0)).
\end{aligned}
$$

The optimal pair has the semigroup property; i.e., $\tilde{x}_{k+s}(x_0) = \tilde{x}_k(\tilde{x}_s(x_0))$, $\tilde{u}_{k+s}(x_0) = \tilde{u}_k(\tilde{x}_s(x_0))$. Therefore,

$$
\begin{aligned}
\sum_{k=0}^{\infty} \mathcal{F}(\tilde{x}_{k+s}(x_0), \tilde{u}_{k+s}(x_0)) \;&=\; \sum_{k=0}^{\infty} \mathcal{F}(\tilde{x}_k(\tilde{x}_s(x_0)), \tilde{u}_k(\tilde{x}_s(x_0))) = \\
&=\; V(\tilde{x}_s(x_0)) = V(\hat{x}_0)
\end{aligned}
$$

Since $J(x,u) \geq V(x_0)$ we shall have

$$
\begin{aligned}
\sum_{t=0}^{s-1} \mathcal{F}(\tilde{x}_t(x_0), \tilde{u}_t(x_0)) + \mathcal{F}(\hat{x}_0, v) + V(\hat{x}) \;&\geq \\
&\geq\; \sum_{t=0}^{s-1} \mathcal{F}(\tilde{x}_t(x_0), \tilde{u}_t(x_0)) + V(\hat{x}_0)
\end{aligned}
$$

which reads

$$
\mathcal{F}(\hat{x}_0, v) + (A\hat{x}_0 + Bv)^* H (A\hat{x}_0 + Bv) - \hat{x}_0^* H \hat{x}_0 \geq 0
$$

for any $v \in \mathbb{C}^m$. If $v = \tilde{u}_s(x_0)$ then u_t defined by (3.134) coincides with $\tilde{u}_t(x_0)$ - the optimal input sequence - and the above inequality becomes equality. We deduce that $\tilde{u}_s(x_0)$ ensures the minimal value of the following quadratic form

$$
\begin{aligned}
\mathcal{G}(\hat{x}_0, v) \;&=\; (A\hat{x}_0 + Bv)^* H (A\hat{x}_0 + Bv) - \hat{x}_0^* H \hat{x}_0 + \mathcal{F}(\hat{x}_0, v) = \\
&=\; v^* (K + B^* H B) v + v^* (B^* H A + L^*) \hat{x}_0 + \hat{x}_0^* (L + A^* H B) v + \\
&\quad + \hat{x}_0^* (A^* H A - H + M) \hat{x}_0 \geq 0.
\end{aligned}
$$

In order to compute v that minimizes $\mathcal{G}(\hat{x}_0, v)$ we check that $K + B^* H B$ is invertible. Since s is arbitrary we may take $s = 0$ to have $\hat{x}_0 = x_0$ which is again arbitrary. Taking $x_0 = 0$ we shall have $\mathcal{G}(0, v) = v^* (K + B^* H B) v \geq 0$ and since v is arbitrary we have $K + B^* H B \geq 0$. To show that $K + B^* H B$ is invertible we construct the following pair

$$
\begin{aligned}
x_0 &= 0, \ u_0 = v; \quad x_1 = A x_0 + B v = B v, \\
u_t &= \tilde{u}_{t-1}\left(x_1\right), \ t = 1, 2, \dots , \\
x_t &= \tilde{x}_{t-1}\left(x_1\right), \ t = 2, 3, \dots
\end{aligned}
$$

Remark that this pair corresponds to a zero initial condition and to an optimal pair for $t \geq 1$. We have as previously using Plancherel equality

$$
\sum_{t=0}^{\infty} \mathcal{F}\left(x_t, u_t\right) = \frac{1}{2\pi} \int_{-\pi}^{\pi} \mathcal{F}\left(\hat{x}\left(\omega\right), \hat{u}\left(\omega\right)\right) d\omega \geq \delta_0 \sum_{0}^{\infty} |u_t|^2 \geq \delta_0 |v|^2
$$

and on the other hand

$$
\sum_{t=0}^{\infty} \mathcal{F}\left(x_t, u_t\right) = \mathcal{F}\left(0, v\right) + x_1^* H x_1 = v^* K v + v^* B H B v \geq \delta_0 |v|^2
$$

for arbitrary v. Therefor $K + B^* H B > 0$ and we may compute the optimal value (according to abstract Lemma 3.9)

$$
\begin{aligned}
\tilde{u}_s\left(x_0\right) &= -\left(K + B^* H B\right)^{-1}\left(L + A^* H B\right)^* \hat{x}_0 = \\
&= -\left(K + B^* H B\right)^{-1}\left(L + A^* H B\right)^* \tilde{x}_s\left(x_0\right)
\end{aligned}
$$

and since s is arbitrary this equality will give a linear dependence of the optimal input on the optimal state (linear feedback)

$$
\tilde{u}_t = -\left(K + B^* H B\right)^{-1}\left(L + A^* H B\right)^* \tilde{x}_t . \tag{3.136}
$$

Since $\tilde{x}_{t+1} = A \tilde{x}_t + B \tilde{u}_t$ we deduce that $\tilde{x}_t$ is a solution of the linear discrete-time system

$$
x_{t+1} = \left[A - B\left(K + B^* H B\right)^{-1}\left(L + A^* H B\right)^*\right] x_t \tag{3.137}
$$

and we deduce that $A - B\left(K + B^* H B\right)^{-1}\left(L + A^* H B\right)^*$ defines an exponentially stable evolution: we know that $\{\tilde{x}_t\}_t$ is in $\ell_+^2\left(\mathbb{C}^n\right)$; hence $\lim_{t \to \infty} \tilde{x}_t = 0$ and the Persidskii type result (Section 2.2) will give exponential stability.

Now, since $K + B^* H B > 0$ there will exist V with $\det V \neq 0$

such that $K + B^* H B = V^* V$. We may write (taking again $s = 0$)

$$\mathcal{G}(x_0, v) =$$
$$\left(V v + (V^*)^{-1} (L + A^* H B)^* x_0 \right)^* \left(V v + (V^*)^{-1} (L + A^* H B)^* x_0 \right) +$$
$$+ x_0^* \left(A^* H A - H + M - (L + A^* H B)(V^* V)^{-1} (L + A^* H B)^* \right) x_0 \,. \tag{3.138}$$

Denote $W = (L + A^* H B) V^{-1}$; we shall have

$$\mathcal{G}(x_0, v) = |V v + W^* x_0|^2 + x_0^* (A^* H A - H + M - W W^*) x_0 \,,$$

where x_0 and v are arbitrary. The minimizing value of v corresponds to $V v + W^* x_0$ and we know that the minimal value is 0 for any x_0. We deduce that

$$A^* H A - H + M = W W^* \,.$$

Summarizing we obtained the following result:

Theorem 3.15 *Assume that (A, B) is stabilizable and the inequality*

$$\mathcal{F}(x, u) \geq \delta \left(|x|^2 + |u|^2 \right) > 0 \tag{3.139}$$

holds for some $\delta > 0$ and $x \in \mathbb{C}^n$, $u \in \mathbb{C}^m$ such that $zx = Ax + Bu$, $|z| = 1$ or, if $\det(z\mathbf{I} - A) \neq 0$ for $|z| = 1$, the strong frequency domain inequality

$$\mathcal{F}\left((z\mathbf{I} - A)^{-1} Bu, u \right) \geq \delta_1 |u|^2 > 0 \tag{3.140}$$

holds for some $\delta_1 > 0$ and $u \in \mathbb{C}^m$, $|z| = 1$. Then there exist matrices V, W, H with $\det V \neq 0$ and $H = H^*$ such that the following hold:

$$\begin{aligned} K + B^* H B &= V^* V \\ L + A^* H B &= W V \\ A^* H A - H + M &= W W^* \end{aligned} \tag{3.141}$$

Corollary 3.16 *Under the assumptions of Theorem 3.15 the index*
$$\eta_T = \sum_{t=0}^{T-1} \mathcal{F}(x_t, u_t) \text{ may be written as}$$

$$\eta_T = x_0^* H x_0 - x_T^* H x_T + \sum_{t=0}^{T-1} |V u_t + W^* x_t|^2 \tag{3.142}$$

Proof. Since (3.141) hold we may write

$$\mathcal{F}(x,u) =$$
$$= u^*(K + B^*HB)u + u^*(L^* + B^*HA)x + x^*(L + A^*HB)u +$$
$$+x^*(M + A^*HA - H)x - (Ax + Bu)^* H(Ax + Bu) + x^*Hx =$$
$$= u^*V^*Vu + u^*V^*W^*x + x^*WVu + x^*WW^*x -$$
$$- (Ax + Bu)^* H(Ax + Bu) + x^*Hx =$$
$$= |Vu + W^*x|^2 - (Ax + Bu)^* H(Ax + Bu) + x^*Hx$$

If (x_t, u_t) is a pair that satisfies (3.121) we shall have

$$\mathcal{F}(x_t, u_t) = |Vu_t + W^*x_t|^2 - x_{t+1}^* H x_{t+1} + x_t^* H x_t$$

and summing from 0 to $T-1$ will give (3.142).

Corollary 3.17 *Under the assumptions of Theorem 3.15 the characteristic matrix $T(\sigma^{-1}, \sigma)$ can be factorized as follows*

$$T(\sigma^{-1}, \sigma) = F^*(\bar{\sigma}^{-1}) F(\sigma) , \qquad (3.143)$$

where $F(\sigma) = V + W^*(\sigma I - A)^{-1} B$, $\bar{F}(\sigma) = \overline{F(\bar{\sigma})}$.

Proof. Since (3.141) hold we may write, starting from (3.124)

$$T(\lambda, \sigma) = V^*V - B^*HB + (V^*W^* - B^*HA) +$$
$$+ B^*(\lambda I - A^*)^{-1}(WV - A^*HB) +$$
$$+ B^*(\lambda I - A^*)^{-1}(WW^* + H - A^*HA)(\sigma I - A)^{-1} B =$$
$$= \left(V^* + B^*(\lambda I - A^*)^{-1} W\right)\left(V + W^*(\sigma I - A)^{-1} B\right) +$$
$$+ (1 - \lambda\sigma) B^*(\lambda I - A^*)^{-1} H(\sigma I - A)^{-1} B .$$

If $\lambda = \sigma^{-1}$ we obtain (3.143) since we have

$$F(\bar{\sigma}) = V + W^*(\bar{\sigma}I - A)^{-1} B , \quad \overline{F(\bar{\sigma})} = \bar{V} + W^\top(\sigma I - \bar{A})^{-1} \bar{B} ,$$
$$F^*(\sigma) = \left(\overline{F(\bar{\sigma})}\right)^\top = V^* + B^*(\sigma I - A^*)^{-1} W .$$

Corollary 3.18 *Under the assumptions of Theorem 3.15 the linear feedback control law (3.136) exponentially stabilizes system (3.121), minimizes the quadratic functional (3.128) on the set of admissible pairs (x_t, u_t) and the optimal (minimal) value of (3.128) is a* *quadratic form whose matrix H satisfies the discrete-type algebraic Riccati equation*

$$A^* H A - H + M = (L + A^* H B)(K + B^* H B)^{-1}(L + A^* H B)^* .$$
(3.144)

Proof. Since the assumptions of Theorem 3.15 hold there exist V, W, H satisfying (3.141); hence (3.136) reads $\tilde{u}_t = -V^{-1} W^* \tilde{x}_t$. The exponentially stabilizing character of (3.136) has already been proved.

On the other hand (3.142) holds. If (x_t, u_t) is an admissible pair we may let $T \to \infty$ and find

$$\eta_\infty = x_0^* H x_0 + \sum_0^\infty |V u_t + W^* x_t|^2$$

The minimal value of η_∞ obviously corresponds to the optimal pair $(\tilde{x}_t, \tilde{u}_t)$ which makes the sum zero and $\eta_\infty = x_0^* H x_0$. If V, W are eliminated from (3.141) we obtain (3.144). $\blacksquare$

Proposition 3.19 *Assume that (A, B) is a stabilizable pair. In order that there exist a Hermitian matrix $H_0 = H_0^*$ such that*

$$\mathcal{F}(x, u) + (Ax + Bu)^* H_0 (Ax + Bu) - x^* H_0 x \geq \varepsilon \left(|x|^2 + |u|^2 \right)$$
(3.145)

it is necessary and sufficient that the inequality (3.139) holds for some $\delta > 0$ and $x \in \mathbb{C}^n$, $u \in \mathbb{C}^m$ such that $zx = Ax + Bu$, $|z| = 1$ or, if $\det(zI - A) \neq 0$ for $|z| = 1$ the strong frequency domain inequality (3.140) holds for some $\delta_1 > 0$ and $u \in \mathbb{C}^m$, $|z| = 1$.

Proof. Consider the quadratic form

$$\mathcal{F}_1(x, u) = \mathcal{F}(x, u) - \frac{\delta_0}{2} \left(|x|^2 + |u|^2 \right)$$

where $\delta_0 > 0$ is the one of (3.139); we deduce that (3.139) holds for $\mathcal{F}_1(x, u)$ with $\delta_0/2$. Theorem 3.15 will give existence of V_0, $\det V_0 \neq 0$, W_0, $H_0 = H_0^*$ satisfying (3.141) with $K - \frac{\delta_0}{2} I$ and $M - \frac{\delta_0}{2} I$ instead of K and M respectively. We deduce (see the proof to Corollary 3.16) that

$$\mathcal{F}_1(x, u) = |V_0 u + W_0^* x|^2 - (Ax + Bu)^* H_0 (Ax + Bu) + x^* H_0 x$$

and this will give

$$\mathcal{F}(x, u) + (Ax + Bu)^* H_0 (Ax + Bu) - x^* H_0 x =$$
$$= |V_0 u + W_0^* x|^2 + \frac{\delta_0}{2} \left(|x|^2 + |u|^2 \right)$$

what proves sufficiency of (3.139) for (3.145). Necessity is obvious.

3.7.5. The KSPY Lemma under the Weak Frequency Domain Inequality

We shall consider here that only the weak (nonstrict) frequency domain inequality holds

$$\mathcal{F}(x, u) \geq 0, \ x \in \mathbb{C}^n, \ u \in \mathbb{C}^m, \ z \in \mathbb{C}, \ |z| = 1, \ zx = Ax + Bu .$$
$$(3.146)$$

Theorem 3.20 *Assume that (A, B) is a controllable pair. Then the following properties are equivalent:*
1^0 *The weak frequency domain inequality (3.146) holds.*
2^0 *There exist matrices $V, W, H = H^*$ such that (3.141) are satisfied.*
3^0 *The characteristic matrix $T(\lambda, \sigma)$ defined by (3.124) may be factorized as follows*

$$T\left(\sigma^{-1}, \sigma\right) = F^* \left(\bar{\sigma}^{-1}\right) F(\sigma) ,$$

where $F(\sigma) = V + W^* (\sigma I - A)^{-1} B$, $\det(\sigma I - A) \neq 0$.

In order to prove this theorem we shall need some auxiliary results.

Proposition 3.21 *Let $H = H^*$ satisfy the inequality $A^* H A - H \geq G$. If A has its eigenvalues inside the unit disk and $H_1 = H_1^*$ is a solution of $A^* H A - H = G$ then $H \leq H_1$. If A has its spectrum outside the unit disk and $H_2 = H_2^*$ is a solution of $A^* H A - H = G$ then $H \geq H_2$.*

Proof. Let $\Delta H_j = H - H_j$, $j = 1, 2$. We shall have

$$A^* (\Delta H_j) A - \Delta H_j = A^* H A - H - G = \Delta G \geq 0 .$$

Let A have the eigenvalues inside the unit disk. Then

$$\Delta H_1 = -\sum_{t=0}^{\infty} (A^*)^t (\Delta G) A^t \leq 0; \quad \text{hence} \quad H \leq H_1 .$$

Let A have the eigenvalues outside the unit disk. Then

$$\triangle H_2 = \sum_{-\infty}^{-1} (A^*)^t \, (\triangle G) \, A^t \geq 0; \quad \text{hence} \quad H \geq H_2 \,.$$

Proposition 3.22 *Assume that the identity*

$$\mathcal{F}(x,u) + (Ax + Bu)^* H (Ax + Bu) - x^* H x = |Vu + W^* x|^2 \quad (3.147)$$

holds for all $x \in \mathbb{C}^n$, $u \in \mathbb{C}^m$.

   Let F_1 be such that $A + BF_1$ has its eigenvalues inside the unit disk and denote $x^* G_1 x = \mathcal{F}(x, F_1 x)$. If H_1 is the solution of the Liapunov equation

$$(A + BF_1)^* H_1 (A + BF_1) - H_1 = -G_1$$

then $H \leq H_1$.

   Let F_2 be such that $A + BF_2$ has its eigenvalues outside the unit disk and denote $x^* G_2 x = \mathcal{F}(x, F_2 x)$. If H_2 is the solution of the Liapunov equation

$$(A + BF_2)^* H_2 (A + BF_2) - H_2 = -G_2$$

then $H \geq H_2$.

Proof. We take in (3.147) $u = F_i x$, $i = 1, 2$ and find

$$x^* \left[(A + BF_i)^* H (A + BF_i) - H \right] x + \mathcal{F}(x, F_i x) \geq 0$$

and this reads

$$(A + BF_i)^* H (A + BF_i) - H \geq G_i \,.$$

Application of Proposition 3.21 gives $H_2 \leq H \leq H_1$.

Proof of Theorem 3.20. Consider the modified quadratic form $\mathcal{F}_\delta(x, u) = \mathcal{F}(x, u) + \delta \left(|x|^2 + |u|^2 \right)$ where $\delta > 0$. For this quadratic form the strong frequency domain inequality (3.139) holds. The assumptions of Theorem 3.15 being fulfilled there will exist $H_\delta = H_\delta^*, V_\delta, W_\delta$ such that

$$\begin{aligned}
K + \delta \mathbf{I} + B^* H_\delta B &= V_\delta^* V_\delta \\
L + A^* H_\delta B &= W_\delta V_\delta \\
A^* H_\delta A - H_\delta + M + \delta \mathbf{I} &= W_\delta W_\delta^*
\end{aligned} \quad (3.148)$$

or, equivalently

$$\mathcal{F}(x,u) + \delta\left(|x|^2 + |u|^2\right) + (Ax + Bu)^* H_\delta (Ax + Bu) - x^* H_\delta x =$$
$$= |V_\delta u + W_\delta^* x|^2$$

(3.149)

Since (A, B) is controllable there exists F_1 such that $A + BF_1$ has its eigenvalues inside the unit disk and F_2 such that $A + BF_2$ has its eigenvalues outside the unit disk. Let H_δ^1 be the solution of

$$(A + BF_1)^* H_\delta^1 (A + BF_1) - H_\delta^1 = G_1^\delta,$$

where $x^* G_1^\delta x = \mathcal{F}_\delta(x, F_1 x)$. Let H_δ^2 be the solution of

$$(A + BF_2)^* H_\delta^2 (A + BF_2) - H_\delta^2 = G_2^\delta,$$

where $x^* G_2^\delta x = \mathcal{F}_\delta(x, F_2 x)$. According to Proposition 3.22 we shall have

$$H_2^\delta \leq H_\delta \leq H_1^\delta.$$

On the other hand we have for any matrix F

$$\mathcal{F}(x, Fx) \;=\; \mathcal{F}^0(x, Fx) \leq \mathcal{F}(x, Fx) + \delta\left(|x|^2 + |Fx|^2\right) =$$
$$=\; \mathcal{F}^\delta(x, Fx) \leq \mathcal{F}(x, Fx) + |x|^2 + |Fx|^2 = \mathcal{F}^1(x, Fx).$$

We deduce that $G_i^0 \leq G_i^\delta \leq G_i^1$, $i = 1, 2$. Applying Proposition 3.21 it follows that

$$H_2^0 \leq H_2^\delta \leq H_\delta \leq H_1^\delta \leq H_1^1;$$

hence $|H_\delta| \leq \rho$ where ρ does not depend on $\delta \in (0, 1]$. From the last equality of (3.148) we deduce $|W_\delta| \leq \rho_1$ and from the first one we deduce that $|V_\delta| \leq \rho_2$ for $0 < \delta < 1$. We obtained that the families of matrices $(H_\delta)_\delta$, $(W_\delta)_\delta$, $(V_\delta)_\delta$ are bounded for $\delta \in (0, 1]$. Consider a sequence $\{\delta_k\}_k$, $0 < \delta_k \leq 1$, $\lim_{k\to\infty} \delta_k = 0$. Using compactness of bounded and closed sets on finite dimensional spaces we deduce that the sequence $(H_{\delta_k}, W_{\delta_k}, V_{\delta_k})$ contains a convergent subsequence: there will exist the subsequence $\{\delta_{k_l}\}_l$ and the matrices H, W, V satisfying

$$\lim_{l\to\infty} H_{\delta_{k_l}} = H \;, \; \lim_{l\to\infty} W_{\delta_{k_l}} = W \;, \; \lim_{l\to\infty} V_{\delta_{k_l}} = V$$

From (3.148) it follows, by letting $\delta \to 0$ that H, W, V satisfy (3.141). We proved that $1^0 \Rightarrow 2^0$.

Assuming that 2^0 holds we may repeat the proof of Corollary 3.17 and obtain factorization (3.143): we obtained that $2^0 \Rightarrow 3^0$.

Assume now that 3^0 holds. If σ is such that $|\sigma| = 1$ and $\det(\sigma\mathbf{I} - A) \neq 0$ then $u^*T(\sigma^{-1}, \sigma)u = |F(\sigma)u|^2 \geq 0$. On the other hand if $x \in \mathbb{C}^n$ and $u \in \mathbb{C}^m$ are such that $\sigma x = Ax + Bu$, $|\sigma| = 1$, $\det(\sigma\mathbf{I} - A) \neq 0$, $\mathcal{F}(x, u) = u^*T(\sigma^{-1}, \sigma)u \geq 0$. If we consider now the case of A having eigenvalues on the unit disk, these values of σ, $|\sigma| = 1$, for which $\det(\sigma\mathbf{I} - A) = 0$ are a finite number and isolated.

Therefore, $\mathcal{F}(x, u) \geq 0$ will hold for *all* complex σ, x, u such that $\sigma x = Ax + Bu$, $|\sigma| = 1$. The proof is complete.

3.8. APPLICATIONS OF THE MULTI-INPUT KSPY LEMMA

We shall give below some applications of the KSPY lemma which are connected with stability and stabilization.

3.8.1. Linear Feedback Disturbance Attenuation

A. We shall start here with a version of *discrete time bounded real lemma* (e.g., Haddad and Bernstein (1994)) which is called *theorem on the γ -contraction* (Halanay and Ionescu (1994), p. 65).

Theorem 3.23 *Let $G(z)$ be a matrix function with rational proper entries and with its poles inside the unit disk and assume that $\|G(z)\|_\infty = \sup\limits_{|z|=1} \bar{\sigma}(G(z)) < \gamma$ where $\bar{\sigma}(G)$ is the upper singular eigenvalue of G. Then if (A, B, C, D) is a minimal (i.e., with (A, B) controllable and (C, A) observable) state representation of $G(z)$ - $D + C(z\mathbf{I} - A)^{-1}B \equiv G(z)$ - then there exist matrices V with $\det V \neq 0$, W, $P = P^* > 0$ such that the following hold:*

$$\gamma^2\mathbf{I} - D^*D - B^*PB = V^*V$$
$$-C^*D - A^*PB = WV \qquad (3.150)$$
$$A^*PA - P + C^*C = -WW^*$$

and the matrix $A - BV^{-1}W^*$ has its eigenvalues inside the unit disk.

Conversely, assume that there exist V, W, P with the above properties i.e. satisfying $\det V \neq 0$, $P > 0$ and the condition on the eigenvalues of $A - BV^{-1}W^*$ verifying (3.150). Then $\|G(z)\|_\infty < \gamma$.

Proof.   We define $\mathcal{F}(x,u) = \gamma^2 |u|^2 - |Cx + Du|^2$. Since $\det(z\mathbf{I} - A) \neq 0$ for $|z| = 1$ and $\|G(z)\|_\infty < \gamma$ condition (3.140) holds for $(A, B, \gamma^2\mathbf{I} - D^*D, -C^*D, -C^*C)$ and we may apply Theorem 3.15 to deduce existence of V with $\det V \neq 0$, W and H satisfying

$$\begin{aligned}
\gamma^2\mathbf{I} - D^*D + B^*HB &= V^*V \\
-C^*D + A^*HB &= WV \\
A^*HA - H - C^*C &= WW^*.
\end{aligned}$$

Taking $P = -H$ we obtain (3.150). The property of the matrix $A - BV^{-1}W^*$ follows from Corollary 3.18, with $H = -P$. We then use Corollary 3.16 to obtain

$$\begin{aligned}
\eta_T &= \sum_{t=0}^{T-1} \left(\gamma^2 |u_t|^2 - |Cx_t + Du_t|^2\right) = \\
&= x_T^* P x_T - x_0^* P x_0 + \sum_{t=0}^{T-1} |V u_t + W^* x_t|^2
\end{aligned}$$

for all pairs (x_t, u_t) satisfying $x_{t+1} = Ax_t + Bu_t$. Since A defines an exponentially stable evolution (due to the assumption about the poles of $G(z)$) we may take $u_t \equiv 0$ and let $T \to \infty$ obtaining

$$x_0^* P x_0 \geq \sum_0^\infty |Cx_t|^2 \; ;$$

hence $P \geq 0$. Assume there exists $x_0 \neq 0$ such that $x_0^* P x_0 = 0$. Then $Cx_t = CA^t x_0 = 0$, $t = 0, 1, 2, \ldots$ But (C, A) is observable since we assumed a minimal state representation; this gives $x_0 = 0$; hence $P > 0$.

Assume now existence of V, W, P with the properties stated in Theorem 3.23. We may write as in the proof of Corollary 3.16

$$\mathcal{F}(x.u) = |Vu + W^* x|^2 + (Ax + Bu)^* P (Ax + Bu) - x^* P x.$$

Take now $\hat{x}$ and $\hat{u}$ such that $A\hat{x} + B\hat{u} = z\hat{x}$ with $|z| = 1$. Since A has no eigenvalues with $|z| = 1$ we may write $\hat{x} = (z\mathbf{I} - A)^{-1} B\hat{u}$ and

$$\mathcal{F}(\hat{x}, \hat{u}) = \gamma^2 |\hat{u}|^2 - |G(z)\hat{u}|^2 = \left|V\hat{u} + W^* (z\mathbf{I} - A)^{-1} B\hat{u}\right|^2 \geq 0.$$

We deduce that $|G(z)|^2 \leq \gamma^2$ hence $\|G(z)\|_\infty \leq \gamma$. Assume that $\|G(z)\|_\infty = \gamma$: since the unit circle is a compact set and $G(z)$

has no poles in it there will exist some z_0 with $|z_0| = 1$ such that $\bar{\sigma}\left(G\left(z_0\right)\right) = \gamma$ hence $\det\left(\gamma^2\mathbf{I} - G^*\left(z_0\right)G\left(z_0\right)\right) = 0$. We may thus find some $\hat{u} \neq 0$ such that $\left(\gamma^2\mathbf{I} - G^*\left(z_0\right)G\left(z_0\right)\right)\hat{u} = 0$. Taking $\hat{x} = \left(z_0\mathbf{I} - A\right)^{-1}B\hat{u}$ we shall have for this pair $(\hat{x}, \hat{u})$ the equality $\mathcal{F}\left(\hat{x}, \hat{u}\right) = 0$ hence $V\hat{u} + W^*\left(z_0\mathbf{I} - A\right)^{-1}B\hat{u} = 0$. Since $\hat{u} \neq 0$ we deduce that $\det\left(V + W^*\left(z_0\mathbf{I} - A\right)^{-1}B\right) = 0$. But a well known identity (e.g., Popov (1973), p. 61) shows that

$$\det\left(z_0\mathbf{I} - A + BV^{-1}W^*\right) = \det\left(z_0\mathbf{I} - A\right) \times$$
$$\times \det\left(V + W^*\left(z_0\mathbf{I} - A\right)^{-1}B\right)$$

and we deduce $\det\left(z_0\mathbf{I} - A + BV^{-1}W^*\right) = 0$. But $A - BV^{-1}W^*$ has its eigenvalues inside the unit disk and this contradiction shows that $\|G\left(z\right)\|_\infty < \gamma$ what ends the proof.

Remarks. a) System (3.150) is equivalent to the discrete time algebraic Riccati equation for P. We have indeed

$$-W = \left(C^*D + A^*PB\right)V^{-1}$$
$$WW^* = \left(C^*D + A^*PB\right)\left(\gamma^2\mathbf{I} - D^*D - B^*PB\right)\left(D^*C + B^*PA\right)$$

and we deduce that P satisfies the equality

$$\left(C^*D + A^*PB\right)\left(\gamma^2\mathbf{I} - D^*D - B^*PB\right)^{-1}\left(D^*C + B^*PA\right) +$$
$$+A^*PA - P + C^*C = 0$$

$$(3.151)$$

and also the inequality $\gamma^2\mathbf{I} - D^*D - B^*PB > 0$.

b) The γ-contraction property still holds if we take an additional output $y_\varepsilon = \varepsilon x$ for ε small enough which leads to replace C^*C by $C^*C + \varepsilon^2\mathbf{I}$ in (3.150). In this case $\mathcal{F}_\varepsilon\left(x, u\right) = \gamma^2|u|^2 - |Cx + Du|^2 - \varepsilon^2|x|^2$ and we have to show that if $\|G\left(z\right)\|_\infty < \gamma$ condition (3.140) holds for the modified set $\left(A, B, \gamma^2\mathbf{I} - D^*D, -C^*D, -C^*C - \varepsilon^2\mathbf{I}\right)$; in this case (3.140) reads

$$u^*\left(\gamma^2\mathbf{I} - G^*\left(z\right)G\left(z\right)\right)u - \varepsilon^2u^*B^*\left(\bar{z}\mathbf{I} - A^*\right)^{-1}\left(z\mathbf{I} - A\right)^{-1}Bu > \delta_1|u|^2$$

Since the unit circle is compact and $\det\left(z\mathbf{I} - A\right) \neq 0$ on the circle there will exist z_1, $|z_1| = 1$ such that $\sup_{|z|=1}\bar{\sigma}\left(G\left(z\right)\right) = \bar{\sigma}\left(G\left(z_1\right)\right)$ and z_2, $|z_2| = 1$ such that $\sup_{|z|=1}u^*B^*\left(\bar{z}\mathbf{I} - A^*\right)^{-1}\left(z\mathbf{I} - A\right)^{-1}Bu =$

$u^* B^* (z_2 \mathbf{I} - A^*)^{-1} (z_2 \mathbf{I} - A)^{-1} Bu$. The above inequality is fulfilled if $\gamma^2 - (\bar{\sigma}(G(z_1)))^2 - \varepsilon^2 \left| (z_2 \mathbf{I} - A)^{-1} B \right|^2 > 0$ and we may choose

$$\varepsilon^2 < \frac{1}{2 \left| (z_2 \mathbf{I} - A)^{-1} B \right|^2} \left[\gamma^2 - (\bar{\sigma}(G(z_1)))^2 \right].$$

It will then follow that P satisfies the discrete time algebraic Riccati inequality

$$(C^* D + A^* PB) \left(\gamma^2 \mathbf{I} - D^* D - B^* PB \right)^{-1} (D^* C + B^* PA) + \\ + A^* PA - P + C^* C < 0$$

$$(3.152)$$

as well as the inequality $\gamma^2 \mathbf{I} - D^* D - B^* PB > 0$.

B. We consider here a simple version of the so-called *disturbance attenuation problem*. Consider the linear system

$$\begin{aligned} x_{t+1} &= Ax_t + B_1 w_t + B_2 u_t \\ y_t &= Cx_t + Du_t \end{aligned}$$

$$(3.153)$$

which has two input signals: u - called control signal - and w - called disturbance. The difference between the two signals is the following: while the control signal u is available to the designer for the improvement of systems properties, e.g., by feedback, the disturbance has, generally speaking, undesirable effects on system's behavior. These effects should be maintained at lowest possible level. We may state now the disturbance attenuation problem. Given system (3.153), find a state feedback control law $u = Fx$ such that $A + B_2 F$ defines an exponentially stable evolution and the attenuation of the disturbances by the new system as felt in its output is less than γ - a given constant.

In fact the designer has to find a matrix F such that the closed loop system

$$\begin{aligned} x_{t+1} &= (A + B_2 F) x_t + B_1 w_t \\ y_t &= (C + DF) x_t \end{aligned}$$

$$(3.154)$$

is internally stable; i.e., $x_{t+1} = (A + B_2 F) x_t$ is exponentially stable and the matrix transfer function from the disturbance input w to the output y namely $G_F(z) = (C + DF)(z\mathbf{I} - A - B_2 F)^{-1} B_1$ has the γ-*attenuation property* $\|G_F(z)\|_\infty < \gamma$ (γ-contracting input/output operator).

We may state now

Theorem 3.24 *Consider system (3.153) with (A, B_1) controllable and (C, A) observable and assume that $C^* D = 0$, $D^* D = \mathbf{I}$. If there* *exists a matrix F such that the control law $u = Fx$ achieves exponential stability of the free system and also the γ -attenuation property then the following nonlinear matrix inequality of the discrete time algebraic game - Riccati type*

$$A^*PA - P + C^*C + A^*PB_1 \left(\gamma^2 I - B_1^*PB_1\right)^{-1} B_1^*PA -$$
$$- A^*\Phi_1(P) B_2 \left(I + B_2^*\Phi_1(P) B_2\right)^{-1} B_2^*\Phi_1(P) A \leq 0, \qquad (3.155)$$

where $\Phi_1(P) = P + PB_1 \left(\gamma^2 I - B_1^*PB_1\right)^{-1} B_1^*P$ has a strictly positive solution $P > 0$ which is stabilizing; i.e., the matrix

$$A - B_2 \left(I + B_2^*\Phi_1(P) B_2\right)^{-1} B_2^*\Phi_1(P) A$$

defines an exponentially stable evolution; also $P > 0$ is such that $\gamma^2 I - B_1^*PB_1 > 0$.

    Conversely, if $P > 0$ is a solution of (3.155) such that $\gamma^2 I - B_1^*PB_1 > 0$ then the control law

$$u = -B_2 \left(I + B_2^*\Phi_1(P) B_2\right)^{-1} B_2^*\Phi_1(P) A$$

is stabilizing and γ -attenuating.

Proof. Assume first existence of F with the required properties. We may apply Theorem 3.23 for the set $(A + B_2F, B_1, \gamma^2 I, 0, -C^*C - F^*F)$ since $A + B_2F$ defines an exponentially stable evolution and $\|G_F(z)\|_\infty < \gamma$. Accordingly there will exist $P > 0$ satisfying the following discrete-time algebraic Riccati equation

$$(A + B_2F)^* P(A + B_2F) - P + C^*C + F^*F +$$
$$(A + B_2F)^* PB_1 \left(\gamma^2 I - B_1^*PB_1\right)^{-1} B_1^*P(A + B_2F) = 0 \qquad (3.156)$$

and also $\gamma^2 I - B_1^*PB_1 > 0$. After a simple manipulation the above Riccati equation becomes

$$A^*PA - P + C^*C + A^*PB_1 \left(\gamma^2 I - B_1^*PB_1\right)^{-1} B_1^*PA +$$
$$+ F^*B_2^*\Phi_1(P) A + A^*\Phi_1(P) B_2F + F^* \left(I + B_2^*\Phi_1(P) B_2\right) F = 0,$$

where $\Phi_1(P) = P + PB_1 \left(\gamma^2 I - B_1^*PB_1\right)^{-1} B_1^*P > 0$ hence

$\mathbf{I} + B_2^* \Phi_1(P) B_2 > 0$. Completion of the square leads to

$$A^* PA - P + C^* C + A^* PB_1 \left(\gamma^2 \mathbf{I} - B_1^* PB_1 \right)^{-1} B_1^* PA -$$
$$- A^* \Phi_1(P) B_2 \left(\mathbf{I} + B_2^* \Phi_1(P) B_2 \right)^{-1} B_2^* \Phi_1(P) A +$$
$$+ \left[F + \left(\mathbf{I} + B_2^* \Phi_1(P) B_2 \right)^{-1} \Phi_1(P) A \right]^* \left(\mathbf{I} + B_2^* \Phi_1(P) B_2 \right) \times$$
$$\times \left[F + \left(\mathbf{I} + B_2^* \Phi_1(P) B_2 \right)^{-1} \Phi_1(P) A \right] = 0$$

$$(3.157)$$

Obviously the last term in (3.157) is nonnegative hence $P > 0$ satisfies (3.155). We already mentioned that from Theorem 3.23 it follows that $\gamma^2 \mathbf{I} - B_1^* PB_1 > 0$.

Next we choose in (3.156)

$$F = \hat{F} = - \left(\mathbf{I} + B_2^* \Phi_1(P) B_2 \right)^{-1} \Phi_1(P) A$$

obtaining

$$\left(A + B_2 \hat{F} \right)^* P \left(A + B_2 \hat{F} \right) - P \leq -C^* C - \hat{F}^* \hat{F}.$$

Since $P > 0$ we may take $V(x) = x^* Px$ as a Liapunov function to obtain that $A + B_2 \hat{F}$ defines a stable evolution and $\hat{F} x_t \equiv 0$ hence $x_{t+1} = A x_t$. Since (C, A) is observable we deduce that in the set where $V(x_t) \equiv const$ we shall have $x_t \equiv 0$. This gives the exponential stability (the invariance principle of Section 2.6 plus the fact that the system is linear).

Conversely, let $P > 0$ be a solution of (3.155) satisfying also $\gamma^2 \mathbf{I} - B_1^* PB_1 > 0$. We already know that the control

$$u = -B_2 \left(\mathbf{I} + B_2^* \Phi_1(P) B_2 \right)^{-1} B_2^* \Phi_1(P) A$$

is stabilizing. We may take in (3.156)

$$F = \hat{F} = - \left(\mathbf{I} + B_2^* \Phi_1(P) B_2 \right)^{-1} B_2^* \Phi_1(P) A.$$

Since $\gamma^2 \mathbf{I} - B_1^* PB_1 > 0$ there exists V with $\det V \neq 0$ such that $\gamma^2 \mathbf{I} - B_1^* PB_1 = V^* V$. Denoting $\hat{W} = - \left(A + B_2 \hat{F} \right) PB_1 V^{-1}$ we obtain from (3.156) that

$$\left(A + B_2 \hat{F} \right)^* P \left(A + B_2 \hat{F} \right) - P + C^* C + \hat{F}^* \hat{F} + \hat{W} \hat{W}^* = 0.$$

Summarizing, a set of Kalman - Szegö - Popov - Yakubovich equalities

has been obtained for the set $\left(A + B_2 \hat{F}, B_1, \gamma^2 \mathbf{I}, 0, -C^*C - \hat{F}^* \hat{F} \right)$.
We may apply Theorem 3.23 (its second part) to obtain that
$\left\| G_{\hat{F}}(z) \right\|_\infty < \gamma$; i.e., the control $u = \hat{F}x$ is γ-attenuating what
ends the proof.

Remark. Stabilization and γ-attenuation by linear feedback falls in
the larger class of problems called discrete time H_∞-control. With
respect to this the reader is sent to the results of Gu et al. (1989)
and Stoorvogel (1990).

3.8.2. A Small Gain Result

We shall consider here the following nonlinear system with several
nonlinear elements, each of them being function of several scalar arguments and time-varying

$$x_{t+1} = Ax_t - B\Phi_t\left(Cx_t\right), \qquad (3.158)$$

where we assume that (A, B, C) is a minimal state representation of
the matrix function $G(z)$ with rational strictly proper entries and its
poles located inside the open unit disk. The sequence $\{\Phi_t(y)\}_{t\in\mathbb{N}}$ of
functions $\Phi_t : \mathbb{R}^p \to \mathbb{R}^m$ is assumed to satisfy $\Phi_t^*(y)\Phi_t(y) \leq \gamma^2 y^* y$
i.e. $|\Phi_t(y)| \leq \gamma|y|$ for any $t \in \mathbb{N}$. From this inequality it follows that
$\Phi_t(0) \equiv 0$ hence (3.158) has the zero solution.

We are interested in sufficient conditions for the global asymp-
totic stability of the zero solution for all nonlinear functions of the
considered class that is in absolute stability for a system with several
nonlinear elements.

Consider the quadratic form $V(x) = x^* P x$ with $P > 0$ a candidate for Liapunov function and compute its variation along the
solutions of (3.158)

$$W(x) = V\left(Ax - B\Phi\left(Cx\right)\right) - V(x) =$$
$$= \left(x^* A^* - \Phi^*\left(Cx\right) B^*\right) P\left(Ax - B\Phi\left(Cx\right)\right) - x^* P x =$$

$$= x^*\left(A^* P A - P\right)x - \Phi^*\left(Cx\right) B^* P A x - x^* A^* P B \Phi\left(Cx\right) +$$
$$+ \Phi^*\left(Cx\right) B^* P B \Phi\left(Cx\right) + \gamma^2 x^* C^* C x - \Phi^*\left(Cx\right)\Phi\left(Cx\right) +$$
$$+ \Phi^*\left(Cx\right)\Phi\left(Cx\right) - \gamma^2\left(Cx\right)^*\left(Cx\right) =$$

$$= -\Phi^*\left(Cx\right)\left(\mathbf{I} - B^* P B\right)\Phi\left(Cx\right) - \Phi^*\left(Cx\right) B^* P A x - \gamma^2 |Cx|^2 -$$
$$- x^* A^* P B \Phi\left(Cx\right) - x^*\left(P - A^* P A - \gamma^2 C^* C\right)x + |\Phi\left(Cx\right)|^2 \, .$$

We shall now apply Theorem 3.23 (bounded real lemma) to the system

$$x_{t+1} = Ax_t + Bu_t$$
$$y_t = \gamma C x_t$$

If $\|G(z)\|_\infty < \gamma^{-1}$ then there exist matrices V with $\det V \neq 0$, W, $P = P^* > 0$ such that

$$\mathbf{I} - B^* PB = V^* V, \quad -A^* PB = WV, \quad A^* PA - P + \gamma^2 C^* C + \varepsilon^2 \mathbf{I} = -WW^*$$

with $\varepsilon > 0$ sufficiently small (see the second remark to the bounded real lemma).

We choose $P > 0$ whose existence follows from Theorem 3.23 as the matrix of $\mathcal{V}(x) = x^* P x$ considered above. Then

$$\begin{aligned}
\mathcal{W}(x) &= -|-V\Phi(Cx) + W^* x|^2 + |\Phi(Cx)|^2 - \gamma^2 |Cx|^2 - \varepsilon^2 |x|^2 \leq \\
&\leq -\varepsilon^2 |x|^2 .
\end{aligned}$$

We deduce asymptotic stability for the zero solution of (3.158) by applying the theorem of Liapunov of asymptotic stability (Theorem 2.17). Since $\mathcal{V}(x) = x^* P x$ with $P > 0$ we have $\mathcal{V}(x) \geq \lambda_0 |x|^2$ with $\lambda_0 > 0$ and asymptotic stability is global. We obtained in fact the following result

Theorem 3.25 *Consider system* (3.158) *under the following assumptions: i)* (A, B, C) *is a minimal state representation of matrix function* $G(z)$ *with rational strictly proper entries and all its poles inside the open unit disk; ii)* $\Phi_t : \mathbb{R}^p \to \mathbb{R}^m$ *satisfies* $|\Phi_t(y)| \leq \gamma |y|$ *for all* $t \in \mathbb{N}$; *iii) the small gain condition* $\|G(z)\|_\infty < \gamma^{-1}$ *holds. Then the zero solution of* (3.157) *is globally asymptotically stable for all nonlinear elements of the class defined above.*

3.8.3. Absolute Stability for Systems with Several Sector and Slope Restricted Nonlinearities

We shall consider here global asymptotic stability of the zero equilibrium for the following system

$$x_{t+1} = Ax_t - \sum_{j=1}^m b_j \varphi_j \left(c_j^* x_t \right) , \qquad (3.159)$$

where each nonlinear function $\varphi_j(\sigma)$ is subject to the following conditions

$$\varphi_j(0) = 0, \ 0 \le \varphi_j(\sigma) \le \bar{\varphi}_j \sigma^2, \ -\alpha_j \le \varphi'_j(\sigma) \le \beta_j, \ \bar{\varphi}_j \le \beta_j, \ \alpha_j \ge 0. \tag{3.160}$$

In order to state and prove the main result of this section denote by B the matrix having b_j as columns, by C^* the matrix having c_j^* as rows and by $\bar{F}$, F_α, F_β the diagonal matrices having as diagonal elements $\bar{\varphi}_j$, α_j, β_j $(j = \overline{1,m})$ respectively. Let $G(z) = C^*(z\mathbf{I} - A)^{-1}B$ be the matrix transfer function associated to (A, B, C^*).

Theorem 3.26 *Consider system* (3.159) *under the following assumptions: i) matrix A has its eigenvalues inside the unit disk; ii) the pair (A, B) is controllable; iii) the nonlinear functions belong to the class defined by* (3.160) *and are such that for each $j = \overline{1,m}$, $\varphi_j(\sigma)/\sigma$ is either nonincreasing for $\sigma > 0$ and nondecreasing for $\sigma < 0$ or nondecreasing for $\sigma > 0$ and nonincreasing for $\sigma < 0$; iv) there exists a set of parameters $\tau_{1j} > 0$, $\tau_{2j} \ge 0$, $\theta_{1j} \ge 0$, $\theta_{2j} \ge 0$, $j = \overline{1,m}$ such that the following matrix frequency domain inequality holds for $|z| = 1$:*

$$D_1 + \mathrm{Re}\left[(D_1\bar{F} + (D'_2 - D''_2)(1 - z^{-1}))G(z)\right] +$$
$$+ |1 - z|^2 [D_3 + \mathrm{Re}(D_3(F_\beta - F_\alpha)G(z)) - \tag{3.161}$$
$$- \tfrac{1}{2}G^*(z)(F_\alpha D'_2 + F_\beta D''_2 + 2F_\alpha D_3 F_\beta)G(z)] > 0$$

and also

$$(\theta_{1j} - \theta_{2j})\left(\tfrac{1}{2}\sigma_j\varphi_j(\sigma_j) - \int_0^{\sigma_j} \varphi_j(\lambda)\,\mathrm{d}\lambda\right) \le 0, \quad j = \overline{1,m} \tag{3.162}$$

where we denoted by D_1, D'_2, D''_2, D_3 the diagonal matrices having as diagonal elements the parameters τ_{1j}, θ_{1j}, θ_{2j}, τ_{2j}, $j = \overline{1,m}$ respectively and, for a complex matrix M, $\mathrm{Re}\,M = \tfrac{1}{2}(M + M^*)$. Then the zero equilibrium of (3.159) *is globally asymptotically stable.*

Proof. A. We associate to (3.159) the controlled system

$$x_{t+1} = Ax_t + \sum_{j=1}^{m} b_j \xi_t^j$$
$$\xi_{t+1}^j = \xi_t^j + \mu_t^j \,, \ j = \overline{1, m} \qquad (3.163)$$
$$\sigma_t^j = c_j^* x_t \,,$$

where μ^j are the inputs and σ^j the outputs of the system. Let $\hat{x}_t$, $t \geq 0$, be a solution of (3.159). By choosing $\mu_t^j = \varphi_j \left(c_j^* \hat{x}_t \right) - \varphi_j \left(c_j^* \hat{x}_{t+1} \right)$ let us consider the solution of (3.163) corresponding to the initial conditions $x_0 = \hat{x}_0$, $\xi_0^j = -\varphi_j \left(c_j^* \hat{x}_0 \right)$, $j = \overline{1, m}$ and to the above defined input sequences; denote it by x_t, ξ_t^j, $j = \overline{1, m}$ and let us study the evolution of the sequences $x_t - \hat{x}_t$, $\xi_t^j + \varphi_j \left(c_j^* \hat{x}_t \right)$:

$$x_{t+1} - \hat{x}_{t+1} = A \left(x_t - \hat{x}_t \right) + \sum_{j=1}^{m} b_j \left(\xi_t^j + \varphi_j \left(c_j^* \hat{x}_t \right) \right)$$
$$\xi_{t+1}^j + \varphi_j \left(c_j^* \hat{x}_{t+1} \right) = \xi_t^j + \varphi_j \left(c_j^* \hat{x}_t \right) - \varphi_j \left(c_j^* \hat{x}_{t+1} \right) + \varphi_j \left(c_j^* \hat{x}_{t+1} \right) \,,$$
$$j = \overline{1, m} \,.$$

The second equality shows that $\xi_t^j + \varphi_j \left(c_j^* \hat{x}_t \right)$ is constant for all $t \geq 0$ and since $\xi_0^j + \varphi_j \left(c_j^* \hat{x}_0 \right) = 0$ this constant is zero. Then the first equality becomes

$$x_{t+1} - \hat{x}_{t+1} = A \left(x_t - \hat{x}_t \right)$$

and since $x_0 = \hat{x}_0$ we have $x_t \equiv \hat{x}_t$, $\xi_t^j = -\varphi_j \left(c_j^* x_t \right)$.

For $\mu_t^j, \xi_t^j, \sigma_t^j$ defined as above, i.e., starting from a solution of (3.159) we may obtain, using (3.160) and as in the paper of Yakubovich (1968), the following quadratic constraints

$$\xi^j \left(\xi^j + \bar{\varphi}_j \sigma^j \right) = -\varphi_j \left(\sigma_j \right) \left(\bar{\varphi}_j \sigma^j - \varphi_j \left(\sigma^j \right) \right) \leq 0 \,, \ j = \overline{1, m} \quad (3.164)$$

$$\left(\mu_t^j - \alpha_j \left(\sigma_{t+1}^j - \sigma_t^j \right) \right) \left(\mu_t^j + \beta_j \left(\sigma_{t+1}^j - \sigma_t^j \right) \right) =$$
$$= \left(\varphi_j \left(\sigma_t^j \right) - \varphi_j \left(\sigma_{t+1}^j \right) - \alpha_j \left(\sigma_{t+1}^j - \sigma_t^j \right) \right) \times$$
$$\times \left(\varphi_j \left(\sigma_t^j \right) - \varphi_j \left(\sigma_{t+1}^j \right) + \beta_j \left(\sigma_{t+1}^j - \sigma_t^j \right) \right) \leq 0 \,, \ j = \overline{1, m}$$
$$(3.165)$$

$$\sum_{k=s}^{t-1} \left(\mu_k^j + \xi_k^j - \frac{\alpha_j}{2} \left(\sigma_{k+1}^j - \sigma_k^j \right) \right) \left(\sigma_{k+1}^j - \sigma_k^j \right) = \sum_{k=s}^{t-1} \left(-\varphi_j \left(\sigma_{k+1}^j \right) - \right.$$

$$\left. - \frac{\alpha_j}{2} \left(\sigma_{k+1}^j - \sigma_k^j \right) \right) \left(\sigma_{k+1}^j - \sigma_k^j \right) \le \Phi_j \left(\sigma_s^j \right) - \Phi_j \left(\sigma_t^j \right) , \ j = \overline{1,m}$$

$$(3.166)$$

$$- \sum_{k=s}^{t-1} \left(\mu_k^j + \xi_k^j + \frac{\bar{\varphi}_j}{2} \left(\sigma_{k+1}^j - \sigma_k^j \right) + \frac{\beta_j}{2} \left(\sigma_{k+1}^j - \sigma_k^j \right) \right) \left(\sigma_{k+1}^j - \sigma_k^j \right) =$$

$$= - \sum_{k=s}^{t-1} \left(-\varphi_j \left(\sigma_{k+1}^j \right) + \frac{\bar{\varphi}_j}{2} \left(\sigma_{k+1}^j - \sigma_k^j \right) + \frac{\beta_j}{2} \left(\sigma_{k+1}^j - \sigma_k^j \right) \right) \times$$

$$\times \left(\sigma_{k+1}^j - \sigma_k^j \right) \le$$

$$\le \frac{\bar{\varphi}_j}{2} \left(\sigma_s^j \right)^2 - \Phi_j \left(\sigma_s^j \right) - \frac{\bar{\varphi}_j}{2} \left(\sigma_t^j \right)^2 + \Phi_j \left(\sigma_t^j \right) , \ j = \overline{1,m}$$

$$(3.167)$$

where $\Phi_j \left(\sigma \right) = \displaystyle\int_0^\sigma \varphi_j \left(\lambda \right) d\lambda$.

Remark that inequalities (3.164) and (3.165) follow directly from (3.160) while (3.166) and (3.167) are nothing else than inequalities (3.48) and (3.49) obtained from the slope constraints by using backward differences (see Section 3.2).

To the controlled system (3.163) we now associate the following index

$$\eta_{t,s} = \sum_{j=1}^m \left[\tau_{1j} \sum_{k=s}^{t-1} \xi_k^j \left(\xi_k^j + \bar{\varphi}_j \sigma_k^j \right) + \tau_{2j} \sum_{k=s}^{t-1} \left(\mu_k^j - \alpha_j \left(\sigma_{k+1}^j - \sigma_k^j \right) \right) \right.$$

$$\times \left(\mu_k^j + \beta_j \left(\sigma_{k+1}^j - \sigma_k^j \right) \right) + \theta_{1j} \sum_{k=s}^{t-1} \left(\mu_k^j + \xi_k^j - \frac{\alpha_j}{2} \left(\sigma_{k+1}^j - \sigma_k^j \right) \right) \times$$

$$\times \left(\sigma_{k+1}^j - \sigma_k^j \right) - \theta_{2j} \sum_{k=s}^{t-1} \left(\mu_k^j + \xi_k^j + \frac{\bar{\varphi}_j}{2} \left(\sigma_{k+1}^j + \sigma_k^j \right) + \right.$$

$$\left. \left. + \frac{\beta_j}{2} \left(\sigma_{k+1}^j - \sigma_k^j \right) \right) \left(\sigma_{k+1}^j - \sigma_k^j \right) \right] - \varepsilon^2 \sum_{k=s}^{t-1} |x_k|^2$$

$$(3.168)$$

where the parameters τ_{ij}, θ_{ij} $\left(i = 1,2, \ j = \overline{1,m} \right)$ are those for which the frequency domain inequality (3.161) and inequalities (3.162) hold. In this way we have defined a Popov system consisting of the controlled system (3.163) and the index (3.168). Denoting by u the input to (3.163), having μ^j, $j = \overline{1,m}$, as entries, by v the vector with

ξ^j as entries, by y the vector with σ^j as entries and using also the previously introduced notations the Popov system (3.163), (3.168) reads

$$
\begin{aligned}
x_{t+1} &= Ax_t + Bv_t \\
v_{t+1} &= v_t + u_t \\
y_t &= C^* x_t
\end{aligned}
\qquad (3.169)
$$

$$
\begin{aligned}
\eta_{t,s} = \sum_{k=s}^{t-1} &\Big[v_k^* D_1 \left(v_k + \bar{F} y_k \right) + \\
&+ \left(u_k - F_\alpha \left(y_{k+1} - y_k \right) \right)^* D_3 \left(u_k + F_\beta \left(y_{k+1} - y_k \right) \right) - \\
&- \left(u_k + v_k + \tfrac{1}{2}\bar{F} \left(y_{k+1} + y_k \right) + \tfrac{1}{2} F_\beta \left(y_{k+1} - y_k \right) \right)^* D_2'' \left(y_{k+1} - y_k \right) + \\
&+ \left(u_k + v_k - \tfrac{1}{2} F_\alpha \left(y_{k+1} - y_k \right) \right)^* D_2' \left(y_{k+1} - y_k \right) - \varepsilon^2 \left| x_k \right|^2 \Big]
\end{aligned}
$$
$$(3.170)$$

Remark that

$$
\begin{aligned}
\left(y_{k+1} + y_k \right)^* \bar{F} D_2'' \left(y_{k+1} - y_k \right) &= y_{k+1}^* \bar{F} D_2'' y_{k+1} - y_k^* \bar{F} D_2'' y_k \\
\sum_{k=s}^{t-1} \left(y_{k+1} + y_k \right)^* \bar{F} D_2'' \left(y_{k+1} - y_k \right) &= y_t^* \bar{F} D_2'' y_t - y_s^* \bar{F} D_2'' y_s
\end{aligned}
$$

This will give

$$
\begin{aligned}
\eta_{t,s} = \tfrac{1}{2} \left(y_s^* \bar{F} D_2'' y_s - y_t^* \bar{F} D_2'' y_t \right) + \sum_{k=s}^{t-1} &\Big[v_k^* D_1 \left(v_k + \bar{F} y_k \right) + \\
&+ \left(u_k - F_\alpha \left(y_{k+1} - y_k \right) \right)^* D_3 \left(u_k + F_\beta \left(y_{k+1} - y_k \right) \right) + \\
&+ \left(u_k + v_k - \tfrac{1}{2} F_\alpha \left(y_{k+1} - y_k \right) \right)^* D_2' \left(y_{k+1} - y_k \right) - \\
&- \left(u_k + v_k + \tfrac{1}{2} F_\beta \left(y_{k+1} - y_k \right) \right)^* D_2'' \left(y_{k+1} - y_k \right) - \varepsilon^2 \left| x_k \right|^2 \Big]
\end{aligned}
$$
$$(3.171)$$

We now take into account that $y_k = C^* x_k$ and $y_{k+1} = C^* x_{k+1} = C^* A x_k + C^* B v_k$. It follows that under the sum in (3.171) there is a quadratic form in u and in the state variables x, v. Therefore we may write

$$
\begin{pmatrix} x_{t+1} \\ v_{t+1} \end{pmatrix} = \begin{pmatrix} A & B \\ 0 & I \end{pmatrix} \begin{pmatrix} x_t \\ v_t \end{pmatrix} + \begin{pmatrix} 0 \\ I \end{pmatrix} u_t
\qquad (3.172)
$$

$$\eta_{t,s} = \frac{1}{2}\left(x_s^* C\bar{F}D_2''C^* x_s - x_t^* C\bar{F}D_2''C^* x_t\right) + \sum_{k=s}^{t-1} \mathcal{F}\left(x_k, v_k, u_k\right)$$

$$(3.173)$$

where $\mathcal{F}(x, v, u)$ is the above mentioned quadratic form that can be obtained from (3.171) by substituting the expressions of y_k and y_{k+1}.

B. We shall apply KSPY lemma to the system defined by (3.172) and (3.173). We check first that controllability of (A, B) implies controllability of the pair $\left(\begin{pmatrix} A & B \\ 0 & I \end{pmatrix}, \begin{pmatrix} 0 \\ I \end{pmatrix}\right)$. With respect to this it is easily seen that this last pair is equivalent to $\left(\begin{pmatrix} I & 0 \\ A & B \end{pmatrix}, \begin{pmatrix} I \\ 0 \end{pmatrix}\right)$, the equivalence transformation being defined by the matrix $\begin{pmatrix} 0 & I \\ I & 0 \end{pmatrix}$. The controllability matrix is given by

$$C_{n+m} = \begin{pmatrix} I & I & I & \cdots & I \\ 0 & B & B + AB & \cdots & \sum_{0}^{n+m-1} A^k B \end{pmatrix}.$$

We deduce that

$$\operatorname{rank} C_{n+m} = \operatorname{rank} \begin{pmatrix} I & 0 & 0 & \cdots & 0 \\ 0 & B & B+AB & \cdots & \sum_{0}^{n+m-1} A^k B \end{pmatrix} =$$

$$= m + \operatorname{rank}\left(B \ AB \ \cdots \ A^{n-1}B\right) = m + n \ ;$$

hence controllability is proved.

We have to check now the fulfillment of the frequency domain inequality. We take u, v, x satisfying the equalities

$$zx = Ax + Bv \ , \ (z-1)v = u \ , \ |z| = 1 \ .$$

Since A has its eigenvalues inside the unit disk we may take $x = (zI - A)^{-1} Bv$, $y = C^* x = C^* (zI - A)^{-1} Bv = G(z) v$. Substituting in $\mathcal{F}(x, v, u)$ we obtain the following form of the frequency domain inequality required by the KSPY lemma in this case

$$v^* \left\{ D_1 + \text{Re} \left[(D_1 \bar{F} + (D_2' + D_2'') (1 - z^{-1})) G(z) \right] + |z - 1|^2 \times \right.$$
$$\times [D_3 + \text{Re} \, D_3 (F_\beta - F_\alpha) G(z) -$$
$$- \tfrac{1}{2} G^*(z) (F_\alpha D_2' + F_\beta D_2'' + 2F_\alpha D_3 F_\beta) G(z)] -$$
$$\left. - \varepsilon^2 B^* (\bar{z}\mathbf{I} - A^*)^{-1} (z\mathbf{I} - A)^{-1} B \right\} v \geq 0 \, , \; v \in \mathbb{C}^m, \; |z| = 1 \, .$$

$$(3.174)$$

Assume now that (3.161) holds; since the unit circle is a compact set and A has no eigenvalues on it, $B^* (\bar{z}\mathbf{I} - A^*)^{-1} (z\mathbf{I} - A)^{-1} B$ has bounded norm hence there will exist some $\varepsilon > 0$ such that (3.174) holds. From KSPY lemma we deduce existence of matrices H_{11}, H_{12}, H_{22} with $H_{11} = H_{11}^*, H_{22} = H_{22}^*, V$ with $\det V \neq 0, W_1, W_2$ such that

$$\eta_{t,s} = x_s^* \left(\tfrac{1}{2} C \bar{F} D_2'' C^* + H_{11} \right) x_s + x_s^* H_{12} v_s + v_s^* H_{12}^* x_s + v_s^* H_{22} v_s -$$
$$- x_t^* \left(\tfrac{1}{2} C \bar{F} D_2'' C^* + H_{11} \right) x_t - x_t^* H_{12} v_t - v_t^* H_{12}^* x_t - v_t^* H_{22} v_t +$$
$$+ \sum_{k=s}^{t-1} |V u_k + W_1^* x_k + W_2^* v_k|^2$$

$$(3.175)$$

Choose now $\mu_t^j = \varphi_j \left(c_j^* x_t \right) - \varphi_j \left(c_j^* x_{t+1} \right)$ where x_t is a solution of (3.159), $\xi_0^j = -\varphi_j \left(c_j^* x_0 \right)$ leading to $\xi_t^j = -\varphi_j \left(c_j^* x_t \right)$. Taking into account (3.164) - (3.167) we shall have

$$\eta_{t,s} \leq \sum_{j=1}^m \left[(\theta_{1j} - \theta_{2j}) \Phi_j \left(\sigma_s^j \right) + \tfrac{1}{2} \theta_{2j} \bar{\varphi}_j \left(\sigma_s^j \right)^2 \right] -$$
$$- \sum_{j=1}^m \left[(\theta_{1j} - \theta_{2j}) \Phi_j \left(\sigma_t^j \right) + \tfrac{1}{2} \theta_{2j} \bar{\varphi}_j \left(\sigma_t^j \right)^2 \right] - \varepsilon^2 \sum_{k=s}^{t-1} |x_k|^2$$

$$(3.176)$$

Using (3.176) in (3.175) we shall have, after a re-ordering of the terms and taking into account that $v_k = -f(C^* x_k)$ where $f(C^* x_k)$ is the vector with the entries $\varphi_j \left(c_j^* x_k \right), \, j = \overline{1, m}$:

$$- x_t^* \left(\frac{1}{2} C \bar{F} D_2'' C^* + H_{11} \right) x_t + x_t^* H_{12} f(C^* x_t) +$$

$$+ f^* (C^* x_t) H_{12}^* x_t - f^* (C^* x_t) H_{22} f^* (C^* x_t) -$$

$$-\sum_{j=1}^{m}\left[\theta_{1j}\Phi_j\left(c_j^*x_t\right)+\theta_{2j}\left(\frac{1}{2}\bar{\varphi}_j\left(c_j^*x_t\right)^2-\Phi_j\left(c_j^*x_t\right)\right)\right]\leq$$

$$\leq-x_s^*\left(\frac{1}{2}C\bar{F}D_2''C^*+H_{11}\right)x_s+x_s^*H_{12}f\left(C^*x_s\right)+$$

$$+f^*\left(C^*x_s\right)H_{12}^*x_s-f^*\left(C^*x_s\right)H_{22}f^*\left(C^*x_s\right)+$$

$$+\sum_{j=1}^{m}\left[\theta_{1j}\Phi_j\left(c_j^*x_s\right)+\theta_{2j}\left(\frac{1}{2}\bar{\varphi}_j\left(c_j^*x_s\right)^2-\Phi_j\left(c_j^*x_s\right)\right)\right]-$$

$$-\sum_{k=s}^{t-1}|V\left(f\left(C^*x_k\right)-f\left(C^*x_{k+1}\right)\right)+W_1^*x_k-W_1^*f\left(C^*x_{k+1}\right)|^2-$$

$$-\varepsilon^2\sum_{k=s}^{t-1}|x_k|^2\ .\tag{3.177}$$

C. We may now introduce the candidate Liapunov function for (3.159)

$$\begin{aligned}V\left(x\right)=&-x^*\left(\tfrac{1}{2}C\bar{F}D_2''C^*+H_{11}\right)x+x^*H_{12}f\left(C^*x\right)+\\&+f^*\left(C^*x\right)H_{12}^*x-f^*\left(C^*x\right)H_{22}f^*\left(C^*x\right)+\\&+\sum_{j=1}^{m}\theta_{1j}\int_0^{c_j^*x}\varphi_j\left(\lambda\right)\mathrm{d}\lambda+\sum_{j=1}^{m}\theta_{2j}\int_0^{c_j^*x}\left(\bar{\varphi}_j\lambda-\varphi_j\left(\lambda\right)\right)\mathrm{d}\lambda\end{aligned}\tag{3.178}$$

(Remember that $\Phi_j\left(\sigma\right)=\int_0^{\sigma}\varphi_j\left(\lambda\right)\mathrm{d}\lambda$.) From (3.177) we deduce that

$$V\left(x_t\right)\leq V\left(x_s\right)-\varepsilon^2\sum_{k=s}^{t-1}|x_k|^2$$

along the solutions of (3.159); by taking $s=t-1$ we obtain

$$V\left(x_t\right)-V\left(x_{t-1}\right)\leq-\varepsilon^2\left|x_{t-1}\right|^2\ .\tag{3.179}$$

In the following we shall show that $V\left(x\right)$ is strictly positive definite by using the properties of (3.159) for linear functions $\varphi_j\left(\sigma\right)=f_j\sigma$, $0\leq f_j\leq\bar{\varphi}_j$; since $\bar{\varphi}_j\leq\beta_j$ we shall have $-\alpha_j\leq0\leq f_j\leq\bar{\varphi}_j\leq$

β_j; hence these functions belong to the class defined by (3.160) and inequality (3.179) holds for them. Remark also that for these functions (3.162) is automatically satisfied without any additional constraint on the parameters θ_{1j}, θ_{2j}.

Denote by F the diagonal matrix with f_j as diagonal elements. For the above linear functions we shall have $f(C^*x) = FC^*x$ and the function $V(x)$ defined by (3.178) becomes

$$V(x) = -x^*H_{11}x + x^*H_{12}FC^*x + x^*CFH_{12}^*x - \\ -x^*CFH_{22}FC^*x + \tfrac{1}{2}x^*C\left(D_2' - D_2''\right)FC^*x = x^*Px \; ; \tag{3.180}$$

hence a quadratic form.

We shall prove now that (3.159) is exponentially stable for all linear functions defined above provided frequency domain inequality (3.161) holds. If this were not true there would exist some F satisfying the sector conditions; i.e., $0 \leq F \leq \bar{F}$ and some z_0 with $|z_0| = 1$ being an eigenvalue of $A - BFC^*$. Using again the known identity (Popov (1973), p. 61) we obtain

$$0 = \det\left(z_0\mathbf{I} - A + BFC^*\right) = \det\left(z_0\mathbf{I} - A\right)\det\left(\mathbf{I} + FG\left(z_0\right)\right) \; .$$

Since A has no eigenvalues on the unit circle we deduce that $\det\left(\mathbf{I} + FG\left(z_0\right)\right) = 0$; hence there will exist some $u \neq 0$ such that $u + FG\left(z_0\right)u = 0$. Assume first that F is invertible hence $G\left(z_0\right)u = -F^{-1}u$. We write down the left-hand side of (3.161) for $z = z_0 = \cos\theta_0 + i\sin\theta_0$ and substitute $G\left(z_0\right)u$ from the above equality to obtain

$$u^*D_1u - u^*D_1\bar{F}F^{-1}u - u^*\left(D_2' - D_2''\right)F^{-1}u\left(1 - \cos\theta_0\right) + \\ +2\left(1 - \cos\theta_0\right)\left[u^*D_3u - u^*D_3\left(F_\beta F^{-1} - F_\alpha F^{-1} + \right. \right. \\ \left. \left. +F^{-1}F_\alpha F_\beta F^{-1}\right)u - \tfrac{1}{2}u^*F^{-1}\left(F_\alpha D_2' + F_\beta D_2''\right)F^{-1}u\right] =$$

$$= u^*D_1\left(\mathbf{I} - \bar{F}F^{-1}\right)u - \left(1 - \cos\theta_0\right)u^*D_2'\left(\mathbf{I} + F_\alpha F^{-1}\right)F^{-1}u + \\ + \left(1 - \cos\theta_0\right)u^*D_2''F^{-1}\left(\mathbf{I} - F_\beta F^{-1}\right)u + \\ +2\left(1 - \cos\theta_0\right)u^*D_3\left(\mathbf{I} - F_\beta F^{-1}\right)\left(\mathbf{I} + F_\alpha F^{-1}\right)u \leq 0$$

The last inequality follows from $\mathbf{I} - \bar{F}F^{-1} < 0$, $\mathbf{I} - F_\beta F^{-1} < 0$, the other matrices being diagonal and positive semidefinite. We obtained a contradiction to the frequency domain inequality (3.161).

Consider now the case when some of the diagonal elements of F may be zero (not all of them since A has all its eigenvalues inside the unit disk). We have $u^k + f_k \sum_{j=1}^m c_k^*\left(z_0\mathbf{I} - A\right)^{-1}b_ju^j = 0$, $k = \overline{1, m}$;

for those k for which $f_k = 0$ we have $u^k = 0$ while for the other we have $\sum_{j=1}^m c_k^* (z_0 \mathbf{I} - A)^{-1} b_j u^j = \frac{1}{f_k} u^k$. Since the terms corresponding to the indices k for which $f_k = 0$ do not appear in the quadratic form we deduce that the former inequality holds and the same contradiction occurs.

We obtained that $A - BFC^*$ defines an exponentially stable evolution. We already know that the quadratic form (3.180) satisfies inequality (3.179); hence, the following discrete Liapunov matrix inequality holds

$$(A - BFC^*)\, P\, (A - BFC^*) - P \leq -\varepsilon^2 \mathbf{I}$$

It follows that P is positively definite; i.e., there exists $\delta > 0$ such that $x^* Px \geq \delta |x|^2$ for all x. Let now x be arbitrary and choose the diagonal matrix F as follows: a diagonal element equals $\varphi_j \left(c_j^* x \right) / c_j^* x$ if $c_j^* x \neq 0$ and 0 if $c_j^* x = 0$. With this choice we shall have

$$x^* P\,(x)\, x = -x^* H_{11} x + x^* H_{12} f\,(C^* x) + f^*\,(C^* x)\, H_{12}^* x - \\ - f^*\,(C^* x)\, H_{22} f\,(C^* x) + \tfrac{1}{2} x^* C\,(D_2' - D_2'')\, f\,(C^* x) =$$

$$= -x^* \left(\tfrac{1}{2} C \bar{F} D_2'' C^* + H_{11} \right) x + x^* H_{12} f\,(C^* x) + f^*\,(C^* x)\, H_{12}^* x - \\ - f^*\,(C^* x)\, H_{22} f\,(C^* x) + \tfrac{1}{2} \sum_{j=1}^m \bar{p}_j \theta_{2j} \left(c_j^* x \right)^2 + \\ + \tfrac{1}{2} \sum_{j=1}^m (\theta_{1j} - \theta_{2j}) \left(c_j^* x \right) \varphi_j \left(c_j^* x \right) =$$

$$= \mathcal{V}\,(x) + \sum_{j=1}^m (\theta_{1j} - \theta_{2j}) \left[\tfrac{1}{2} \left(c_j^* x \right) \varphi_j \left(c_j^* x \right) - \int_0^{c_j^* x} \varphi_j\,(\lambda)\, \mathrm{d}\lambda \right] \geq \delta |x|^2$$

The assumption about monotonicity of $\varphi_j\,(\sigma) / \sigma$ implies that $\tfrac{1}{2} \sigma \varphi_j\,(\sigma) - \int_0^\sigma \varphi_j\,(\lambda) \mathrm{d}\lambda$ keeps constant sign for any σ hence (3.162) holds for suitable choice of $\theta_{1j} - \theta_{2j}$. We deduce that $\mathcal{V}\,(x) \geq \delta_0 |x|^2$ and we may apply Liapunov theorem on asymptotic stability (Theorem 2.17). The above inequality satisfied by $\mathcal{V}\,(x)$ shows also that asymptotic stability is global (from the Barbashin - Krasovskii type theorem, Corollary 2.31). Remark that $\delta > 0$ in the estimate for

$x^* P x$ can be chosen independent of F since F is contained in a compact set and the smallest eigenvalue of P depends continuously on F and is strictly positive. The proof is thus complete.

Remarks. 1^0 From (3.178) it follows, taking also into account the sector conditions satisfied by the nonlinear functions $\varphi_j(\sigma)$ that $\mathcal{V}(x) \le \delta_1 |x|^2$. This inequality, combined with the previous one satisfied by $\mathcal{V}(x)$ and with (3.179), shows that the zero equilibrium is globally exponentially stable. Existence of a single globally exponentially stable equilibrium of a nonlinear system with sector and slope restricted nonlinearities may be called *almost linear behavior*.

2^0 The absolute stability criterion obtained above is a fairly general one allowing the rediscovery of other previously known criteria. The reader is sent to the papers of Yakubovich (1967, 1968) for a quite long list of frequency domain inequalities for absolute stability. We shall pay some attention to a stability criterion reported in the paper of Haddad and Bernstein (1994) namely Theorem 6.1 of this paper. Let us assume in (3.160) that $\alpha_j = 0$, $j = \overline{1, m}$. In this case the class of nonlinear functions is the class considered by Haddad and Bernstein.

Assume that (3.161) holds for the following set of parameters: $D_1 = \bar{F}^{-1}$, $D_2' = 0$, $D_3 = 0$. Since $F_\alpha = 0$ inequality (3.161) becomes

$$\bar{F}^{-1} + \mathrm{Re}\left[(\mathbf{I} + (\bar{z} - 1) D_2'') G(z)\right] - \frac{1}{2} |z - 1|^2 G^*(z) F_\beta D_2'' G(z) > 0.$$

This is exactly the frequency domain inequality occurring in the cited paper of Haddad and Bernstein (1994), more precisely in the section called "The Popov criterion" (Theorem 6.1). For the single nonlinearity case it is known as Szegö - Pearson criterion (Szegö (1963 b), Szegö and Pearson (1964), see also the cited paper of Yakubovich (1968)). When slope restrictions are left aside (only monotonicity counts) the term $-\frac{1}{2} |z - 1|^2 G^*(z) F_\beta D_2'' G(z)$, which is "spoiling" the frequency domain inequality, no longer appears and we have the modified Tsypkin criterion (see Section 3.3) for the case of several nonlinearities. The difference with respect to Section 3.3 is that here the frequency domain inequality ensures existence of a Liapunov function (via KSPY lemma) - the tool allowing one to obtain the absolute stability - whereas previously the stability estimates were obtained directly from the frequency domain inequalities using the Plancherel theorem.

CHAPTER 4

STABLE OSCILLATIONS

4.1. PERIODIC SOLUTIONS OF FORCED LINEAR SYSTEMS WITH PERIODIC COEFFICIENTS (FORCED OSCILLATIONS)

Consider the system

$$x_{t+1} = A_t x_t + f_t, \qquad (4.1)$$

where A_t and f_t are T-periodic. We are interested in finding conditions for (4.1) to have T-periodic solutions.

A. A solution $x_t(x_0)$ of (4.1) is periodic if $x_T(x_0) = x_0$. Using the variation of constants formula

$$x_t = X_{t,0} x_0 + \sum_0^{t-1} X_{t,k+1} f_k$$

the necessary and sufficient condition for $x_t(x_0)$ to be periodic is

$$(\mathbf{I} - U) x_0 = \sum_0^{T-1} X_{T,k+1} f_k \qquad (4.2)$$

From (4.2) we deduce the following:

1. *If the free system $x_{t+1} = A x_t$ has no T-periodic solutions except the zero solution, then for any T-periodic sequence $\{f_t\}_t$ the forced system (4.2) has a unique T-periodic solution.*

Indeed the condition for the free system to have a T-periodic solution is $(\mathbf{I} - U) x_0 = 0$. If this equality holds only for $x_0 = 0$ then $(\mathbf{I} - U)$ is nonsingular and this allows a unique x_0 satisfying (4.2);

221

this x_0 determines the periodic solution of (4.1) in a unique way.

2. *Conversely, if system* (4.1) *has a T-periodic solution, whatever is the T-periodic sequence* $\{f_t\}_t$ *then the corresponding free system* $x_{t+1} = A_t x_t$ *has no T-periodic solutions except the zero solution.*

To prove this it suffices to show that whatever vector u is we may choose the sequence f_t such that $\sum_0^{T-1} X_{T,k+1} f_k = u$; we choose $f_{T-1} = u$, $f_k = 0$, $k = \overline{0, T-2}$. In this case we shall have

$$(\mathbf{I} - U) x_0 = u$$

for any u hence $(\mathbf{I} - U) x_0 = 0$ only if $x_0 = 0$. We have proved in fact

Theorem 4.1 *A linear system with T-periodic coefficients and with a T-periodic forcing term has a T-periodic solution for every T-periodic forcing term if and only if the free system has no other T-periodic solution except the identically zero one.*

The necessary and sufficient condition for the free system to have no T-periodic solution is that $\det (\mathbf{I} - U) \neq 0$ that is the system has no multiplier equal to 1. If this condition holds, the initial condition of the periodic solution is given by

$$x_0 = (\mathbf{I} - U)^{-1} \sum_0^{T-1} X_{T,k+1} f_k = \sum_0^{T-1} (\mathbf{I} - U)^{-1} X_{T,k+1} f_k$$

We deduce that the periodic solution may be written as follows

$$
\begin{aligned}
x_t(x_0) &= X_{t,0} x_0 + \sum_0^{t-1} X_{t,k+1} f_k = \\
&= X_{t,0} \sum_0^{T-1} (\mathbf{I} - U)^{-1} X_{T,k+1} f_k + \sum_0^{t-1} X_{t,k+1} f_k = \\
&= \sum_0^{T-1} X_{t,0} (\mathbf{I} - U)^{-1} X_{T,k+1} f_k + \sum_0^{t-1} X_{t,k+1} f_k = \\
&= \sum_0^{T-1} G_{t,k+1} f_k
\end{aligned}
$$

where

$$G_{t,k} = \begin{cases} X_{t,0} \left(\mathbf{I} - X_T\right)^{-1} X_{T,k} f_k + X_{t,k}, & k \leq t \\ X_{t,0} \left(\mathbf{I} - X_T\right)^{-1} X_{T,k} f_k, & t < k \leq T. \end{cases}$$

It follows that the periodic solution allows estimates of the form

$$|x_t| \leq M \sup_t |f_t|, \quad |x_t| \leq \tilde{M} \sum_0^{T-1} |f_k|$$

where the constants M and $\tilde{M}$ depend only on the free system data.

B. If the free system $x_{t+1} = A_t x_t$ admits nonzero T-periodic solutions, the initial conditions for these solutions will verify the system

$$(\mathbf{I} - U) x_0 = 0.$$

If this system has k linearly independent solutions, $rank\,(\mathbf{I} - U) = n - k$ hence the system $w\,(\mathbf{I} - U) = 0$ also has k linearly independent solutions.

Remark that $U = X_{T,0} = Y_{0,T}$ hence $w\,(\mathbf{I} - U) = 0$ reads $w = wY_{0,T}$. The sequence $y_t = wY_{t,T}$ is a solution of the adjoint system satisfying $y_T = w$. Indeed we have

$$y_{t-1} = wY_{t-1,T} = wY_{t,T} A_{t-1} = y_t A_{t-1}.$$

Since $w = wY_{0,T}$ it follows that $y_T = y_0$; hence y_t thus defined is periodic. The converse is also true. We proved in fact the following

Proposition 4.2 *The system $x_t = A_t x_t$ and its adjoint $y_{t-1} = y_t A_{t-1}$ have the same number of linearly independent periodic solutions.*

C. Let y_t be a periodic solution of the adjoint system $y_{t-1} = y_t A_{t-1}$ and x_t some solution of (4.1). We shall have

$$y_{t+1} x_{t+1} = y_{t+1} \left(A_t x_t + f_t\right) = y_{t+1} A_t x_t + y_{t+1} f_t = y_t x_t + y_{t+1} f_t$$

Therefore,

$$y_{t+1} x_{t+1} - y_t x_t = y_{t+1} f_t, \quad y_T x_T - y_0 x_0 = \sum_0^{T-1} y_{t+1} f_t$$

and since y_t was supposed T-periodic it follows that

$$y_T (x_T - x_0) = \sum_0^{T-1} y_{t+1} f_t . \tag{4.3}$$

From here we deduce that if (4.1) has periodic solutions then

$$\sum_0^{T-1} y_{t+1} f_t = 0 \tag{4.4}$$

for any periodic solution of the adjoint system.

Assume now that (4.4) holds for any periodic solution of the adjoint system and let w be a nonzero solution of $w(\mathbf{I} - U) = 0$ (such solutions exist since the adjoint system has nonzero periodic solutions). For any x_0 we compute

$$' \quad w \left[X_{T,0} x_0 - x_0 + \sum_0^{T-1} X_{T,k+1} f_k \right] = y_T (x_T - x_0) = \sum_0^{T-1} y_{t+1} f_t = 0 \tag{4.5}$$

since if $w(\mathbf{I} - U) = 0$ then $y_T = w$, where y_t is a periodic solution of the adjoint system defined by $y_t = w Y_{t,T}$ (see above); we used also (4.3) and (4.4). From (4.5) we deduce that for any w such that $w = w X_{T,0}$ it follows also that $w \sum_0^{T-1} X_{T,k+1} f_k = 0$. This shows that the system

$$w(\mathbf{I} - X_{T,0}) = 0$$

and the augmented system

$$w(\mathbf{I} - X_{T,0}) = 0 , \quad w \sum_0^{T-1} X_{T,k+1} f_k = 0$$

have the same number of linearly independent solutions, i.e., the matrices of the two systems have equal rank. Consequently, system (4.2) has solutions hence there exist periodic solutions for (4.1). We have proved

Proposition 4.3   *A necessary and sufficient condition for (4.1) to have periodic solutions is that (4.4) holds for any periodic solution of the adjoint system.*

D. Consider again system (4.1) and let $x_t (t_0, x)$ be the solution satisfying $x_t (t_0, x) = x$.

Proposition 4.4 *If system (4.1) has no T-periodic solutions then* $\lim\limits_{t \to \infty} |x_t (t_0, x)| = \infty$ *uniformly with respect to* $|x| \leq \gamma$.

Proof. Since (4.1) has no periodic solutions, there exists a T-periodic solution y_t of the adjoint system such that $\sum\limits_0^{T-1} y_{k+1} f_k = \mu \neq 0$. We have

$$
y_{t+T} x_{t+T} - y_t x_t = \sum_t^{t+T-1} y_{k+1} f_k = \sum_0^{T-1} y_{k+1} f_k - \sum_0^{t-1} y_{k+1} f_k =
$$

$$
= \sum_0^{T-1} y_{k+1} f_k + \sum_T^{t+T-1} y_{k+1} f_k - \sum_0^{t-1} y_{k+1} f_k =
$$

$$
= \sum_0^{T-1} y_{k+1} f_k + \sum_0^{t-1} y_{k+T+1} f_{k+T} - \sum_0^{t-1} y_{k+1} f_k = \sum_0^{T-1} y_{k+1} f_k
$$

the last equality following from the periodicity of both y_t and f_t. Since y_t is periodic the above equality becomes

$$
y_t (x_{t+T} - x_t) = \sum_0^{T-1} y_{k+1} f_k = \mu
$$

But if x_t is a solution of (4.1) then x_{t+kT} is also a solution and we may write

$$
y_t \left(x_{t+(k+1)T} - x_{t+kT} \right) = \mu \quad , \quad k = \overline{0, p-1}
$$

Summing the above equalities we shall have

$$
y_t (x_{t+pT} - x_t) = p\mu
$$

which gives $|y_t x_{t+pT}| \geq p\mu - |y_t x_t|$. But $|y_t x_{t+pT}| \leq |y_t| |x_{t+pT}|$ hence $|y_t| |x_{t+pT}| \geq p |\mu| - |y_t x_t|$. The sequence y_t does not contain any zero vector since it is a nonzero periodic solution of a free linear system. Denoting $\tilde{\mu} = \inf\limits_t |\mu| / |y_t|$ we shall have

$$
|x_{t+pT}| \geq p\tilde{\mu} - |x_t|
$$

Let $t \geq t_0$; we may write $t = t_0 + pT + k$ where $0 \leq k \leq T$; hence $|x_t| = |x_{t_0+k+pT}| \geq p\tilde{\mu} - |x_{t_0+k}|$. But $x_{t_0+k} = X_{t_0+k,t_0}x$ and if $|x| \leq \gamma$ then $|x_{t_0+k}| \leq M\gamma$; hence $|x_t| \geq p\tilde{\mu} - M\gamma$. If $\dfrac{t - t_0 - T}{T} > \dfrac{M\gamma + L}{\tilde{\mu}}$ that is if $\dfrac{t}{T} > 1 + \dfrac{t_0}{T} + \dfrac{M\gamma + L}{\tilde{\mu}}$ then $p > \dfrac{M\gamma + L}{\tilde{\mu}}$; hence $|x_t| > L$ where L may be arbitrarily large which ends the proof.

Remark. The meaning of this result is the following. Existence of T-periodic solutions for the free system (that follows from nonexistence of T-periodic solutions for the forced system - Theorem 4.1) implies, in the general situation when the forcing term does not satisfy the orthogonality condition (4.4), unboundedness of the forced oscillations (*resonance*).

E. Assume now that the multipliers of (4.1) are inside the unit disk of the complex plane. Since no multiplier equals 1 system (4.1) has a T-periodic solution. *This solution is exponentially stable.* Indeed let $\hat{x}_t$ be the T-periodic solution and x_t any other solution of (4.1). The deviation $x_t - \hat{x}_t$ is a solution of the free system

$$x_{t+1} - \hat{x}_{t+1} = A(x_t - \hat{x}_t)$$

and since the multipliers are inside the unit disk any solution of the free system tends exponentially to zero (see Section 2.4). Therefore,

$$|x_t - \hat{x}_t| \leq \beta_0 \rho^{t-t_0}|x_{t_0} - \hat{x}_{t_0}|, \quad 0 < \rho < 1, \; t > t_0.$$

We proved in fact

Proposition 4.5 *If the multipliers of A_t are located inside the unit disk then system (4.1) has a unique T-periodic solution that is exponentially stable.*

We may thus call the periodic solution a *stable forced oscillation.*
Remark also that if A_t defines an exponentially stable evolution and f_t is bounded then (4.1) will have a bounded solution for all integers t, expressed by

$$x_t = \sum_{-\infty}^{t-1} X_{t,k+1} f_k.$$

(One may see that this is a solution by direct check; boundedness follows from the facts that f_k are bounded and $|X_{t,k+1}| \leq \beta_0 \rho^{t-k}$.

$0 < \rho < 1$.) On the other side, if A_t and f_t are T-periodic the above solution is exactly the unique T-periodic solution of (4.1) since

$$x_{t+T} = \sum_{-\infty}^{t+T-1} X_{t+T,k+1} f_k = \sum_{-\infty}^{t-1} X_{t+T,k+T+1} f_{k+T} = \sum_{-\infty}^{t-1} X_{t,k+1} f_k = x_t$$

(we used the basic definition of $X_{t,k}$ and the periodicity of A_t and f_t).

4.2. ALMOST PERIODIC SEQUENCES

There is no need to advocate the significance of periodic functions. It is also clear that functions as $f(t) = \sin t + \sin \sqrt{2} t$ or $f(t) = \sin t + \sin \pi t$ are no longer periodic although such functions arise in a natural way, e.g., in mechanics or in electronic communications. For instance, an *amplitude modulated signal* occurring in AM (amplitude modulation) radio is described by $v(t) = (A_0 + u(t)) \cos \omega_c t$, where ω_c is the so-called *carrier frequency* (e.g., Kamen (1990), p. 462-466). If $u(t)$ is the so-called *tone basic signal* then $u(t) = u_0 \cos \omega_0 t$ and the AM signal takes the form

$$
\begin{aligned}
v(t) &= (A_0 + u_0 \cos \omega_0 t) \cos \omega_c t = \\
&= A_0 \left[\frac{1}{2} \frac{u_0}{A_0} \cos (\omega_c - \omega_0) t + \cos \omega_c t + \frac{1}{2} \frac{u_0}{A_0} \cos (\omega_c + \omega_0) t \right] = \\
&= A_0 \left[m \cos (\omega_c - \omega_0) t + \cos \omega_c t + m \cos (\omega_c + \omega_0) t \right] ,
\end{aligned}
$$

where m denotes the so-called *modulation depth*. Generally speaking the frequency ratio ω_0/ω_c is not rational hence the AM signal is no longer periodic.

The general theory for such class of functions has been developed by Harald Bohr (1923): a function $f : \mathbb{R} \to X$, where (X, d) is a metric space is *almost periodic* if it is continuous and for every $\varepsilon > 0$ there exists $L(\varepsilon) > 0$ such that in each interval of length $L(\varepsilon)$ there exists a number τ such that $d(f(t + \tau), f(t)) < \varepsilon$ for all $t \in \mathbb{R}$. It is easy to see that if f is periodic with period T then for every $\varepsilon > 0$ we may take $L(\varepsilon) = T$ hence a periodic function is almost periodic.

The next step was done by S. Bochner (1927) who proved that the definition of Bohr was equivalent to the following one: f is almost periodic if it is continuous and if for every sequence $\{\tau_k\}_k$ there exists a subsequence $\{\tau_{k_l}\}_l$ such that the sequence $\{f(t + \tau_{k_l})\}_l$ converges uniformly with respect to $t \in \mathbb{R}$.

From this definition it follows directly that if X is a Banach space and $f_i : \mathbb{R} \rightarrow X$, $i = 1, 2$, are almost periodic, then $\alpha f_1 + \beta f_2$ is almost periodic for all $\alpha, \beta \in \mathbb{R}$. Hence the functions as $f(t) = \alpha \sin t + \beta \sin \sqrt{2} t$ are almost periodic. It is also true that if $f(t) = \sum_1^\infty (\alpha_k \sin \omega_k t + \beta_k \cos \omega_k t)$ with arbitrary $\omega_k \in \mathbb{R}$ is uniformly convergent, it defines an almost periodic function. A sort of converse is also true: to each almost periodic function we may associate a sort of Fourier series, but we shall not insist upon such facts. In the following we shall consider in some detail their discrete time analogues.

4.2.1. Motivation

We shall start with some examples. Consider first the discrete time linear equation

$$y_{t+1} + 2\alpha y_t + y_{t-1} = 0 , \quad t = 2, 3, \ldots$$

Its characteristic function is $\lambda^2 + 2\alpha\lambda + 1 = 0$. If α is a real number such that $|\alpha| < 1$ then the characteristic equation has two complex roots on the unit circle. Let $e^{i\theta}, e^{-i\theta}$ be these roots where $\cos \theta = -\alpha$. If the initial conditions y_0, y_1 are real then the solution of the discrete time equation is given by

$$y_t = \frac{y_1 + \alpha y_0}{\sqrt{1 - \alpha^2}} \sin \theta t + y_0 \cos \theta t , \quad t = 2, 3, \ldots$$

where $\sin \theta t$ and $\cos \theta t$ are, generally speaking, not periodic for $t \in \mathbb{Z}$ unless θ is a rational multiple of π. As we shall see, for all real θ the sequences $\{\sin \theta t\}_t$, $\{\cos \theta t\}_t$, $t \in \mathbb{Z}$ are almost periodic.

A most natural way of introducing almost periodic sequences is in connection with sampling (see Section 1.3 or the introductory part to Chapter 3). Assume we have a periodic signal, say $f(t) = \sin t$; we measure the signal at $t_k = k\delta$, where δ is a given sampling period. We may ask if in such case the sequence $\{f(t_k)\}_k = \{\sin k\delta\}_k$ will be periodic. The answer is yes only if there exists integers N and p such that $N\delta = 2p\pi$ (see also the previous example). Since such situation is rather unlike to occur, the next question to be asked is: what properties does the sequence $\{f(t_k)\}_k$ inherit from the periodicity of f? It is a simple matter to show that such sequence will be almost periodic, with a natural definition for almost periodicity. Moreover, the following is true: a sequence $\{x_k\}_k$ is almost periodic if and only if there exists an almost periodic function f such that $x_k = f(k)$. Since for every almost periodic f and every $\delta > 0$ the function $t \rightarrow f(t\delta)$ is almost periodic, we see that almost periodic sequences are generated by sampling from almost periodic functions.

The first study of almost periodic sequences seems to be that of Walther (1928), an important early reference in the field being that of Ky Fan (1943). More recent references are Corduneanu (1982, 1989).

4.2.2. Standard Theory of Almost Periodic Sequences

A. We give the following

Definition 4.1 A sequence $(x_t)_t$, $t \in \mathbb{Z}$, $x_t \in X$, X being a Banach space, is *almost periodic* if for every $\varepsilon > 0$ there exists $T(\varepsilon)$ such that among $T(\varepsilon)$ consecutive integers there is one, call it p, such that $\|x_{t+p} - x_t\| < \varepsilon$ for all $t \in \mathbb{Z}$. Integers p with the above property are called ε - *almost periods of the sequence*. If a sequence is periodic, all multiples of the period are ε - almost periods for all $\varepsilon > 0$.

Proposition 4.6 *An almost periodic sequence is bounded.*

Proof. Let $s \in \mathbb{Z}$ be arbitrary and let $\varepsilon > 0$; according to the above definition there exists p, $-s \leq p \leq -s + T(\varepsilon)$, an ε -almost period. We deduce that $0 \leq s + p \leq T(\varepsilon)$ and

$$\|x_s\| = \|x_s - x_{s+p} + x_{s+p}\| \leq \|x_s - x_{s+p}\| + \|x_{s+p}\| < \varepsilon + \max_{0 \leq t \leq T(\varepsilon)} \|x_t\|$$

and since s was arbitrary the proposition is proved.

For an almost periodic sequence $\{x_t\}_t$, denoted x, we define $\|x\| = \sup_t \|x_t\|$ which is finite, according to Proposition 4.6; for two sequences we define the distance $\rho(x, y) = \|x - y\| = \sup_t \|x_t - y_t\|$.

Given a sequence $\{x^l\}_l$ of almost periodic sequences, $x^l = \{x_t^l\}_t$, $t \in \mathbb{Z}$, $l \in \mathbb{N}$, it converges to a sequence x if $\lim_{l \to \infty} \rho(x^l, x) = 0$; hence if for every $\varepsilon > 0$ there exists $\lambda(\varepsilon) \in \mathbb{N}$ such that $\sup_t \|x_t^l - x_t\| < \varepsilon$, $l > \lambda(\varepsilon)$.

Proposition 4.7 *Let $\{x^l\}_l$ be a sequence of almost periodic sequences converging to a sequence x in the sense of the above definition; then sequence x is almost periodic.*

Proof. Let $\varepsilon > 0$ and $\lambda(\varepsilon)$ such that $\left\| x_t^{\lambda(\varepsilon)} - x_t \right\| < \varepsilon/3$ for all $t \in \mathbb{Z}$; let p be an $(\varepsilon/3)$ -almost period of $x^{\lambda(\varepsilon)}$. We deduce

$$\left\| x_{t+p} - x_t \right\| \leq \left\| x_{t+p} - x_{t+p}^{\lambda(\varepsilon)} \right\| + \left\| x_{t+p}^{\lambda(\varepsilon)} - x_t^{\lambda(\varepsilon)} \right\| + \left\| x_t^{\lambda(\varepsilon)} - x_t \right\| <$$
$$< \varepsilon/3 + \varepsilon/3 + \varepsilon/3 = \varepsilon$$

Therefore p is an ε -almost period of x since it may be found among every $T^\varepsilon(\varepsilon/3)$ consecutive integers where $T^\varepsilon(\cdot)$ is associated to the almost periodic sequence $x^{\lambda(\varepsilon)}$.

Corollary 4.8 *The metric space of almost periodic sequences is complete.*

Proof. Let $\left(x^l \right)_l$ be a Cauchy sequence of almost periodic sequences: there exists $\lambda(\varepsilon)$ such that $\left\| x_t^{l+m} - x_t^l \right\| < \varepsilon$ if $l \geq \lambda(\varepsilon)$, $m \in \mathbb{N}$, $t \in \mathbb{Z}$; for fixed t we have a Cauchy sequence in X and this Cauchy sequence has a limit x_t. For the sequence $\{ x_t \}_t$ thus defined we shall have

$$\left\| x_t^l - x_t \right\| \leq \left\| x_t^l - x_t^{l+m} \right\| + \left\| x_t^{l+m} - x_t \right\| < \varepsilon + \left\| x_t^{l+m} - x_t \right\|$$

and by letting $m \to \infty$ we deduce $\left\| x_t^l - x_t \right\| < \varepsilon$ for $l > \lambda(\varepsilon)$. We have $x^l \to x$ in the sense of the metric previously defined; according to Proposition 4.7 x is an almost periodic sequence.

    B. In the following we shall give a necessary and sufficient condition of almost periodicity.

Theorem 4.9 *A necessary and sufficient condition for a sequence $\{x_t\}_t$ to be almost periodic is that for every sequence of integers $\{t_j\}_j$, $j \in \mathbb{N}$, there exists a subsequence $\{ t_{j_i} \}_i$, $i \in \mathbb{N}$ such that $\{ x_{t+t_{j_i}} \}_i$, $i \in \mathbb{N}$, converges, uniformly with respect to $t \in \mathbb{Z}$.*

Proof. a) Assume that $\{x_t\}_t$ is almost periodic and let $\{t_j\}_j$, $j \in \mathbb{N}$, be a sequence of integers $(t_j \in \mathbb{Z})$. For $\varepsilon > 0$ there exists $T(\varepsilon) > 0$ such that between $t_j - T(\varepsilon)$ and t_j there is an ε -almost period p_j; hence $0 \leq t_j - p_j \leq T(\varepsilon)$. Denoting $q_j = t_j - p_j$, the sequence $\{q_j\}_j$ can take only a finite number (at most $T(\varepsilon) + 1$) values; hence there is some q, $0 \leq q \leq T(\varepsilon)$, such that $q_j = q$ for an infinite number of j's; let these indices be numbered as j_i. We have

$$\left\| x_{t+t_j} - x_{t+q_j} \right\| = \left\| x_{t+q_j+p_j} - x_{t+q_j} \right\| < \varepsilon;$$

hence $\left\|x_{t+t_j} - x_{t+q_j}\right\| < \varepsilon$ for all $t \in \mathbb{Z}$. Consider now a sequence $\{\varepsilon_l\}_l$, $\varepsilon_l \to 0$, for instance $\varepsilon_l = 1/l$; from $\left\{x_{t+t_j}\right\}_j$ we take a subsequence chosen as above in order that $\left\|x_{t+t_{j_i^1}} - x_{t+q^1}\right\| < \varepsilon_1$; from this last subsequence we take a new subsequence $\left\{x_{t+t_{j_i^2}}\right\}_i$ such that $\left\|x_{t+t_{j_i^2}} - x_{t+q^2}\right\| < \varepsilon_2$; we go on in the same way and for each $r \in \mathbb{N}$ we obtain a subsequence $\left\{x_{t+t_{j_i^r}}\right\}_i$ such that $\left\|x_{t+t_{j_i^r}} - x_{t+q^r}\right\| < \varepsilon_r$. Take now the "diagonal" subsequence $\left\{x_{t+t_{j_i^i}}\right\}_i$; for $\varepsilon > 0$ take $k(\varepsilon) \in \mathbb{N}$ such that $\varepsilon_{k(\varepsilon)} < \varepsilon/2$ where ε_l belong to the previously defined sequence. For $r \geq k(\varepsilon)$, $s \geq k(\varepsilon)$ we shall have

$$
\begin{aligned}
\left\|x_{t+t_{j_r^r}} - x_{t+t_{j_s^s}}\right\| &\leq \left\|x_{t+t_{j_r^r}} - x_{t+q^k}\right\| + \left\|x_{t+q^k} - x_{t+t_{j_s^s}}\right\| < \\
&< \varepsilon_{k(\varepsilon)} + \varepsilon_{k(\varepsilon)} < \varepsilon
\end{aligned}
$$

since $\left\{t_{j_r^r}\right\}_r$ and $\left\{t_{j_s^s}\right\}_s$ are both subsequences of $\left\{t_{j_i^{k(\varepsilon)}}\right\}_i$. We obtained in fact that the sequence $\left\{x_{t+t_{j_i^i}}\right\}_i$ is a Cauchy sequence hence it converges uniformly (to an almost periodic sequence).

b) Let now the condition in the statement be fulfilled and assume that the sequence is not almost periodic; there will exist then some $\varepsilon_0 > 0$ such that for every $T \in \mathbb{N}$ there are T consecutive integers among which there is no ε_0 -almost period for $\{x_t\}_t$. Let L_T be such a group of T consecutive integers and let t_1 be an arbitrary integer; choose t_2 such that $t_2 - t_1$ is in L_1 (for instance, if $k \in L_1$ we may choose $t_2 = t_1 - k$); denote $L_1 = L_{\nu_1}$. Choose now $\nu_2 > |t_1 - t_2|$ and t_3 such that $t_3 - t_1$ and $t_3 - t_2$ are in L_{ν_2}; to do that let $l, l+1, ..., l+\nu_2-1$ be the numbers in L_{ν_2} and assume $t_2 \leq t_1$. Choose $t_3 = l + t_1$; hence $t_3 - t_1 \in L_{\nu_2}$ and $t_3 - t_2 \geq l$, $t_3 - t_2 \leq l + t_1 - t_2 < l + \nu_2 - 1$; hence $t_3 - t_2 \in L_{\nu_2}$. We may go on in the same way to obtain $\nu_j \geq \max_{1 \leq \mu < \nu \leq j} |t_\nu - t_\mu|$ then t_{j+1} such that $t_{j+1} - t_\mu \in L_{\nu_j}$ for $1 \leq \mu \leq j$; we may take $t_{j+1} = \min\{l : l \in L_{\nu_j}\} + \max(t_\mu : 1 \leq \mu \leq j)$.

For the sequence $\{t_j\}_j$ constructed in this way we have

$$
\sup_t \|x_{t+t_r} - x_{t+t_s}\| = \sup_t \|x_{t+t_r-t_s} - x_t\|, \quad t_r - t_s \in L_{\nu_{r-1}} \quad \text{(if } r \geq s\text{)}.
$$

According to the definition of L_T we deduce that

$$
\sup_t \|x_{t+t_r} - x_{t+t_s}\| \geq \varepsilon_0.
$$

By assumption we may take a subsequence $\{t_{j_i}\}_i$ from the sequence $\{t_j\}_j$ constructed above such that $\{x_{t+t_{j_i}}\}_i$ converges uniformly with respect to $t \in \mathbb{Z}$, i.e., there exists j_0 such that if $r \geq j_0$, $s \geq j_0$; then $\|x_{t+t_{j_r}} - x_{t+t_{j_s}}\| \geq \varepsilon_0/2$ and this contradicts the property of sequence $\{t_j\}_j$ constructed above.

This theorem has two important consequences.

Corollary 4.10 *If x and y are almost periodic sequences and $\alpha \in \mathbb{R}$, $\beta \in \mathbb{R}$ then $\alpha x + \beta y$ is an almost periodic sequence.*

Indeed if $\{x_t\}_t$ and $\{y_t\}_t$ are almost periodic we may take from the sequence $\{t_k\}_k$ a subsequence $\left\{t_{k'_j}\right\}_j$ such that $\left\{x_{t+t_{k'_j}}\right\}_j$ converges uniformly and from the sequence $\left\{t_{k'_j}\right\}_j$ the subsequence $\left\{t_{k_j}\right\}_j$ such that $\left\{y_{t+t_{k_j}}\right\}_j$ converges uniformly; it follows that from the sequence $\{t_k\}_k$ we may take a subsequence $\left\{t_{k_j}\right\}_j$ such that $\left\{\alpha x_{t+t_{k_j}} + \beta y_{t+t_{k_j}}\right\}_j$ converges uniformly.

Corollary 4.11 *If $\{x_t\}_t$ is an almost periodic sequence and $\{\alpha_t\}_t$ is an almost periodic sequence of real numbers then the sequence $\{\alpha_t x_t\}_t$ is almost periodic.*

The proof of this corollary goes as previously. From the two corollaries and the previous result we deduce that the set of almost periodic sequences with their values in a Banach space is also a Banach space with the sup norm defined previously.

Corollary 4.12 *If X is an algebra and if x and y are almost periodic sequences then xy defined by $x_t y_t$ is an almost periodic sequence.*

Corollary 4.13 *If X is a finite dimensional space and if $x = \{x_t\}_t$ is an almost periodic sequence then the coordinate sequences $x^i = \{x_t^i\}_t$, $i = \overline{1, n}$, are almost periodic.*

This result follows directly from Theorem 4.9 since in finite dimension uniform convergence of the coordinates follows from the uniform convergence of the vector.

Corollary 4.14 *If $\{x_t^i\}_t$, $i = \overline{1, n}$, are almost periodic sequences then the sequence $x = \{x_t\}_t$ in the product space with $x = \left(x^1, ..., x^n\right)$ is almost periodic.*

Corollary 4.15 *If $\left\{x_t^i\right\}_t$, $i = \overline{1,n}$, are almost periodic then for every $\varepsilon > 0$ there exists $T(\varepsilon)$ such that among $T(\varepsilon)$ consecutive integers there exists an integer p such that $\left\|x_{t+p}^i - x_t^i\right\| < \varepsilon$ for all $t \in \mathbb{Z}$ and $i = \overline{1,n}$.*

Here the numbers $T(\varepsilon)$ and p are the ones associated to the sequence $x = \{x_t\}_t$ in the product space with the induced norm.

We shall give further the discrete counter-part of a well known result of Bochner.

Theorem 4.16 *A necessary and sufficient condition for the sequence $\{x_t\}_t$ to be almost periodic is the existence of an almost periodic function $f : \mathbb{R} \to X$ such that $x_t = f(t)$, $t \in \mathbb{Z}$.*

Proof. Assume $f : \mathbb{R} \to X$ be almost periodic and let $\{t_j\}_j$ be a sequence of integers; consider the sequence $x_{t+t_j} = f(t + t_j)$, $j \in \mathbb{N}$, $t \in \mathbb{Z}$. From almost periodicity of f we deduce the existence of a subsequence $\{t_{j_i}\}_i$ such that $\{f(s + t_{j_i})\}_i$ converges uniformly with respect to $s \in \mathbb{R}$; hence $\left\{x_{t+t_{j_i}}\right\}_i$ converges uniformly with respect to $t \in \mathbb{Z}$; hence $\{x_t\}_t$ is almost periodic.

To see the converse one takes for a sequence $\{x_k\}_k$, $k \in \mathbb{Z}$ the function $f(t) = x_k + (t - k)(x_{k+1} - x_k)$, $k \le t < k + 1$; one can easily see that an $\varepsilon/3$-almost period of x is also an ε-almost period for f.

C. It is a known fact that for an almost periodic function the limit $\lim\limits_{T \to \infty} \frac{1}{T} \int\limits_{t}^{t+T} f(s)\mathrm{d}s$ exists uniformly with respect to $t \in \mathbb{R}$ and it is called the *mean value of the function*. A similar property holds for almost periodic sequences too.

Proposition 4.17 *If $\{x_t\}_t$, $t \in \mathbb{Z}$ is an almost periodic sequence then the limit*

$$\hat{x} = \lim_{j \to \infty} \frac{x_{t+1} + x_{t+2} + \cdots + x_{t+j}}{j}$$

exists uniformly with respect to $t \in \mathbb{Z}$ and is called the mean value of the sequence x.

Proof. Let f be the almost periodic function $f(t) = x_t$ defined in

Theorem 4.16. We shall have that $\lim\limits_{j\to\infty} \frac{1}{j} \int\limits_{t}^{t+j} f(s)\mathrm{d}s$ exists since f is almost periodic. Taking into account the form of f we shall have

$$\frac{1}{j}\int\limits_{t}^{t+j} f(s)\,\mathrm{d}s = \frac{1}{j}\sum_{i=0}^{j-1} \int\limits_{t+i}^{t+i+1} f(s)\,\mathrm{d}s$$

$$f(s) = x_{t+i} + (s-k-i)(x_{t+i+1} - x_{t+i})$$

$$\int\limits_{t+i}^{t+i+1} f(s)\,\mathrm{d}s = x_{t+i} + \frac{1}{2}(x_{t+i+1} - x_{t+i}) = \frac{1}{2}(x_{t+i} + x_{t+i+1})$$

$$\frac{1}{j}\int\limits_{t}^{t+j} f(s)\,\mathrm{d}s = \frac{x_{t+1} + x_{t+2} + \cdots + x_{t+j}}{j} + \frac{x_t - x_{t+j}}{2j}.$$

Since $\{x_t\}_t$ is a bounded sequence the second term in the right-hand side converges uniformly to zero with respect to $t \in \mathbb{Z}$ as $j \to \infty$. The proposition is proved.

4.2.3. A New Bochner Theory for Almost Periodic Sequences

The purpose of this section is to show that the results of Bochner (1962) on almost periodic functions are also valid for almost periodic sequences and may be applied to obtain existence of almost periodic solutions for discrete systems. We shall use the following notation: if $l = \{l_j\}_j$ is a sequence and $\{m_i\}_i$ is a subsequence of it, i.e., $m_i = l_{j_i}$ we shall write $m \subset l$; if $l = \{l_j\}_j$ and $\{m_i\}_i$ are sequences then $l + m$ is the sequence $\{l_j + m_j\}_j$. If $l = \{l_j\}_j$, $m = \{m_j\}_j$ are sequences then the subsequences $\{l_{j_i}\}_i$ and $\{m_{j_i}\}_i$ will be termed common subsequences. Notation $T_l x = y$ means $y_k = \lim\limits_{j\to\infty} x_{k+l_j}$ for each fixed $k \in \mathbb{Z}$. If further $T_m y = z$ we shall naturally write $T_m T_l x = z$. With the above notation the property in Theorem 4.9 (equivalent to Definition 4.1) is: x is almost periodic if and only if for every sequence l of integers there exists a subsequence $m \subset l$ such that $T_m x$ exists uniformly on $\mathbb{Z}$.

Theorem 4.18 Let x be a sequence. Assume for every pair l', m' of sequences of integers there exist common subsequences $l, m, l \subset l'$,

$m \subset m'$, such that $T_l T_m x = T_{l+m} x$ pointwise in $\mathbb{Z}$. Then x is almost periodic.

Proof. Let n' be a sequence of integers: if we choose $l' = 0$, $m' = n'$ we deduce, by using the property in the statement, existence of a subsequence $n \subset n'$ such that $T_n x = y$ exists pointwise. Assume $\lim_{j \to \infty} x_{k+n_j} = y_k$ is not uniform with respect to $k \in \mathbb{Z}$. Then there exists $\varepsilon_0 > 0$ such that for every J there exist $j' \geq J$, $j'' \geq J$, k_J such that $\left\| x_{k_J + n_{j'}} - x_{k_J + n_{j''}} \right\| \geq \varepsilon_0$. Take $J = i$, denote $n_{j'} = \hat{n}'_i$, $n_{j''} = \hat{n}''_i$ and write $\left\| x_{k_i + \hat{n}'_i} - x_{k_i + \hat{n}''_i} \right\| \geq \varepsilon_0$. From the property in the statement we deduce existence of subsequences $\tilde{n}'$, $\tilde{k}$, $\tilde{n}' \subset \hat{n}'$, $\tilde{k} \subset k$ such that pointwise $T_{\tilde{k} + \tilde{n}'} x = T_{\tilde{k}} T_{\tilde{n}'} x$. Again from the property in the statement we deduce existence of subsequences $\check{n}'' \subset \tilde{n}''$, $\check{k} \subset \tilde{k}$ such that $T_{\check{k} + \check{n}''} x = T_{\check{k}} T_{\check{n}''} x$; for the corresponding subsequence $\check{n}'$ we have also $T_{\check{k} + \check{n}'} x = T_{\check{k}} T_{\check{n}'} x$. But $\check{n}'$, $\check{n}''$ are subsequences of n hence $T_{\check{n}'} x = T_{\check{n}''} x = y$ pointwise. We deduce that $\lim_{j \to \infty} x_{\check{k}_j + \check{n}'_j} = \lim_{j \to \infty} x_{\check{k}_j + \check{n}''_j}$; hence $\lim_{j \to \infty} \left\| x_{\check{k}_j + \check{n}'_j} - x_{\check{k}_j + \check{n}''_j} \right\| = 0$ contradicting $\left\| x_{\check{k}_j + \check{n}'_j} - x_{\check{k}_j + \check{n}''_j} \right\| \geq \varepsilon_0$; this contradiction proves the theorem.

A *useful remark* is that if x is an almost periodic sequence then for every pair of sequences of integers l', m' there exist common subsequences $l \subset l'$, $m \subset m'$ such that $T_l T_m x = T_{l+m} x$ uniformly on $\mathbb{Z}$. We choose $l \subset l'$, $m \subset m'$ successively in order to obtain common subsequences such that $T_l x = y$, $T_{l+m} x = z$ uniformly on $\mathbb{Z}$; choose $\varepsilon > 0$ and J_ε such that $\left\| x_{k+m_i+l_j} - z_k \right\| < \varepsilon/2$ for all $k \in \mathbb{Z}$ and all $j \geq J_\varepsilon$, $\left\| x_{k+m_i+l_j} - y_{k+m_i} \right\| < \varepsilon/2$ for all $i \geq 1$, $j \geq J_\varepsilon$. We deduce that for every $k \in \mathbb{Z}$ and $j \geq J_\varepsilon$ $\left\| y_{k+m_i} - z_k \right\| < \varepsilon$ hence $T_m y = z$ uniformly on $\mathbb{Z}$.

4.2.4. Asymptotic Almost Periodic Sequences

We give the following

Definition 4.2 A sequence u_t, $t \in \mathbb{N}$ is called *asymptotic almost periodic* if for any $\varepsilon > 0$ there exist $T(\varepsilon) > 0$ and $B(\varepsilon) > 0$ such that among T consecutive integers there should exist an integer p with the property that $t \geq B(\varepsilon)$ and $t + p \geq B(\varepsilon)$ implies $\left\| u_{t+p} - u_t \right\| < \varepsilon$.

The significance of this definition appears from the following fact:

if $u_t = \alpha_t + \omega_t$, where α is an almost periodic sequence and $\lim\limits_{t \to \infty} \omega_t = 0$ then u is an asymptotic almost periodic sequence. Indeed let $\varepsilon > 0$; there will exist $T \geq B(\varepsilon)$ such that $\|\omega_t\| < \varepsilon/3$. On the other hand, since α is almost periodic there exists $T(\varepsilon) > 0$ such that there exists an integer p among $T(\varepsilon)$ consecutive integers such that $\|\alpha_{t+p} - \alpha_t\| < \varepsilon/3$ for any t. Let now $t \geq B(\varepsilon)$, $t + p \geq B(\varepsilon)$; we shall have

$$
\begin{aligned}
\|u_{t+p} - u_t\| &= \|\alpha_{t+p} + \omega_{t+p} - \alpha_t - \omega_t\| \leq \\
&\leq \|\alpha_{t+p} - \alpha_t\| + \|\omega_{t+p}\| + \|\omega_t\| < \varepsilon \, ;
\end{aligned}
$$

hence u is asymptotic almost periodic in the sense of the above definition.

The converse of this proposition is the following

Theorem 4.19 *If u is a $\mathbb{R}^n$ -valued asymptotic almost periodic sequence, there exist an almost periodic sequence α and a sequence ω tending to zero such that $u_t = \alpha_t + \omega_t$; the sequences α and ω are determined in a unique way.*

The proof of this theorem requires some preliminaries.

Proposition 4.20 *Any asymptotic almost periodic sequence is bounded.*

Proof. Let $s \geq B(1)$ be an arbitrary integer; choose p between $-s + B(1)$ and $-s + B(1) + T(1)$; it will follow that $s + p > B(1)$; hence $\|u_{s+p} - u_s\| < 1$. We deduce that $\|u_s\| \leq \|u_{s+p} - u_s\| + \|u_{s+p}\| < 1 + \|u_{s+p}\|$ hence $\|u_s\| < 1 + M_1$, where $M_1 = \max\limits_{B(1) \leq t \leq B(1)+T(1)} \|u_t\|$; if $M_2 = \max\limits_{0 \leq t \leq B(1)} \|u_t\|$ we deduce

$$
\|u_t\| < 1 + \max(M_1, M_2) = 1 + M \, , \quad M = \max_{0 \leq t \leq B(1)+T(1)} \|u_t\| \, .
$$

Prop sition 4.21 *If u is a $\mathbb{R}^n$ -valued asymptotic almost periodic sequence then from any sequence $\{h_s\}_s$ of integers, $h_s \to \infty$, we can select a subsequence k_s such that the sequence $\{u_{t+k_s}\}_s$ converges uniformly with respect to $t \geq -j$, j being an arbitrary positive integer. Moreover $\alpha_t = \lim\limits_{s \to \infty} u_{t+k_s}$ is an almost periodic sequence.*

Proof. Let $v_{t_0} = u_{t+h_s}$: since u is bounded there exists M such that $|v_{t_s}| < M$ and using Cesaro lemma we may select a subsequence of indices $\{s_k\}_k$ such that $\left\{v_{t_{*k}}\right\}_k$ is convergent; moreover, if j is an arbitrary positive integer we may choose the subsequence such that $h_{s_k} \geq j$ (since $\lim_{s \to \infty} h_s = \infty$); hence $t + h_{s_k} \geq 0$ if $t \geq -j$; in this way $v_{t_{*k}} = u_{t+h_{*k}}$ is defined for $t \geq -j$.

Let now r be a positive integer: $r = 1, 2, ...$; for each r we may find a subsequence $h_{s_k^r}$ such that $\left\{u_{t+h_{*k}^r}\right\}_k$ is convergent for $-r \leq t \leq r$ and $\left\{h_{s_k^r}\right\}_k$ is contained in $\left\{h_{s_k^{r-1}}\right\}_k$. From the 2-index sequence $h_{s_k^i}$ we select the "diagonal" sequence $h_{s_r^r}$. For an arbitrary t we take $R > 0$ such that $|t| \leq R$. For $r \geq R$ $\{h_{s_r^r}\}_r$ is contained in $\left\{h_{s_k^R}\right\}_k$: hence $\left\{u_{t+h_{*r}^r}\right\}_r$ is convergent for $-R \leq t \leq R$. We deduce that $\left\{u_{t+h_{*r}^r}\right\}_r$ is convergent for any t and this convergence is uniform with respect to t when t goes a finite number of indices. Denote further $h_{s_r^r} = h_r'$. Let $\varepsilon > 0$ and p_r chosen between $h_r' - T(\varepsilon)$ and h_r' from the definition of asymptotic almost periodic sequences; for r large enough we shall have $h_r' > T(\varepsilon)$ hence $p_r > 0$; if also $t > B(\varepsilon)$ we shall have $t + p_r > B(\varepsilon)$ hence $|u_{t+p_r} - u_t| < \varepsilon$. We denote $m_r = h_r' - p_r$; we shall have $0 < m_r \leq T(\varepsilon)$; if $t \geq B(\varepsilon)$ we have $t + m_r > B(\varepsilon)$, $t + m_r + p_r > B(\varepsilon)$ and $|u_{t+m_r+p_r} - u_{t+m_r}| < \varepsilon$ hence $|u_{t+h_r'} - u_{t+m_r}| < \varepsilon$. But m_r can take only a finite number of distinct values hence there exists m^* corresponding to an infinite sequence of indices r_j : we shall thus have $\left|u_{t+h_{r_j}'} - u_{t+m^*}\right| < \varepsilon$. But sequence $\left\{u_{t+h_{r_j}'}\right\}_j$ is a subsequence of the convergent sequence $\left\{u_{t+h_r'}\right\}_r$; this sequence is convergent for any t and let α_t be the limit. We have $|\alpha_t - u_{t+m^*}| < \varepsilon$ for $t \geq B(\varepsilon)$ and $\left|u_{t+h_{r_j}'} - \alpha_t\right| \leq \left|u_{t+h_{r_j}'} - u_{t+m^*}\right| + |u_{t+m^*} - \alpha_t| \leq 2\varepsilon$ for $t \geq B(\varepsilon)$.

Let now l be an arbitrary positive integer: consider those t such that $-l \leq t \leq B(\varepsilon)$; from the selection of h_r' it follows that $\{u_{t+h_r'}\}_r$ converges uniformly on finite sets of indices to α_t and this is true for the subsequence $\left\{u_{t+h_{r_j}'}\right\}_j$; this last subsequence converges uniformly to α_t for $-l \leq t \leq B(\varepsilon)$. We have already shown uniform convergence for $t > B(\varepsilon)$; hence we have obtained uniform convergence for all $t \geq -l$, i.e., $\left|u_{t+h_{r_j}'} - \alpha_t\right| \leq 2\varepsilon$ for all $t \geq l$ provided $j \geq \rho(\varepsilon)$.

Take now $\varepsilon = \frac{1}{3s}$, $l = s$; since h'_{r_j} depends on ε it will depend on s; $\rho(\varepsilon)$ depends on ε but also on l hence it depends on s. Denote $k_s = h'_{r_{\rho_s}}$; we shall have $|u_{t+k_s} - \alpha_t| < \frac{2}{3s} < \frac{1}{s}$ for $t \geq -s$.

Therefore for $\varepsilon > 0$ and j an arbitrary positive integer we choose $K(\varepsilon) > \max\left(\frac{1}{\varepsilon}, j\right)$; for $s \geq K(\varepsilon)$ we shall have $|u_{t+k_s} - \alpha_t| < \varepsilon$ for any $t \geq -j$, i.e., uniform convergence for $t \geq -j$ is proved.

It remains to show that $\{\alpha_t\}_t$, defined for any integer t is almost periodic. Let $\varepsilon > 0$: we have $|u_{t+p} - u_t| < \varepsilon/2$ for $t \geq B(\varepsilon/2)$, $t + p \geq B(\varepsilon/2)$ hence $|u_{t+p+k_s} - u_{t+k_s}| < \varepsilon/2$ for $t + k_s \geq B(\varepsilon/2)$, $t + p + k_s \geq B(\varepsilon/2)$. Consider now an arbitrary t; we choose s large enough to have $t + k_s \geq B(\varepsilon/2)$, $t + p + k_s \geq B(\varepsilon/2)$, this choice being possible since $\lim_{s \to \infty} k_s = \infty$; we deduce from the above inequality that $|\alpha_{t+p} - \alpha_t| \leq \varepsilon/2 < \varepsilon$ what shows that p is ε-almost period for α_t.

Proof of Theorem 4.19. a) Consider a sequence $\{\varepsilon_s\}_s$, $\varepsilon_s \to 0$ and p_s a ε_s-asymptotic almost period located between s and $s+T(\varepsilon_s)$: we have $|u_{t+p_s} - u_t| < \varepsilon_s$ for $t \geq B(\varepsilon_s)$. But $p_s \geq s$ implies $p_s \to \infty$ and, according to Proposition 4.21 we can select a subsequence p_{s_k} such that $\left\{u_{t+p_{s_k}}\right\}_k$ converges uniformly for $t \geq -j$ to an almost periodic sequence α_t. Denote $\omega_t = u_t - \alpha_t$: we have $|\omega_t| \leq \left|u_t - u_{t+p_{s_k}}\right| + \left|u_{t+p_{s_k}} - \alpha_t\right| < \varepsilon_{s_k} + \left|u_{t+p_{s_k}} - \alpha_t\right|$ for $t \geq B(\varepsilon_{s_k})$. For $k > K(\varepsilon)$ we shall have $\varepsilon_{s_k} < \varepsilon/2$ and $\left|u_{t+p_{s_k}} - \alpha_t\right| < \varepsilon/2$ (Proposition 4.21); denoting $B\left(\varepsilon_{s_{K(\varepsilon)}}\right) = \hat{B}(\varepsilon)$ we deduce $|\omega_t| < \varepsilon$ for $T \geq \hat{B}(\varepsilon)$ what shows that $\lim_{t \to \infty} \omega_t = 0$.

b) Assume we have also $u_t = \alpha'_t + \omega'_t$ where α'_t is almost periodic and $\omega'_t \to 0$. We deduce $\alpha_t - \alpha'_t = \omega'_t - \omega_t$ and since $\omega'_t - \omega_t \to 0$ we deduce that $\alpha_t - \alpha'_t \to 0$. Since $\alpha_t - \alpha'_t$ is almost periodic it remains to show that *an almost periodic sequence tending to zero is identically zero*.

Let β_t be such an almost periodic sequence, $\varepsilon_s \to 0$ and p_s a ε_s-almost period between s and $s + T(\varepsilon_s)$; $\{\beta_{t+p_s}\}_s$ contains a subsequence that converges uniformly with respect to t (Theorem 4.9); let it be $\left\{\beta_{t+p_{s_k}}\right\}_k$. From $p_s \geq s$ we deduce $p_{s_k} \to \infty$ hence $\lim_{k \to \infty} \beta_{t+p_{s_k}} = 0$ uniformly with respect to t. On the other hand we deduce from $\left|\beta_t - \beta_{t+p_{s_k}}\right| < \varepsilon_{s_k}$ by letting $k \to \infty$ that $|\beta_t| \leq 0$; hence $\beta_t = 0$ what ends the proof.

4.3. ALMOST PERIODIC LINEAR OSCILLATIONS

A. We shall consider first the case of the almost periodic solutions for free (autonomous) linear systems.

Proposition 4.22 *Consider the linear system with constant coefficients*

$$x_{t+1} = Ax_t \tag{4.6}$$

under the assumption that A has all its eigenvalues on the unit circle, with simple elementary divisors. Then all solutions of the system are almost periodic.

Proof. The solutions of (4.6) have the form

$$x_t = \sum_{k=1}^{n} c_k e^{i\theta_k t} u_k \ , \ t \in \mathbb{Z}$$

where the eigenvalues $\lambda_k = e^{i\theta_k}$, $k = \overline{1,n}$, are not necessarily distinct, u_k, $k = \overline{1,n}$, are the eigenvectors associated to the eigenvalues λ_k and c_k, $k = \overline{1,n}$, are arbitrary constants.

Since $f : \mathbb{R} \to \mathbb{R}^n$ defined by $f(t) = \sum_{k=1}^{n} c_k e^{i\theta_k t} u_k$ is almost periodic, it follows from Theorem 4.16 that x_t is an almost periodic sequence.

An alternative proof is the following. Since the eigenvalues of A have simple elementary divisors, A can be given by the diagonal form: there exists T with $\det T \neq 0$ such that $A = TDT^{-1}$, $D = diag\left(e^{i\theta_k}\right)$. We have $A^t = TDT^{-1} = T \, diag\left(e^{i\theta_k t}\right) T^{-1}$. Since $e^{i\theta_k t}$ are almost periodic, D^t is almost periodic hence A^t is almost periodic.

As an application of this result we shall consider the equation

$$y_{t+1} - Ay_t + y_{t-1} = 0 \tag{4.7}$$

that has been already considered in Section 2.4 (in fact the multi-dimensional counter-part of the equation considered in Subsection 4.2.1). We may state

Proposition 4.23 *Assume that A is symmetric and has its eigenvalues distinct and located inside the interval $|\lambda| < 2$. Then all solutions of the system are almost periodic.*

Proof. We have already seen in Section 2.4 that equation (4.7) may be written as a system under the form

$$\begin{aligned} x_{t+1} &= y_t \\ y_{t+1} &= -x_t + Ay_t \end{aligned} \tag{4.8}$$

and this system is canonical. Its matrix is symplectic and has distinct eigenvalues that are located on the unit circle (see Proposition 2.15). We apply Proposition 4.22 and the conclusion follows.

    B. We shall consider now the case of the forced almost periodic linear oscillations. Our object of study will be the system

$$x_{t+1} = A_t x_t + f_t , \ t \in \mathbb{Z} \tag{4.9}$$

under the assumption that A_t and f_t are almost periodic. We shall need first a preliminary result.

Lemma 4.24 *If the zero solution of the free system $x_{t+1} = A_t x_t$ is exponentially stable, i.e., there exist $\beta_0 > 0$ and $0 < \rho < 1$ such that $|X_{t,k}| \le \beta_0 \rho^{t-k}$, $t \ge k$, $X_{t,k}$ being the state transition matrix associated to A_t, then for any $\alpha > 0$ such that $0 < \rho + \alpha < 1$ and $\eta > 0$ there exists $\varepsilon_{\eta,\alpha}$ such that if p is an $\varepsilon_{\eta,\alpha}$ -almost period of A_t then $|X_{t+p,k+p} - X_{t,k}| \le \eta (\rho + \alpha)^{t-k}$ for $t \ge k$.*

Proof. Denote $Z_t = X_{t+p,k+p} - X_{t,k}$; we shall have

$$\begin{aligned} Z_{t+1} &= X_{t+p+1,k+p} - X_{t+1,k} = A_{t+p} X_{t+p,k+p} - A_t X_{t,k} = \\ &= A_t Z_t + (A_{t+p} - A_t) X_{t+p,k+p} , \ Z_k = 0 . \end{aligned}$$

Using the formula of variations of constants we obtain

$$Z_t = \sum_{j=k}^{t-1} X_{t,j+1} (A_{t+p} - A_t) X_{j+p,k+p} , \ t \ge k+1 .$$

Taking into account the estimate for $X_{t,k}$ we have

$$\begin{aligned} |Z_t| &\le \left(\sup_j |A_{j+p} - A_j| \right) \sum_{j=k}^{t-1} \beta_0^2 \rho^{t-j-1} \rho^{j-k} = \\ &= \beta_0^2 (t-k) \rho^{t-k-1} \sup_j |A_{j+p} - A_j| . \end{aligned}$$

If p is an $\varepsilon_{\eta,\alpha}$ -almost period it follows that $|Z_t| \le \varepsilon_{\eta,\alpha} \beta_0^2 (t-k) \rho^{t-k-1}$

and $\varepsilon_{\eta,\alpha}$ may be chosen from the inequality

$$\varepsilon_{\eta,\alpha} < \frac{\eta(\rho+\alpha)^{t-k}}{\beta_0^2(t-k)\rho^{t-k-1}}, \quad t \geq k+1$$

and this choice is possible since the sequence $\dfrac{(\rho+\alpha)^t}{t\rho^t}$ tends to infinity and has a positive lower bound.

We are now in position to state and prove the main result on forced almost periodic linear oscillations.

Theorem 4.25 *If the zero solution of the free system $x_{t+1} = A_t x_t$ is uniformly asymptotically stable then for any forcing almost periodic sequence f_t system (4.9) has a unique almost periodic solution satisfying an estimate of the form*

$$|x_t| \leq M \sup_t |f_t| \qquad (4.10)$$

where M is determined by the free system only. Moreover, this solution is exponentially stable.

Remark that this is the almost periodic analogue of Proposition 4.5 of the periodic case.

Proof. Uniform asymptotic stability implies in this case exponential stability hence $|X_{t,k}| \leq \beta_0 \rho^{t-k}$ with $0 < \rho < 1$, $X_{t,k}$ being the state transition matrix associated to A_t. We consider the sequence $\hat{x}_t = \sum_{k=-\infty}^{t-1} X_{t,k+1} f_k$ that is well defined due to the above exponential estimate for $X_{t,k}$. Moreover, this sequence is bounded

$$|\hat{x}_t| \leq \sum_{k=-\infty}^{t-1} |X_{t,k+1}||f_k| \leq$$

$$\leq \beta_0 (\sup_k |f_k|) \sum_{k=-\infty}^{t-1} \rho^{t-k-1} = \frac{\beta_0}{1-\rho} \sup_k |f_k|$$

and is a solution of (4.9)

$$\hat{x}_{t+1} = \sum_{k=-\infty}^{t} X_{t+1,k+1} f_k = \sum_{k=-\infty}^{t-1} X_{t+1,k+1} f_k + f_t =$$

$$= \sum_{k=-\infty}^{t-1} A_t X_{t,k+1} f_k + f_t = A_t \hat{x}_t + f_t .$$

We have to check now that $\hat{x}_t$ is almost periodic. Let $\varepsilon > 0$ and p be an η_ε-almost period for both A_t and f_t, where η_ε will be specified below. We shall have

$$\hat{x}_{t+p} - \hat{x}_t = \sum_{k=-\infty}^{t+p-1} X_{t+p,k+1} f_k - \sum_{k=-\infty}^{t-1} X_{t,k+1} f_k =$$

$$= \sum_{k=-\infty}^{t-1} (X_{t+p,k+p+1} f_{k+p} - X_{t,k+1} f_k)$$

It follows that

$$|\hat{x}_{t+p} - \hat{x}_t| \leq \sum_{k=-\infty}^{t-1} |X_{t+p,k+p+1} - X_{t,k+1}| |f_{k+p}| +$$

$$+ \sum_{k=-\infty}^{t-1} |X_{t,k+1}| |f_{k+p} - f_k|$$

$$\leq \left(\sup_k |f_k| \right) \sum_{k=-\infty}^{t-1} |X_{t+p,k+p+1} - X_{t,k+1}| + \eta_\varepsilon \frac{\beta_0}{1-\rho} .$$

Consider now some $\alpha > 0$ such that $0 < \rho + \alpha < 1$. Since p is an η_ε-almost period for A_t we may obtain the following estimate

$$|X_{t+p,k+p+1} - X_{t,k+1}| \leq \eta_\varepsilon \beta_0^2 (t-k) \rho^{t-k-1}$$

and choosing $\eta_\varepsilon < \dfrac{\tilde{\eta} (\rho + \alpha)^{t-k}}{\beta_0^2 (t-k) \rho^{t-k-1}}, t \geq k+1$ we obtain

$$|X_{t+p,k+p+1} - X_{t,k+1}| \leq \tilde{\eta} (\rho + \alpha)^{t-k}$$

(we applied Lemma 4.24). Therefore,

$$|\hat{x}_{t+p} - \hat{x}_t| \leq \tilde{\eta} \left(\sup_k |f_k| \right) \sum_{k=-\infty}^{t-1} (\rho + \alpha)^{t-k} + \frac{\beta_0}{1-\rho} \eta_\varepsilon \leq$$

$$\leq \frac{\tilde{\eta}}{1-\rho-\alpha} \sup_k |f_k| + \frac{\beta_0}{1-\rho} \eta_\varepsilon .$$

Choose now $\tilde{\eta}_\varepsilon = \dfrac{\varepsilon}{2} \dfrac{1 - \rho - \alpha}{\sup_k |f_k|}$, $\eta_\varepsilon < \min \left\{ \dfrac{\varepsilon (1 - \rho)}{\beta_0}, \min_{t \geq 1} \dfrac{\tilde{\eta}_\varepsilon (\rho + \alpha)^t}{\beta_0^2 t \rho^{t-1}} \right\}$.

This choice, which is possible according to Lemma 4.24 will give $|\hat{x}_{t+p} - \hat{x}_t| < \varepsilon$ and almost periodicity is proved.

Uniqueness follows at once from the fact that an almost periodic sequence with the limit zero is identically zero (see the final part of the proof of Theorem 4.19) in the previous section).

Exponential stability of the unique almost periodic solution follows as in the case of the unique periodic solution (see Section 4.1 E). The proof is complete.

A *second proof* relies on Bochner theory for almost periodic sequences presented in Section 4.2.3. It is considered again the solution $\hat{x}_t = \sum\limits_{k=-\infty}^{t-1} X_{t,k+1} f_k$ that we showed previously to be bounded. We shall show that it is almost periodic using Bochner theory.

Let l' be an arbitrary sequence of integers; from almost periodicity of A_t we deduce existence of a subsequence $l \subset l'$ such that $\tilde{A} = T_l A$ exists uniformly on $\mathbb{Z}$ $\left(\tilde{A} = \lim\limits_{j \to \infty} A_{t+l_j} \right)$. From exponential stability we have

$$|X_{t,k}| = |A_{t-1} A_{t-2} \cdots A_k| \leq \beta_0 \rho^{t-k}, \quad 0 < \rho < 1$$

and therefore

$$\left| X_{t+l_j, k+l_j} \right| = \left| A_{t+l_j-1} \cdots A_{k+l_j} \right| \leq \beta_0 \rho^{t-k}.$$

By letting $j \to \infty$ and denoting $\tilde{X}_{t,k} = \tilde{A}_{t-1} \tilde{A}_{t-2} \cdots \tilde{A}_k$ we obtain $\left| \tilde{X}_{t,k} \right| \leq \beta_0 \rho^{t-k}$ i.e. $\tilde{A}_t$ also defines an exponentially stable evolution.

Let now l', m' be arbitrary sequences of integers and consider common subsequences $l \subset l'$, $m \subset m'$ such that $T_{l+m} A = T_l T_m A$, $T_{l+m} f = T_l T_m f$. From

$$\hat{x}_{t+l_i} = \sum_{k=-\infty}^{t+l_i-1} X_{t+l_i, k+1} f_k = \sum_{s=-\infty}^{t-1} X_{t+l_i, s+l_i+1} f_{s+l_i} =$$

$$= \sum_{s=-\infty}^{t-1} A_{t+l_i-1} \cdots A_{s+l_i} f_{s+l_i}$$

we deduce by letting $i \to \infty$ that

$$\lim_{i \to \infty} \hat{x}_{t+l_i} = (T_l \hat{x})_t = \sum_{s=-\infty}^{t-1} \tilde{A}_{t-1} \cdots \tilde{A}_s \tilde{f}_s =$$

$$= \sum_{s=-\infty}^{t-1} (T_l A)_{t-1} \cdots (T_l A)_s (T_l f)_s$$

and, further

$$(T_m T_l \hat{x})_t = \sum_{s=-\infty}^{t-1} (T_m T_l A)_{t-1} \cdots (T_m T_l A)_s (T_m T_l f)_s =$$

$$= \sum_{s=-\infty}^{t-1} (T_{m+l} A)_{t-1} \cdots (T_{m+l} A)_s (T_{m+l} f)_s = (T_{m+l} \hat{x})_t$$

Applying Theorem 4.18 we obtain almost periodicity of $\hat{x}_t$. Uniqueness and exponential stability follows as previously.

A *third proof* is based on the properties of Liapunov functions for exponentially stable systems.

We consider the free system $x_{t+1} = A_t x_t$ whose zero solution is exponentially stable. In this case a converse Liapunov type theorem (Section 2.5) ensures existence of a sequence of functions $\{V_t(x)\}_t$ such that $|V_t(x) - V_t(\tilde{x})| \leq L|x - \tilde{x}|$, $|x| \leq V_t(x) \leq L|x|$, $V_{t+1}(x_{t+1}) - V_t(x_t) \leq -(1-\rho)V_t(x_t)$ where $0 < \rho < 1$.

Let x_t be some solution of (4.9) and let $y_{k,t}$ be the solution of the free system for $k > t$ satisfying $y_{t,t} = x_t$. From the properties of the Liapunov function we deduce

$$V_{k+1}(y_{k+1,t}) - V_k(y_{k,t}) \leq -(1-\rho)V_k(y_{k,t})$$

and, if $k = t$

$$V_{t+1}(A_t x_t) - V_t(x_t) \leq -(1-\rho)V_t(x_t) .$$

We deduce further

$$V_{t+1}(x_{t+1}) - V_t(x_t) = V_{t+1}(A_t x_t + f_t) - V_t(x_t) =$$
$$= V_{t+1}(A_t x_t + f_t) - V_{t+1}(A_t x_t) + V_{t+1}(A_t x_t) - V_t(x_t) \leq$$
$$\leq L|f_t| - (1-\rho)V_t(x_t) .$$

Assume now that this solution x_t of (4.9) is determined by the initial condition $|x_{t_0}| < K$: for K large enough it will follow that

$V_t(x_t) \leq LK$ for any $t \geq t_0$. Indeed, if this were not true there would exist $t_1 \geq t_0$ such that $V_{t_1}(x_{t_1}) \leq LK$ and $V_{t_1+1}(x_{t_1+1}) > LK$; this will give $V_{t_1+1}(x_{t_1+1}) - V_{t_1}(x_{t_1}) > 0$. On the other hand, we may deduce from the above proved inequality that

$$V_{t_1+1}(x_{t_1+1}) \leq L|f_{t_1}| + \rho V_{t_1}(x_{t_1}) \leq L\sup_t|f_t| + \rho LK .$$

It follows that $K < \sup_t|f_t| + \rho K$ and by choosing $K \geq \dfrac{1}{1-\rho}\sup_t|f_t|$ we obtain a contradiction.

We obtained that every solution starting in the ball $|x| \leq \dfrac{1}{1-\rho}\sup_t|f_t|$ is contained in the ball $|x| \leq \dfrac{L}{1-\rho}\sup_t|f_t|$ for $t \geq t_0$.

Remark that this is valid for linear forced systems that have A_t generating an exponentially stable evolution regardless A_t and f_t are almost periodic or not.

Let now A_t and f_t be almost periodic with A defining an exponentially stable evolution. Let x_t be some solution of (4.9) with $|x_t| \leq \dfrac{1}{1-\rho}\sup_t|f_t|$ (which is bounded as we have just seen). We shall have

$$x_{t+p+1} - x_{t+1} = A_t(x_{t+p} - x_t) + (A_{t+p} - A_t)x_{t+p} + f_{t+p} - f_t$$

Let y_t be the solution of the free system with the initial condition $y_0 = x_p - x_0$ and denote $v_t = x_{t+p} - x_t - y_t$. It follows that $v_0 = 0$ and

$$\begin{aligned} v_{t+1} &= x_{t+p+1} - x_{t+1} - y_{t+1} = \\ &= A_t(x_{t+p} - x_t) + (A_{t+p} - A_t)x_{t+p} + f_{t+p} - f_t - A_t y_t = \\ &= A_t v_t + (A_{t+p} - A_t)x_{t+p} + f_{t+p} - f_t \end{aligned}$$

Using the above estimates we obtain

$$|v_t| \leq \frac{L}{1-\rho}\left(\sup_t|x_t|\sup_t|A_{t+p} - A_t| + \sup_t|f_{t+p} - f_t|\right), \quad t \geq 0$$

Now if p is an η_ε-common almost period for A_t and f_t and η_ε is chosen such that

$$\frac{L}{1-\rho}\left(1 + \frac{L}{1-\rho}\sup_t|f_t|\right)\eta_\varepsilon < \frac{\varepsilon}{2}$$

we obtain $|v_t| < \dfrac{\varepsilon}{2}$ hence $|x_{t+p} - x_t - y_t| < \dfrac{\varepsilon}{2}$ for $t \geq 0$. From exponential stability of the free system we deduce existence of $B(\varepsilon)$ such that $|y_t| < \dfrac{\varepsilon}{2}$ if $t \geq B(\varepsilon)$. It follows that $|x_{t+p} - x_t| \leq |v_t| + |y_t| < \varepsilon$ for $t \geq B(\varepsilon)$ hence x_t *is asymptotic almost periodic*. Accordingly we shall have $x_t = \alpha_t + \omega_t$ where α_t is almost periodic and $\omega_t \to 0$. We shall have $\alpha_{t+1} + \omega_{t+1} = A_t\alpha_t + f_t + A_t\omega_t$. But $A_t\omega_t \to 0$ for $t \to \infty$ and $A_t\alpha_t + f_t$ is almost periodic. It follows that $\alpha_{t+1} = A_t\alpha_t + f_t$; hence α_t is an almost periodic solution of (4.9).

Let now $|x_{t_0}| \leq \dfrac{1}{1-\rho} \sup_t |f_t|$; hence $|x_t| \leq \dfrac{L}{1-\rho} \sup_t |f_t|$ for $t \geq t_0$. Let $\varepsilon > 0$ and take $C(\varepsilon) > 0$ such that if $t > C(\varepsilon)$ $|\omega_t| < \dfrac{\varepsilon}{2}$; hence $|\alpha_t| \leq \dfrac{L}{1-\rho} \sup_t |f_t| + \dfrac{\varepsilon}{2}$.

Consider further an arbitrary integer t and let p be an $\dfrac{\varepsilon}{2}$-almost period of α_t with $p > C(\varepsilon) - t$; it follows that $|\alpha_{t+p}| \leq \dfrac{L}{1-\rho} \sup_t |f_t| + \dfrac{\varepsilon}{2}$ hence $|\alpha_t| < \dfrac{L}{1-\rho} \sup_t |f_t| + \varepsilon$ for any t. Since ε is arbitrary we deduce $|\alpha_t| \leq \dfrac{L}{1-\rho} \sup_t |f_t|$ - the estimate obtained previously. Uniqueness and exponential stability of the almost periodic solution are obtained in the same way as in the other proofs.

C. We shall consider now the case when the free system has almost periodic solutions. Unlike the periodic case there are no longer general results concerning existence of forced almost periodic oscillations. Nevertheless even partial results may be useful.

Proposition 4.26 *If system (4.9) has an almost periodic solution then* $\displaystyle \lim_{T \to \infty} \sum_{t=1}^{T-1} y_t x_t = 0$ *for any almost periodic solution y_t (a row vector) of the adjoint system* $y_{t-1} = y_t A_{t-1}$.

Proof. We obtained in Section 4.1 (while proving Proposition 4.2) the following formula

$$y_T x_T - y_0 x_0 = \sum_{t=1}^{T-1} y_t x_t$$

which holds for any solution x_t of (4.9) and any solution y_t of the adjoint system. If x_t and y_t are almost periodic they are bounded hence $\lim_{T\to\infty} \frac{1}{T}(y_T x_T - y_0 x_0) = 0$ and the proposition is proved.

It is clear that Proposition 4.26 contains a necessary condition of almost periodicity. In some cases this condition is also sufficient. We shall give below such a case, the discrete counter-part of a result due to Mal'kin (1954).

Theorem 4.27 *Consider system (4.9) under the following assumptions: i)* $A_t \equiv A_{t+p}$; *ii)* $f_t = \sum_{k=1}^{m} (z_k)^t \varphi_t^k$ *where* $z_k, k = \overline{1,m}$ *are distinct and located on the unit circle and* $\{\varphi_t^k\}_t$, $k = \overline{1,m}$ *are p -periodic sequences. If for any almost periodic solution of the adjoint system* $y_{t-1} = y_t A_{t-1}$ *condition* $\lim_{T\to\infty} \frac{1}{T} \sum_{t=0}^{T-1} y_{t+1} f_t = 0$ *holds, then system (4.9) has an almost periodic solution of the form* $x_t = \sum_{k=1}^{m} (z_k)^t \psi_t^k$ *where* $\{\psi_t^k\}_t$, $k = \overline{1,m}$ *are p -periodic.*

Proof. Consider first the system

$$x_{t+1} = A_t x_t + (z_k)^t \varphi_t^k . \tag{4.11}$$

This system has a solution $\hat{x}_t^k = (z_k)^t \psi_t^k$ where ψ_t^k is p -periodic. Indeed, if $\{x_t^k\}_t$ is a solution of (4.11) then $y_t^k = (z_k)^{-1} x_t^k$ satisfies

$$y_{t+1}^k = (z_k)^{-t-1} x_{t+1}^k = z_t^{-t-1} [A_t x_t^k + z_k^t \varphi_t^k] = \frac{1}{z_k} A_t y_t^k + \varphi_t^k \tag{4.12}$$

which is periodic. According to Proposition 4.3 this system has a periodic solution if and only if $\sum_{t=0}^{p-1} u_{t+1}^k \varphi_t^k = 0$ for any periodic solution of the adjoint system $u_t = u_{t+1} \frac{1}{z_k} A_t$. But if u_t^k is a periodic solution of this system then the system has a multiplier equal to 1 hence z_k^p has to be a multiplier of the free system $x_{t+1} = A_t x_t$. In this case $v_t^k = (z_k)^{-t} u_t^k$ is a solution of the adjoint system $v_t = v_{t+1} A_t$ that is almost periodic and z_k^{-p} is a multiplier of this system. Indeed we have

$$v_{t+1}^k = (z_k)^{-t-1} u_{t+1}^k; \quad v_{t+1}^k A_t = (z_k)^{-t} u_{t+1}^k z_k^{-1} A_t = z_k^{-1} u_t^k = v_t^k$$

which shows that v_t^k is a solution to the adjoint system and

$$v_{t+p}^k = (z_k)^{-t-p} u_{t+p}^k = (z_k)^{-p} (z_k)^{-t} u_t^k = (z_k)^{-p} v_t^k$$

which shows that z_k^{-p} is a multiplier of $v_t = v_{t+1} A_t$ (we made use of the periodicity of u_t^k).

It is now clear that we have to consider two distinct cases. First, if $(z_k)^p$ is *not* a multiplier of the free system $x_{t+1} = A_t x_t$ then the system $u_t = u_{t+1} z_k^{-1} A_t$ has no periodic solution; hence (4.12) has a periodic solution $\left\{ \psi_t^k \right\}_t$. We deduce that $(z_k)^t \psi_t^k$ is a solution of (4.11). Next, if $(z_k)^p$ *is* a multiplier of the free system $x_{t+1} = A_t x_t$ then $(z_k)^{-p}$ is a multiplier of the adjoint system $v_t = v_{t+1} A_t$ and this system has an almost periodic solution $v_t^k = (z_k)^{-t} u_t^k$, where u_t^k is a periodic solution of the system $u_t = u_{t+1} z_k^{-1} A_t$. We shall have

$$\lim_{T \to \infty} \frac{1}{T} \sum_{t=0}^{T-1} v_{t+1}^k \left[f_t - (z_k)^t \varphi_t^k \right] =$$

$$= \lim_{T \to \infty} \frac{1}{T} \sum_{t=0}^{T-1} (z_k)^{-t-1} u_{t+1}^k \sum_{\substack{r=1 \\ r \neq k}}^{m} (z_k)^t \varphi_t^r =$$

$$= \sum_{\substack{r=1 \\ r \neq k}}^{m} \lim_{T \to \infty} \frac{1}{T} \sum_{t=0}^{T-1} z_k^{-1} \left(\frac{z_r}{z_k} \right)^t u_{t+1}^k \varphi_t^r = 0$$

(each limit under the sum with respect to r equals 0 since u_t^k and φ_t^r are periodic and bounded and $z_r z_k^{-1}$ is a complex number located on the unit circle).

Since $\lim_{T \to \infty} \dfrac{1}{T} \sum_{t=0}^{T-1} v_{t+1}^k f_t = 0$ we deduce that

$$\lim_{T \to \infty} \frac{1}{T} \sum_{t=0}^{T-1} v_{t+1}^k z_k^t \varphi_t^k = 0;$$

hence $\lim_{T \to \infty} \dfrac{1}{T} \sum_{t=0}^{T-1} u_{t+1}^k \varphi_t^k = 0$. But u_t^k and φ_t^k are p-periodic and we may write

$$\lim_{T \to \infty} \frac{1}{T} \sum_{t=0}^{T-1} u_{t+1}^k \varphi_t^k = \sum_{t=0}^{p-1} u_{t+1}^k \varphi_t^k = 0$$

and this condition holds for each periodic solution of the system $u_t = u_{t+1} z_k^{-1} A_t$. Proposition 4.3 gives that (4.12) has a periodic solution ψ_t^k and $\hat{x}_t^k = (z_k)^t \psi_t^k$ is a solution of (4.11). The proof may be repeated for each k and the theorem is proved.

4.4. PERIODIC AND ALMOST PERIODIC OSCILLATIONS IN NONLINEAR SYSTEMS

4.4.1. Periodic Oscillations

A. Consider the quasi-linear system

$$x_{t+1} = A_t x_t + f_t (x_t) \tag{4.13}$$

where both A_t and $f_t (\cdot)$ are T -periodic. We shall assume that the linear system $x_{t+1} = A_t x_t$ has no T -periodic solution except the zero solution. In this case the Green matrix defined in Section 4.1 by

$$G_{t,k} = \begin{cases} X_{t,0} (\mathbf{I} - X_T)^{-1} X_{T,k} + X_{t,k} & , \quad k \le t \\ X_{t,0} (\mathbf{I} - X_T)^{-1} X_{T,k} & , \quad t < k < T \end{cases}$$

will be used to define the equation

$$x_t = \sum_{k=0}^{T-1} G_{t,k+1} f_k (x_k) . \tag{4.14}$$

As follows from the results of Section 4.1, any periodic solution of (4.13) will be a solution of (4.14). Conversely, a solution of (4.14) is a solution of (4.13) and is periodic. Indeed we have

$$x_t = X_{t,0} (\mathbf{I} - X_T)^{-1} \sum_{k=0}^{T-1} X_{T,k+1} f_k (x_k) + \sum_{k=0}^{t-1} X_{t,k+1} f_k (x_k)$$

for any solution of (4.14). We deduce that

$$x_0 = (\mathbf{I} - X_T)^{-1} \sum_{k=0}^{T-1} X_{T,k+1} f_k (x_k)$$

and

$$x_{t+1} = X_{t+1,0} (\mathbf{I} - X_T)^{-1} \sum_{k=0}^{T-1} X_{T,k+1} f_k (x_k) + \sum_{k=0}^{t} X_{t+1,k+1} f_k (x_k) =$$

$$= A_t X_{t,0} (\mathbf{I} - X_T)^{-1} \sum_{k=0}^{T-1} X_{T,k+1} f_k (x_k) +$$

$$+ \sum_{k=0}^{t-1} A_t X_{t,k+1} f_k (x_k) + f_t (x_t) = A_t x_t + f_t (x_t) ;$$

hence x_t, solution of (4.14), is a solution of (4.13) with the above initial condition. We have also

$$x_T = X_T (\mathbf{I} - X_T)^{-1} \sum_{k=0}^{T-1} X_{T,k+1} f_k (x_k) + \sum_{k=0}^{T-1} X_{T,k+1} f_k (x_k) =$$

$$= \left(X_T (\mathbf{I} - X_T)^{-1} + \mathbf{I} \right) \sum_{k=0}^{T-1} X_{T,k+1} f_k (x_k) =$$

$$= (\mathbf{I} - X_T)^{-1} \sum_{k=0}^{T-1} X_{T,k+1} f_k (x_k) = x_0$$

which shows periodicity of the considered solution x_t.
   If $f_t (0) = g_t$, $f_t (x) - f_t (0) = \varepsilon F_t (x)$ then system (4.13) becomes

$$x_{t+1} = A_t x_t + g_t + \varepsilon F_t (x_t) \qquad (4.15)$$

and equation (4.14) reads

$$x_t = \sum_{k=0}^{T-1} G_{t,k+1} g_k + \varepsilon \sum_{k=0}^{T-1} G_{t,k+1} F_k (x_k) . \qquad (4.16)$$

Denote $\tilde{x}_t = \sum_{k=0}^{T-1} G_{t,k+1} g_k$; $\tilde{x}_t$ represents the unique periodic solution of the linear system $x_{t+1} = A_t x_t + y_t$, corresponding to $\varepsilon = 0$; equation

(4.16) becomes

$$x_t = \tilde{x}_t + \varepsilon \sum_{k=0}^{T-1} G_{t,k+1} F_k \left(x_k \right) . \tag{4.17}$$

For this equation a solution may be constructed by iterations as follows

$$x_t^0 = \tilde{x}_t , \ x_t^i = \tilde{x}_t + \varepsilon \sum_{k=0}^{T-1} G_{t,k+1} F_k \left(x_k^{i-1} \right) , \ i = 1, 2, ...$$

For this sequence $\left\{ x_t^i \right\}_i$ of iterations to be well defined we have to assume that $F_t \left(\cdot \right)$ is defined in a neighborhood of $\tilde{x}_t$.

We deduce

$$\left| x_t^i - \tilde{x}_t \right| \leq \varepsilon M L , \ M = \sup_{0 \leq t \leq T-1} \sum_{k=0}^{T-1} \left| G_{t,k+1} \right| , \ L = \sup_{\substack{|x - \tilde{x}_t| \leq K \\ 0 \leq t \leq T-1}} \left| F_t \left(x \right) \right| ,$$

where the closed ball $\{ x : |x - \tilde{x}_t| \leq K \}$ is included in the neighborhood of $\tilde{x}_t$ where $F_t \left(\cdot \right)$ is defined.

If $F_t \left(\cdot \right)$ is Lipschitz then we shall have

$$\left| x_t^{i+1} - x_t^i \right| \leq \left| \varepsilon \right| M L_1 \sup_{0 \leq t \leq T-1} \left| x_t^i - x_t^{i-1} \right| ,$$

where L_1 is the Lipschitz constant. From this inequality we obtain that

$$\left| x_t^{i+1} - x_t^i \right| \leq \left(\left| \varepsilon \right| M L_1 \right)^i \sup_K \left| x_t^1 - \tilde{x}_t \right| \leq K \left(\left| \varepsilon \right| M L_1 \right)^i$$

and convergence is ensured if $\left| \varepsilon \right| < \dfrac{\rho}{M L_1}, 0 < \rho < 1.$

We obtained the following result

Proposition 4.28 *If the free system $x_{t+1} = A_t x_t$ has no other periodic solution than the zero one and $F_t \left(\cdot \right)$ is defined in a neighborhood of $\tilde{x}_t$, the unique periodic solution of the unperturbed linear system $x_{t+1} = A_t x_t + g_t$, and if it is Lipschitz in any compact set contained in this neighborhood then for $|\varepsilon|$ small enough system (4.15) has a unique periodic solution that may be obtained by iterations.*

Remarks. 1. If A_t defines an exponentially stable evolution (system's multipliers are inside the unit disk) then $\tilde{x}_t^\varepsilon$ - the unique periodic solution of (4.15) - is exponentially stable. This result may be obtained by considering perturbed solutions (with respect to the basic one, $\tilde{x}_t^\varepsilon$), writing down the system in perturbations and applying stability by the first approximation.

2. If we do not require uniqueness and construction by iterations then existence conditions for the periodic solutions may be weakened. Assuming, for instance, that $f_t(x)$ *satisfies the estimate* $|f_t(x)| \leq \beta(|x|)$ *with* $\beta(r)$ *monotonically increasing and that there exists* α_0 *with the property* $\beta(\alpha_0)/\alpha_0 \leq M^{-1}$ then $|x_t| \leq \alpha_0$ will imply

$$|f_t(x_t)| \leq \beta(\alpha_0) \leq \alpha_0 M^{-1} , \quad \left| \sum_{k=0}^{T-1} G_{t,k+1} f_k(x_k) \right| \leq M\alpha_0 M^{-1} = \alpha_0$$

We deduce that the operator associating to the vector $x_0, x_1, ..., x_{T-1}$ the vector $y_0, y_1, ..., y_{T-1}$ defined by $y_t = \sum_{k=0}^{T-1} G_{t,k+1} f_k(x_k)$, $t = \overline{0, T-1}$, applies the set defined by $\sup_{0 \leq t \leq T-1} |x_t| \leq \alpha_0$ in itself. If $f_t(\cdot)$ is continuous the operator is continuous; from the fixed point theorem of Brouwer we deduce existence of a fixed point of the operator; for this fixed point we have

$$x_t = \sum_{k=0}^{T-1} G_{t,k+1} f_k(x_k)$$

and as we have seen previously this equation defines a periodic solution of (4.13).

B. We shall consider now a more general class of nonlinear systems described by

$$x_{t+1} = f_t(x_t) + \varepsilon F_t(x_t, \varepsilon) , \tag{4.18}$$

where $f_t(x)$ and $F_t(x)$ are T-periodic with respect to t. If the system corresponding to $\varepsilon = 0$ has a periodic solution $\tilde{x}_t$ we may ask whether this property still holds for $|\varepsilon|$ sufficiently small. The result is the following

Theorem 4.29 *If the unperturbed system* $x_{t+1} = f_t(x_t)$ *with* $\{f_t(x)\}_t$ *being* T-*periodic has a periodic solution* $\{\tilde{x}_t\}_t$ *that is asymptotically stable then for* $|\varepsilon|$ *sufficiently small system* (4.18) *has a periodic solution* $\{x_t^\varepsilon\}_t$; *moreover* $\lim_{\varepsilon \to 0} x_t^\varepsilon = \tilde{x}_t$.

Proof. We may assume $\tilde{x}_t \equiv 0$ since a suitable change of variable will always reduce the problem to this case. Uniform asymptotic stability will give existence of $\delta_0 > 0$ and $\delta(\eta)$, $K(\eta)$ such that $|y_0| < \delta(\eta)$ implies $|y_t| < \eta$ and $|y_0| < \delta_0$, $t \geq K(\eta)$ imply $|y_t| < \eta$, where y_t is the solution of the unperturbed system $y_{t+1} = f_t(y_t)$. Let $\delta_1(\alpha) = \delta\left(\frac{1}{2}\alpha\right)$, $0 < \zeta < \min\left\{\delta_0, \delta\left(\frac{1}{2}\delta_1\right)\right\}$, $K_0 = K\left(\frac{1}{2}\delta\left(\frac{1}{2}\zeta\right)\right)$, m the smallest positive integer such that $mT > K_0$. Let $B = \{x : |x| < \delta_1(\alpha)\}$ - the ball of radius $\delta_1(\alpha)$; if $x_0 \in B$ then $|y_t(x_0)| < \frac{1}{2}\alpha$. If $|\varepsilon|$ is sufficiently small then it follows that $|x_t^\varepsilon(x_0)| < \alpha$ for $x_0 \in B$, $0 \leq t \leq mT$, where $x_t^\varepsilon(x_0)$ is the solution of the perturbed system (4.18) which equals x_0 at $t = 0$. If f_t and F_t are defined only in a neighborhood of $x = 0$ for α sufficiently small then the solution $x_t^\varepsilon(x_0)$ will remain in this neighborhood for $0 \leq t \leq mT$ hence it may be constructed up to $t = mT$. Denote $\varphi^\varepsilon(x_0) = x_T^\varepsilon(x_0)$; the mapping $\varphi^\varepsilon(\cdot)$ and its iterates $\varphi_j^\varepsilon(\cdot)$, $j = \overline{2, m}$ are defined on B. Let B_1 be the ball $B_1 = \{x : |x| < \zeta\}$ and B_0 the ball $B_0 = \left\{x : |x < \frac{3}{4}\delta\left(\frac{1}{2}\zeta\right)|\right\}$. For $x_0 \in B_1$ and $t \geq 0$ we have $|y_1(x_0)| < \frac{1}{2}\delta_1(\alpha)$ since $\zeta < \delta\left(\frac{1}{2}\delta_1(\alpha)\right)$; for $|\varepsilon|$ small enough it follows that $|x_t^\varepsilon(x_0)| < \delta_1(\alpha)$, $0 \leq t \leq mT$ hence $\varphi_j^\varepsilon(B_1) \subset B$, $j = \overline{1, m}$.

For $x_0 \in B_0$ we deduce that $|y_t(x_0)| < \frac{1}{2}\zeta$, $t \geq 0$; for $|\varepsilon|$ small enough it follows that $|x_t^\varepsilon(x_0)| < \zeta$, $0 \leq t \leq mT$ hence $\varphi_j^\varepsilon(B_0) \subset B$, $j = \overline{0, m}$. From $\zeta < \delta_0$ it follows that $x_0 \in B_1$, $t > T$ imply $|y_t(x_0)| < \frac{1}{2}\delta\left(\frac{1}{2}\zeta\right)$ hence for $|\varepsilon|$ small enough $|x_{mT}^\varepsilon(x_0)| \leq \frac{3}{4}\delta\left(\frac{1}{2}\zeta\right)$ i.e. $\varphi_m^\varepsilon(B_1) \subset B_0$.

We may now use the fixed-point Theorem of Browder (1959) and deduce that φ^ε has a fixed point in B_0. This fixed point is the initial condition of a periodic solution. Since the radius of B_0 may be taken arbitrarily small it follows that the periodic solution tends to zero for $\varepsilon \to 0$ and the proof is complete.

4.4.2. Almost Periodic Oscillations

A. We shall prove first a perturbation result which is analogous to Theorem 4.29 but for a linear unperturbed system.

Theorem 4.30 *If the zero solution of the almost periodic system $x_{t+1} = A_t x_t$ is exponentially stable then for $|\varepsilon|$ sufficiently small system (4.15) - with F_t and g_t almost periodic - has a unique almost periodic solution and this solution may be constructed by iterations provided $F_t(x)$ is Lipschitz uniformly with respect to t. Moreover this almost periodic solution is exponentially stable.*

Proof. We shall apply Theorem 4.25 to obtain for $\varepsilon = 0$ existence of a unique almost periodic solution $\tilde{x}_t$ of the linear system $x_{t+1} = A_t x_t + g_t$. For the iterations we take $x_t^0 = \tilde{x}_t$ and x_t^j the unique almost periodic solution of the system

$$x_{t+1} = A_t x_t + g_t + \varepsilon F_t \left(x_t^{j-1} \right),$$

We deduce that $x_t^j - \tilde{x}_t$ is the unique almost periodic solution of the system

$$x_{t+1} = A_t x_t + \varepsilon F_t \left(x_t^{j-1} \right).$$

We show further that for $|\varepsilon|$ small enough the iterations do not leave a given neighborhood of $\tilde{x}_t$. Indeed we have from the estimate satisfied by the unique almost periodic solution of a forced linear almost periodic system (Section 4.3)

$$\left| x_t^j - \tilde{x}_t \right| \le M\varepsilon \sup_t \left| F_t \left(x_t^{j-1} \right) \right|,$$

where M depends on the free system $x_{t+1} = A_t x_t$ only. Let $M_1 = \sup_t |F_t(x)|$ on the compact set $|x - \tilde{x}_t| \le K$. If we assume $\left| x_t^{j-1} - \tilde{x}_t \right| \le K$ then $\left| x_t^j - \tilde{x}_t \right| \le \varepsilon M M_1 < K$ if $|\varepsilon| < \dfrac{K}{M_1 M}$; by induction we deduce $\left| x_t^j - \tilde{x}_t \right| \le K$ for all $j \ge 0$.

If $F_t(x)$ is uniformly Lipschitz then we deduce that $\left| x_t^{j+1} - x_t^j \right| \le \varepsilon M L \sup_k \left| x_t^j - x_t^{j-1} \right|$ and this will give uniform convergence of the iterations in the same way as in the periodic case; the limit of the sequence $\left\{ x_t^j \right\}_j$ is an almost periodic solution of the perturbed system. The same argument as in the periodic case shows that this solution is exponentially stable.

B. We shall obtain now a rather general result for nonlinear systems reducing existence of almost periodic solutions to existence of a bounded solution with certain stability properties.

Theorem 4.31 *Consider the system $x_{t+1} = f_t(x_t)$ with $f_t(\cdot)$ continuous and $\{f_t(x)\}_t$ being T-periodic. If this system has a bounded uniformly stable solution, this solution is asymptotic almost periodic and the system has an almost periodic solution.*

Proof. Let $\tilde{x}_t$ be the bounded uniformly stable solution of the system: there exists $\delta(\varepsilon)$ such that $|x_{t_0} - \tilde{x}_{t_0}| < \delta(\varepsilon)$ implies $|x_t - \tilde{x}_t| < \varepsilon$ for $t \geq t_0$. Taking $t_0 = kT$ where T is the period of the system we deduce that $|x_{kT} - \tilde{x}_{kT}| < \delta(\varepsilon)$ implies $|x_{t+kT} - \tilde{x}_{t+kT}| < \varepsilon$ for $t \geq 0$. The given system being T-periodic it follows that $x_{t+kT} = x_t(x_{kT})$ hence $|y - \tilde{x}_{kT}| < \delta(\varepsilon)$ implies $|x_t(y) - \tilde{x}_{t+kT}| < \varepsilon$ for $t \geq 0$.

Consider now the mapping $Tx_0 = x_T(x_0)$; we deduce that $x_t(Tx_0) = x_t(x_T(x_0)) = x_{t+T}(x_0)$ and further that $x_t(T^k x_0) = x_{t+kT}(x_0)$, $T^k x_0 = x_{kT}(x_0)$. From $|x_t(y) - \tilde{x}_{t+kT}| < \varepsilon$ it follows that $|x_{lT}(y) - \tilde{x}_{(l+k)T}| < \varepsilon$ hence $|y - T^k \tilde{x}_0| < \delta(\varepsilon)$ implies $|T^l y - T^{l+k} \tilde{x}_0| < \varepsilon$.

Let now $\{h_k\}_k$, $\lim_{k \to \infty} h_k = \infty$, be a sequence of positive integers. The set $\{T^{h_k} \tilde{x}_0\}_k$ is relatively compact since the solution $\tilde{x}_t$ is bounded. Let k_m be a subsequence such that $\lim_{m \to \infty} T^{k_m} \tilde{x}_0 = \hat{x}_0$; for m sufficiently large we have $|T^{k_m} \tilde{x}_0 - \hat{x}_0| < \delta(\varepsilon)$ hence $|T^l \hat{x}_0 - T^{l+k_m} \tilde{x}_0| < \varepsilon$, i.e., $\lim_{m \to \infty} T^{l+k_m} \tilde{x}_0 = T^l \hat{x}_0$. We proved that mapping T has the property that from any sequence $\{T^{l+h_m} \tilde{x}_0\}_m$ we can select a sequence $\{T^{l+k_m} \tilde{x}_0\}_m$ that is uniformly convergent with respect to l hence it is asymptotic almost periodic at $\tilde{x}_0$. We obtained that sequence $\{\tilde{x}_{kT}\}_k$ is asymptotic almost periodic, i.e., for any $\eta > 0$ there exists $l(\eta)$, $K(\eta)$ such that among $l + 1$ consecutive integers there exists an integer p such that, for $t \geq K(\eta)$, $|\tilde{x}_{(k+p)T} - \tilde{x}_{kT}| < \eta$. Consider a sufficiently large t, $t = t' + kT$, $0 \leq t' < T$; we have $\tilde{x}_t = \tilde{x}_{t'+kT} = \tilde{x}_{t'}(\tilde{x}_{kT}) = \tilde{x}_{t'}(T^k \tilde{x}_0)$.

Let now $\varepsilon > 0$; there exists $\eta > 0$ such that $|x - y| < \eta$ implies $|x_t(x) - x_t(y)| < \varepsilon$ for $0 \leq t \leq T$. Let $t > K(\eta)$; we have

$$|\tilde{x}_t - \tilde{x}_{t+pT}| = |\tilde{x}_{t'}(\tilde{x}_{kT}) - \tilde{x}_{t'}(\tilde{x}_{(k+p)T})| < \varepsilon$$

In fact for $\varepsilon > 0$ there exist L_ε and l_ε such that between $(l_\varepsilon + 1)T$ consecutive integers one can find an integer pT such that if $t \geq L_\varepsilon$ then $|\tilde{x}_t - \tilde{x}_{t+pT}| < \varepsilon$ nothing else but asymptotic almost periodicity of $\tilde{x}_t$.

We have further $\tilde{x}_t = \alpha_t + \omega_t$ with α_t almost periodic and $\lim_{t \to \infty} \omega_t = 0$; also $\alpha_{t+1} + \omega_{t+1} = f_t(\alpha_t + \omega_t) = f_t(\alpha_t) + [f_t(\alpha_t + \omega_t) - f_t(\alpha_t)]$. Remark that $\lim_{t \to \infty} [f_t(\alpha_t + \omega_t) - f_t(\alpha_t)] = 0$ hence $\alpha_{t+1} = f_t(\alpha_t)$ and the system has an almost periodic solution.

4.5. AVERAGING

A. We already know that both periodic and almost periodic sequences allow a mean (average) value that is well defined. The method of averaging consists in describing some results on periodic and almost periodic systems using the properties of the associated "averaged" system whose right hand side is the average of the basic system's right hand side. Such approach is sometimes convenient, e.g., when the system is linear and the associated "averaged" system is linear with constant coefficients. In the following we shall give some applications of the averaging. With respect to this we prove first a preliminary result - the discrete analogue of a lemma belonging to Bogoliubov.

Lemma 4.32   *Let* $\{f_k(x)\}_k$ *be a sequence of functions* $f_k : E \to \mathbb{R}$. *E being a compact set from the metric space* (X, ρ). *Assume that:*

i) for any $x \in E$ *we have* $\displaystyle \lim_{N \to \infty} \frac{1}{N} \sum_{i=k}^{k+N-1} f_i(x) = 0$ *uniformly with respect to* k; *ii) the functions are bounded and Lipschitz on* E. *i.e..* $|f_k(x)| < M$, $|f_k(x') - f_k(x'')| \le \lambda \rho(x', x'')$. *Let* $f_k^\eta(x) = \displaystyle \sum_{i=-\infty}^{k} \eta^{k-i} f_i(x)$, $0 < \eta < 1$. *Then* $|f_k^\eta(x)| \le \dfrac{\zeta(\eta)}{1-\eta}$ *where* $\zeta(\eta)$ *is such that* $\displaystyle \lim_{\eta \to 1} \zeta(\eta) = 0$.

Proof.   Under the assumption *ii)* the limit in the lemma will be uniform with respect to k and $x \in E$. This fact will be useful in what follows. We have

$$
\begin{aligned}
f_k^\eta(x) &= \sum_{l=0}^{\infty} \eta^l f_{k-l}(x) = \sum_{j=0}^{\infty} \sum_{l=jN}^{(j+1)N-1} \eta^l f_{k-l}(x) = \\
&= \sum_{j=0}^{\infty} \eta^{jN} \sum_{l=jN}^{(j+1)N-1} \eta^{l-jN} f_{k-l}(x) = \\
&= \sum_{j=0}^{\infty} \eta^{jN} \sum_{l=jN}^{(j+1)N-1} \left[f_{k-l}(x) + f_{k-l}(x) \left(\eta^{l-jN} - 1 \right) \right].
\end{aligned}
$$

It follows that

$$
|f_k^\eta (x)| \;\leq\; \sum_{j=0}^{\infty} \eta^{jN} \left| \sum_{l=jN}^{(j+1)N-1} f_{k-l}(x) \right| +
$$

$$
+ M \sum_{j=0}^{\infty} \eta^{jN} \sum_{l=jN}^{(j+1)N-1} \left(1 - \eta^{l-jN}\right)
$$

$$
= \sum_{j=0}^{\infty} \eta^{jN} \left| \sum_{l=jN}^{(j+1)N-1} f_{k-l}(x) \right| + M \sum_{j=0}^{\infty} \eta^{jN} \sum_{i=0}^{N-1} \left(1 - \eta^{i}\right) =
$$

$$
= \sum_{j=0}^{\infty} \eta^{jN} \left| \sum_{l=jN}^{(j+1)N-1} f_{k-l}(x) \right| + \frac{MN}{1-\eta^N} - \frac{M\eta}{1-\eta} \leq
$$

$$
\leq \sum_{j=0}^{\infty} \eta^{jN} N \varepsilon (N) + \frac{MN}{1-\eta^N} - \frac{M\eta}{1-\eta} ,
$$

where $\varepsilon (N) = \left| \dfrac{1}{N} \sum_{l=k}^{k+N-1} f_k (x) \right|$; hence $\lim\limits_{N \to \infty} \varepsilon (N) = 0$ according to the assumption.

Let $\xi = 1 - \eta$, $N(\xi) = 1 + \left[\frac{1}{\sqrt{\xi}}\right]$, $[\cdot]$ being the entire part of the quantity under braces; it follows that $\lim\limits_{\xi \to 0} N(\xi) = \infty$ hence $\lim\limits_{\xi \to 0} \varepsilon (N(\xi)) = 0$. On the other hand $\frac{N-1}{2}\xi = \frac{1}{2}\xi \left[\frac{1}{\sqrt{\xi}}\right] < \frac{1}{2}\sqrt{\xi}$ hence $\lim\limits_{\xi \to 0} \frac{N-1}{2}\xi = 0$; it follows that $\frac{N(\xi)-1}{2}\xi < 1$ if ξ is sufficiently small. From the inequality

$$
1 - \eta^N = 1 - (1 - \xi)^N \geq N\xi \left(1 - \frac{N-1}{2}\xi\right)
$$

we deduce that

$$
\frac{MN}{1-\eta^N} - \frac{M\eta}{1-\eta} \;\leq\; \frac{M}{\xi\left(1 - \frac{N-1}{2}\xi\right)} - \frac{M(1-\xi)}{\xi} =
$$

$$
= \frac{M(N+1) - (N-1)\xi}{2} \cdot \frac{1}{1 - \frac{N-1}{2}\xi}
$$

and therefore

$$|f_k^{\eta}(x)| \leq \frac{1}{\xi\left(1 - \frac{N(\xi)-1}{2}\xi\right)} \left[\varepsilon\left(N(\xi)\right) + \frac{M\left(N(\xi)+1\right)\xi}{2}\right]$$

Denoting $\zeta(\eta) = \frac{\varepsilon(N(1-\eta))}{1 - \frac{N(1-\eta)-1}{2}(1-\eta)} + \frac{M(N(1-\eta)+1)(1-\eta)}{2\left(1 - \frac{N(1-\eta)-1}{2}(1-\eta)\right)}$ we shall obvi-

ously have $\lim_{\eta \to 1} \zeta(\eta) = 0$ and $|f_k^{\eta}(x)| \leq \frac{\zeta(\eta)}{1-\eta}$ and the lemma is proved.

We shall give now, using the averaging, a simple criterion of uniform asymptotic stability.

Proposition 4.33 *Consider the system*

$$x_{t+1} = x_t + \varepsilon A_t x_t \qquad (4.19)$$

and let $\tilde{A} = \lim_{T \to \infty} \frac{1}{T} \sum_{t=1}^{T} A_t$ be the mean value of the almost periodic

sequence $\{A_t\}_t$. If the eigenvalues of $\tilde{A}$ have negative real parts then system (4.19) is exponentially stable provided $\varepsilon > 0$ is small enough.

Proof. Let $B_t = A_t - \tilde{A}$, $B_t^{1-\varepsilon} = \sum_{k=-\infty}^{t-1} (1-\varepsilon)^{t-k-1} B_k$. Applying

the lemma we obtain that $|B_t^{1-\varepsilon}| \leq \frac{\zeta(1-\varepsilon)}{\varepsilon}$. If ε is sufficiently small

then $\mathbf{I} + \varepsilon B_t^{1-\varepsilon}$ is invertible and v_t from $x_t = \left(\mathbf{I} + \varepsilon B_t^{1-\varepsilon}\right) v_t$ is well defined. We shall have

$$\left(\mathbf{I} + \varepsilon B_{t+1}^{1-\varepsilon}\right) v_{t+1} = \left(\mathbf{I} + \varepsilon B_t^{1-\varepsilon}\right) v_t + \varepsilon A_t \left(\mathbf{I} + \varepsilon B_t^{1-\varepsilon}\right) v_t$$

On the other hand

$$B_{t+1}^{1-\varepsilon} = \sum_{k=-\infty}^{t} (1-\varepsilon)^{t-k} B_k = (1-\varepsilon) B_t^{1-\varepsilon} + B_t$$

We have further

$$\begin{aligned}\left(\mathbf{I} + \varepsilon B_{t+1}^{1-\varepsilon}\right) v_{t+1} &= \left(\mathbf{I} + \varepsilon B_t^{1-\varepsilon} + \varepsilon B_t - \varepsilon^2 B_t^{1-\varepsilon}\right) v_{t+1} = \\ &= \left(\mathbf{I} + \varepsilon B_t^{1-\varepsilon}\right) v_t + \varepsilon A_t \left(\mathbf{I} + \varepsilon B_t^{1-\varepsilon}\right) v_t\end{aligned}$$

For ε small enough the matrix $\mathbf{I} + \varepsilon B_t + \left(\varepsilon - \varepsilon^2\right) B_t^{1-\varepsilon}$ is invertible. Since $\left|\varepsilon B_t^{1-\varepsilon}\right| \leq \zeta(1-\varepsilon)$ with $\lim_{\varepsilon \to 0} \zeta(1-\varepsilon) = 0$ (Lemma 4.32) we may write down the following expansion

$$\left(\mathbf{I} + \varepsilon B_t + \varepsilon B_t^{1-\varepsilon} - \varepsilon^2 B_t^{1-\varepsilon}\right)^{-1} = \mathbf{I} + D_t^1 \varepsilon + D_t^2 \varepsilon^2 + \cdots + D_t^k \varepsilon^k + \cdots$$

hence

$$\left(\mathbf{I} + D_t^1 \varepsilon + \cdots + D_t^k \varepsilon^k + \cdots\right)\left(\mathbf{I} + \varepsilon B_t + \varepsilon B_t^{1-\varepsilon} - \varepsilon^2 B_t^{1-\varepsilon}\right) \equiv \mathbf{I}.$$

We deduce

$$D_t^1 + B_t + B_t^{1-\varepsilon} = 0$$
$$D_t^2 + D_t^1\left(B_t + B_t^{1-\varepsilon}\right) - B_t^{1-\varepsilon} = 0$$
$$\cdots\cdots\cdots\cdots$$
$$D_t^k + D_t^{k-1}\left(B_t + B_t^{1-\varepsilon}\right) - D_t^{k-2}B_t^{1-\varepsilon} = 0.$$
$$\cdots\cdots\cdots\cdots\cdots$$

Taking into account the simple estimates that may be deduced from the above equalities we find

$$\left(\mathbf{I} + \varepsilon\left(B_t + B_t^{1-\varepsilon}\right) - \varepsilon^2 B_t^{1-\varepsilon}\right)^{-1} = \mathbf{I} - \varepsilon\left(B_t + B_t^{1-\varepsilon}\right) + \varepsilon^2 R_t\left(\varepsilon\right)$$

with $|\varepsilon R_t\left(\varepsilon\right)| \leq \beta_0 \zeta\left(1 - \varepsilon\right)$; remark that $\varepsilon\zeta\left(1 - \varepsilon\right) = o\left(\varepsilon\right)$.

We obtain further

$$\begin{aligned}
v_{t+1} &= \left(\mathbf{I} - \varepsilon B_t - \varepsilon B_t^{1-\varepsilon} + \varepsilon^2 R_t\left(\varepsilon\right)\right)\left(\mathbf{I} + \varepsilon A_t\right)\left(\mathbf{I} + \varepsilon B_t^{1-\varepsilon}\right)v_t = \\
&= \left(\mathbf{I} + \varepsilon\left(A_t + B_t\right) + \tilde{R}_t\left(\varepsilon\right)\varepsilon^2\right)v_t,
\end{aligned}$$

where $\left|\varepsilon\tilde{R}_t\left(\varepsilon\right)\right| \leq \tilde{\beta}_0 \zeta\left(1 - \varepsilon\right)$ and $A_t - B_t = \tilde{A}$. Therefore,

$$v_{t+1} = \left(\mathbf{I} + \varepsilon\tilde{A}\right)v_t + \varepsilon C_t\left(\varepsilon\right)v_t, \quad \left(C_t\left(\varepsilon\right) = \varepsilon\tilde{R}_t\left(\varepsilon\right)\right), \qquad (4.20)$$

If $\tilde{A}$ has distinct eigenvalues there exists S nonsingular such that $\tilde{A} = SA_dS^{-1}$ where $A_d = diag\left(\lambda_1, ...\lambda_n\right)$, where λ_i are the eigenvalues of $\tilde{A}$. We have also $\mathbf{I} + \varepsilon\tilde{A} = S\left(\mathbf{I} + \varepsilon A_d\right)S^{-1}$ hence $\left(\mathbf{I} + \varepsilon\tilde{A}\right)^t = S\left(\mathbf{I} + \varepsilon A_d\right)^t S^{-1}$. The eigenvalues of $\mathbf{I} + \varepsilon\tilde{A}$ are the diagonal elements of $\mathbf{I} + \varepsilon A_d$ having the form $1 + \varepsilon\left(-\mu + i\nu\right)$, $\mu > 0$, since $\tilde{A}$ has its eigenvalues with negative real parts. For ε small enough all these eigenvalues are inside the unit disk and we choose the one with the largest modulus to represent the spectral radius

$$\left|\mathbf{I} + \varepsilon\tilde{A}\right|^t \leq \beta_1 \rho_\varepsilon^t, \quad \rho_\varepsilon = |1 - \varepsilon\mu + i\varepsilon\nu|, \quad 0 < \rho_\varepsilon < 1.$$

The variation of constants formula will give

$$v_t = \left(\mathbf{I} + \varepsilon \tilde{A}\right)^t v_0 + \sum_{k=0}^{t-1} \left(\mathbf{I} + \varepsilon \tilde{A}\right)^{t-1-k} \varepsilon C_k(\varepsilon) v_k \; ;$$

hence

$$|v_t| \leq \beta_1 \rho_\varepsilon^t |v_0| + \sum_{k=0}^{t-1} \varepsilon \tilde{\beta}_0 \zeta (1 - \varepsilon) \beta_1 \rho_\varepsilon^{t-1-k} |v_k|$$

Applying Proposition 2.9 we obtain the estimate

$$|v_t| \leq \beta_1 \left(1 + \tilde{\beta}_0 \varepsilon \zeta (1 - \varepsilon) \rho_\varepsilon^{-1}\right) \left(\rho_\varepsilon + \tilde{\beta}_0 \beta_1 \varepsilon \zeta (1 - \varepsilon)\right)^t |v_0|$$

and this gives exponential stability provided $\tilde{\beta}_0 \beta_1 \varepsilon \zeta (1 - \varepsilon) < 1$. A sufficient condition for the fulfillment of this inequality is $2 \tilde{\beta}_0 \beta_1 \varepsilon \zeta (1 - \varepsilon) < 1 - \rho_\varepsilon^2$ which may be given the form $(\mu^2 + \nu^2) \varepsilon + 2 \beta_1 \tilde{\beta}_0 \zeta (1 - \varepsilon) < 2\mu$; since $\lim_{\varepsilon \to 0} \zeta (1 - \varepsilon) = 0$ the inequality holds for $\varepsilon > 0$ small enough.

The result holds nevertheless without the assumption of distinct eigenvalues. Indeed since $\tilde{A}$ has its eigenvalues with negative real parts there exists a positive definite matrix P_0 satisfying the Liapunov equation $\tilde{A}^* P_0 + P_0 \tilde{A} = -\mathbf{I}$.

For the discrete time system (4.20) we consider the Liapunov function $V(x) = x^* P_0 x$. We shall have

$$V(x_{t+1}) - V(x_t) =$$
$$= v_t^* \left(\mathbf{I} + \varepsilon \tilde{A}^* + \varepsilon C_t^*(\varepsilon)\right) P_0 \left(\mathbf{I} + \varepsilon \tilde{A} + \varepsilon C_t(\varepsilon)\right) v_t - v_t^* P_0 v_t =$$

$$= \varepsilon v_t^* \left[\tilde{A}^* P_0 + P_0 \tilde{A} + C_t^*(\varepsilon) P_0 + P_0 C_t(\varepsilon) + \varepsilon \left(\tilde{A}^* P_0 \tilde{A} + \right.\right.$$
$$\left.\left. + C_t^*(\varepsilon) P_0 \tilde{A} + \tilde{A}^* P_0 C_t(\varepsilon) + C_t^*(\varepsilon) P_0 C_t(\varepsilon)\right)\right] v_t =$$

$$= \varepsilon \left[-|v_t|^2 + v_t^* (C_t^*(\varepsilon) P_0 + P_0 C_t(\varepsilon)) v_t + \right.$$
$$\left. + \varepsilon v_t^* \left(\tilde{A} + C_t(\varepsilon)\right)^* P_0 \left(\tilde{A} + C_t(\varepsilon)\right) v_t\right] \leq$$
$$\leq \varepsilon \left[-|v_t|^2 + 2 |P_0| \tilde{\beta}_0 \zeta (1 - \varepsilon) |v_t|^2 + \varepsilon |P_0| \gamma_1 |v_t|^2\right] .$$

We may choose $\varepsilon > 0$ small enough to fulfil the following inequality

$$\varepsilon |P_0| \gamma_1 + 2\tilde{\beta}_0 |P_0| \zeta (1 - \varepsilon) < \frac{1}{2}$$

which gives

$$V(x_{t+1}) - V(x_t) \leq -\frac{\varepsilon}{2} |v_t|^2$$

and the proof ends.

B. We consider now the following nonlinearly perturbed system

$$x_{t+1} = Ax_t + \varepsilon F_t(x_t, \varepsilon) \tag{4.21}$$

where A has its eigenvalues located on the unit circle and with simple elementary divisors, $F_t(\cdot, \cdot)$ is C^1 and almost periodic with respect to t, uniformly with respect to the other two arguments. We already know (Proposition 4.22) that all solutions of the unperturbed system $y_{t+1} = Ay_t$ are almost periodic hence A^t - the state transition matrix - is almost periodic. Since A is nonsingular we may consider the change of variable $x_t = A^t z_t$ to obtain

$$z_{t+1} = z_t + \varepsilon A^{-t-1} F_t \left(A^t z_t, \varepsilon \right)$$

that is a system of the form

$$z_{t+1} = z_t + \varepsilon Z_t(z_t, \varepsilon) , \tag{4.22}$$

where $Z_t(z, \varepsilon) = A^{-t-1} F_t(A^t z, \varepsilon)$ is almost periodic uniformly with respect to z and ε due to the facts that the product of two almost periodic sequences is almost periodic and the sequence $f_t(\varphi_t)$ is almost periodic if φ_t is almost periodic, $f_t(\cdot)$ is continuous and the sequence $\{f_t(\cdot)\}_t$ is almost periodic. Denote by $\tilde{Z}(z, \varepsilon)$ the mean value of $Z_t(z, \varepsilon)$

$$\tilde{Z}(z, \varepsilon) = \lim_{T \to \infty} \frac{1}{T} \sum_{t=0}^{T-1} Z_t(z, \varepsilon) .$$

We may state now

Theorem 4.34 *Let z^0 be a solution of the equation $\tilde{Z}(z, 0) = 0$ such that the Jacobian matrix $J(z^0, 0) = \dfrac{\partial \tilde{Z}}{\partial z}(z^0, 0)$ has its eigenvalues with negative real parts. Then system (4.21) has an almost* *periodic solution* $\hat{x}_t\left(\varepsilon\right)$ *provided* $\varepsilon > 0$ *is small enough; moreover* $\lim_{\varepsilon \to 0} \hat{x}_t\left(\varepsilon\right) = A^t z^0$.

Proof. We shall construct a solution of (4.22) using iterations. Let

$$z_t^0 = z^0 + \varepsilon \sum_{k=-\infty}^{t-1} (1-\varepsilon)^{t-1-k} Z_k \left(z^0, 0\right) .$$

Since Z_k is almost periodic and $\varepsilon > 0$ we deduce that z_t^0 is an almost periodic sequence. From Lemma 4.32 it follows that $\left|z_t^0 - z^0\right| < \eta\left(\varepsilon\right)$ with $\lim_{\varepsilon \to 0} \eta\left(\varepsilon\right) = 0$. Remark also that

$$
\begin{aligned}
z_{t+1}^0 &= z^0 + \varepsilon \sum_{k=-\infty}^{t} (1-\varepsilon)^{t-k} Z_k \left(z^0, 0\right) = \\
&= z^0 + \varepsilon \sum_{k=-\infty}^{t-1} (1-\varepsilon)(1-\varepsilon)^{t-1-k} Z_k \left(z^0, 0\right) + \varepsilon Z_t \left(z^0, 0\right)
\end{aligned}
$$

which reads

$$z_{t+1}^0 = z_t^0 + \varepsilon \left(Z_t \left(z^0, 0\right) - \left(z_t^0 - z^0\right) \right) . \qquad (4.23)$$

Define z_t^{j+1} as the unique almost periodic solution of the system

$$z_{t+1} = z_t + \varepsilon \frac{\partial \tilde{Z}_t}{\partial z} \left(z^0, 0\right) z_t + \varepsilon Z_t \left(z_t^j, \varepsilon\right) - \varepsilon \frac{\partial Z_t}{\partial z} \left(z^0, 0\right) z_t^j \qquad (4.24)$$

Indeed denote $A_t = \dfrac{\partial Z_t}{\partial z} \left(z^0, 0\right)$: we have

$$\tilde{A} = \lim_{T \to \infty} \frac{1}{T} \sum_{t=0}^{T-1} A_t = \lim_{T \to \infty} \frac{1}{T} \sum_{t=0}^{T-1} \frac{\partial Z_t}{\partial z} \left(z^0, 0\right) = \frac{\partial \tilde{Z}}{\partial z} \left(z^0, 0\right)$$

and $\tilde{A}$ has its eigenvalues with negative real parts; system (4.24) has the form

$$z_{t+1} = \left(\mathbf{I} + \varepsilon A_t\right) z_t + \varepsilon \psi_t^j ,$$

where ψ_t^j is almost periodic. If $\varepsilon > 0$ is small enough the free system is exponentially stable (Proposition 4.33) and there exists a unique almost periodic solution of the forced system.

We have further

$$
z_{t+1}^{j+1} - z_{t+1}^0 =
$$

$$
= z_t^{j+1} - z_t^0 + \varepsilon \frac{\partial Z_t}{\partial z} \left(z^0, 0\right) \left(z_t^j - z_t^0\right) + \varepsilon \frac{\partial Z_t}{\partial z} \left(z^0, 0\right) z_t^0 -
$$

$$
- \varepsilon Z_t \left(z_t^0, 0\right) + \varepsilon \left(z_t^0 - z^0\right) + \varepsilon Z_t \left(z_t^j, \varepsilon\right) - \varepsilon \frac{\partial Z_t}{\partial z} \left(z^0, 0\right) z_t^j =
$$

$$
= \left[\mathbf{I} + \varepsilon \frac{\partial Z_t}{\partial z} \left(z^0, 0\right)\right] \left(z_t^{j+1} - z_t^0\right) +
$$

$$
+ \varepsilon \left[Z_t \left(z_t^j, \varepsilon\right) - Z_t \left(z_t^0, 0\right) - \frac{\partial Z_t}{\partial z} \left(z^0, 0\right) \left(z_t^j - z_t^0\right)\right] + \varepsilon \left(z_t^0 - z^0\right)
$$

On the other hand

$$
Z_t \left(z_t^j, \varepsilon\right) - Z_t \left(z^0, 0\right) - \frac{\partial Z_t}{\partial z} \left(z^0, 0\right) \left(z_t^j - z_t^0\right) =
$$

$$
= \left(\int_0^1 \frac{\partial Z_t}{\partial z} \left(z^0 + \lambda \left(z_t^j - z^0\right), \lambda \varepsilon\right) d\lambda\right) \left(z_t^j - z^0\right) +
$$

$$
+ \varepsilon \int_0^1 \frac{\partial Z_t}{\partial \varepsilon} \left(z^0 + \lambda \left(z_t^j - z^0\right), \lambda \varepsilon\right) d\lambda - \frac{\partial Z_t}{\partial z} \left(z^0, 0\right) \left(z_t^j - z_t^0\right) =
$$

$$
= \left(\int_0^1 \left[\frac{\partial Z_t}{\partial z} \left(z^0 + \lambda \left(z_t^j - z^0\right), \lambda \varepsilon\right) - \frac{\partial Z_t}{\partial z} \left(z^0, 0\right)\right] d\lambda\right) \left(z_t^j - z_t^0\right) +
$$

$$
+ \left(\int_0^1 \frac{\partial Z_t}{\partial z} \left(z^0 + \lambda \left(z_t^j - z^0\right), \lambda \varepsilon\right) d\lambda\right) \left(z_t^j - z^0\right) +
$$

$$
+ \varepsilon \int_0^1 \frac{\partial Z_t}{\partial \varepsilon} \left(z^0 + \lambda \left(z_t^j - z^0\right), \lambda \varepsilon\right) d\lambda.
$$

Since $F_t \left(\cdot, \cdot\right)$ is $\mathcal{C}^1$ we deduce that $Z_t \left(\cdot, \cdot\right)$ is $\mathcal{C}^1$; we take into account that $\left|z_t^0 - z^0\right| \le \eta\left(\varepsilon\right)$ and assume by induction that $\left|z_t^j - z^0\right| \le$

$\eta_1\left(\varepsilon\right),$ where $\lim\limits_{\varepsilon\to 0}\eta_1\left(\varepsilon\right)=0.$ We deduce that

$$\left| Z_t\left(z_t^j,\varepsilon\right)-Z_t\left(z^0,0\right)-\frac{\partial Z_t}{\partial z}\left(z^0,0\right)\left(z_t^j-z_t^0\right)+z_t^0-z^0\right| \le$$

$$\le \frac{1}{2M}\sup_t\left|z_t^j-z_t^0\right|+\eta_2\left(\varepsilon\right)\quad,\quad \lim_{\varepsilon\to 0}\eta_2\left(\varepsilon\right)=0.$$

It follows that

$$\left|z_{t+1}^{j+1}-z_t^0\right|\le M\left(\frac{1}{2M}\sup_t\left|z_t^j-z_t^0\right|+\eta_2\left(\varepsilon\right)\right);$$

hence if $\left|z_t^j-z^0\right|\le 2M\eta_2\left(\varepsilon\right)$ this inequality holds for $j+1$ too; this gives also that $z_{t+1}^{j+1}-z_t^0\to 0.$
    We have further

$$z_{t+1}^{j+1}-z_{t+1}^j=z_t^{j+1}-z_t^j+\varepsilon\frac{\partial Z_t}{\partial z}\left(z^0,0\right)\left(z_t^{j+1}-z_t^j\right)+\varepsilon Z_t\left(z_t^j,\varepsilon\right)-$$
$$-\varepsilon Z_t\left(z_t^{j-1},\varepsilon\right)-\varepsilon\frac{\partial Z_t}{\partial z}\left(z^0,0\right)\left(z_t^j-z_t^{j-1}\right)$$

and

$$Z_t\left(z_t^j,\varepsilon\right)-Z_t\left(z_t^{j-1},\varepsilon\right)-\frac{\partial Z_t}{\partial z}\left(z^0,0\right)\left(z_t^j-z_t^{j-1}\right)=$$
$$=\int\limits_0^1\left[\frac{\partial Z_t}{\partial z}\left(z_t^{j-1}+\lambda\left(z_t^j-z_t^{j-1}\right),\varepsilon\right)-\frac{\partial Z_t}{\partial z}\left(z^0,0\right)\right]d\lambda\cdot\left(z_t^j-z_t^{j-1}\right)$$

what gives

$$\left|z_t^{j+1}-z_t^j\right|\le \eta_3\left(\varepsilon\right)\sup_t\left|z_t^j-z_t^{j-1}\right|,\quad \lim_{\varepsilon\to 0}\eta_3\left(\varepsilon\right)=0$$

hence the iterations sequence converges and the proof ends.

Remarks. The above theorem justifies the averaging for discrete time systems. It remains valid if instead of A we had an invertible almost periodic matrix B_t such that the free system $y_{t+1}=B_ty_t$ has all its solutions almost periodic. In the same way we may prove existence of almost periodic solutions for systems in a product space.

Theorem 4.35 *Consider the system*

$$x_{t+1} = x_t + \varepsilon F_t(x_t, y_t, \varepsilon)$$
$$y_{t+1} = A_t y_t + f_t + \varepsilon G_t(x_t, y_t, \varepsilon) \tag{4.25}$$

Assume that: i) $F_t(\cdot,\cdot,\cdot)$ and $G_t(\cdot,\cdot,\cdot)$ are C^1 and $\{F_t(x,y,\varepsilon)\}_t$, $\{G_t(x,y,\varepsilon)\}_t$ are almost periodic uniformly with respect to x, y, ε; ii) $\{A_t\}_t$, $\{f_t\}_t$ are almost periodic; iii) the zero solution of the system $y_{t+1} = A_t y_t$ is exponentially stable. Let φ_t be the unique almost periodic solution of the system $y_{t+1} = A_t y_t + f_t$ and denote $\tilde{F}(x,\varepsilon) = \lim_{T\to\infty} \frac{1}{T} \sum_{t=0}^{T-1} F_t(x,\varphi_t,\varepsilon)$; let x^0 be a solution of $\tilde{F}(x,0) = 0$; assume that the Jacobian matrix $\dfrac{\partial \tilde{F}}{\partial x}(x^0,0)$ has its eigenvalues with negative real parts. Then for $\varepsilon > 0$ small enough system (4.25) has a unique almost periodic solution $\hat{x}_t(\varepsilon), \hat{y}_t(\varepsilon)$ such that $\lim_{\varepsilon\to 0} \hat{x}_t(\varepsilon) = x^0$, $\lim_{\varepsilon\to 0} \hat{y}_t(\varepsilon) = \varphi_t$.

The proof follows by iterations with

$$x_t^0 = x^0 + \varepsilon \sum_{k=-\infty}^{t-1} (1-\varepsilon)^{t-1-k} F_k(x^0, \varphi_k, 0) \ , \quad y_t^0 = \varphi_t$$

and x_t^{j+1}, y_t^{j+1} is the unique almost periodic solution of the system

$$x_{t+1} = x_t + \varepsilon \frac{\partial F_t}{\partial x}(x^0, \varphi_t, 0) x_t + \varepsilon F_t\left(x_t^j, y_t^j, \varepsilon\right) -$$
$$\qquad -\varepsilon \frac{\partial F_t}{\partial x}(x^0, \varphi_t, 0) x_t^j$$
$$y_{t+1} = A_t y_t + f_t + \varepsilon G_t\left(x_t^j, y_t^j, \varepsilon\right).$$

REFERENCES

Aizerman, M.A. (1949). On a problem concerning "stability in the large" of dynamical systems, *Uspekhi Matem. Nauk*, **14**(4), 187-188 (in Russian).

Anderson, B.D.O. and J.B. Moore (1969). New Results in Linear System Stability, *SIAM Journ. Control*, **7**(3), 398-414.

Antonov, V.G., A.L. Likhtarnikov and V.A. Yakubovich (1975). A discrete frequency theorem for the case of Hilbert spaces of states and controls I, *Vestnik Leningrad Univ. (Mat.)*, 1, 22-31 (in Russian).

Barabanov, N.E. and V.A. Yakubovich (1979). Absolute stability of control systems with a hysteresis nonlinearity, *Avtom. i Telemekh.* 12, 5-12 (in Russian).

Barabanov, N.E. (1988). About the problem of Kalman, *Sib. Mat. Ž.* **29**(3), 3-11 (in Russian).

Barbălat, I. and A. Halanay (1971). Nouvelles applications de la méthode fréquentielle dans la théorie des oscillations, *Rev. Roum. Sci. Techn. Série Electrotechn. et Energ.* **16**(3), 689-702.

Barbălat, I. and A. Halanay (1974). Conditions de comportement "presque linéaire" dans la théorie des oscillations, *Rev. Roumaine Sci. Techn. Série Electrotechn. et Energ.* **19**(2), 321-341.

Bochner, S. (1927). Beiträge zur Theorie der fastperiodischen Funktionen einer Variablen, *Math. Ann.* **96**, 119-147.

Bochner, S. (1962). A new approach to almost periodicity, *Proc. Nat. Acad. Sci. USA*, **48**, 2039-2043.

Bohr, H. (1923). Sur les fonctions presque-périodiques, *C. R. Acad. Sci. Paris*, **117**, 737-739.

Brockett, R.W. and J.L. Willems (1965). Frequency domain stability criteria I, II, *IEEE Trans. on Aut. Contr.* **AC-10**, 255-261, 407-413.

Browder, F.E. (1959). On a generalisation of the Schauder fixed-point theorem, *Duke Math. Journal*, **26**(2), 291-303.

Brusin, V.A. (1976a). Lurie equations in Hilbert spaces and their solvability, *Prikl. Math. & Mekh.* **40**(5) (in Russian).

Brusin, V.A. (1976b). Existence of a global Liapunov functional for some classes of nonlinear systems with distributed parameters, *Prikl. Math. & Mekh.* **40**(6) (in Russian).

Byrnes, C.J. and W. Lin (1994). Passivity and absolute stabilization for a class of discrete-time nonlinear systems, *IEEE Trans. Aut. Contr.* **AC-39**, 83-98.

Caianiello, E.R. (1966). Decision equations and reverberations, *Kybernetik*, **3**(3), 98-100.

Caianiello, E.R. (1967). Reverberations and control of neural networks, *Kybernetik*, **4**(1), 10-17.

Četaev, N.G. (1936) About the Stable Trajectories of Dynamics *Coll. Sci. Papers Kazan Av. Inst.*, **1**. Reproduced in *Stability of Motion. Papers in Rational Mechanics*, USSR Science Academy Publ. House, Moscow, 1962, 250-268 (in Russian).

Corduneanu, C. (1982). Almost periodic discrete processes, *Libertas Mathematica*, **2**, 159-169.

Corduneanu, C. (1989). *Almost Periodic Functions* (2nd edition) Chelsea Publ. House,

De Wilde, Ph. (1996). *Neural Network Models. An Analysis*, Springer Verlag, Berlin.

Elaydi, S. (1996), *An Introduction to Difference Equations*, Springer Verlag, Berlin.

Fan, Ky (1943) Les fonctions asymptotiquement presque-périodiques d'une variable entière et leur application à l'étude de l'itération des transformations continues, *Math. Z.* **48**, 658-711.

Frank-Kamenetskii, D.A. (1987). *Diffusion and Heat Conduction in Chemical Kinetics*, Nauka, Moscow (in Russian).

Gelig, A.Kh., G.A. Leonov and V.A. Yakubovich (1978). *Stability of Nonlinear Systems with Non-unique Equilibria*, Nauka Publ. House, Moscow (in Russian).

Gelig, A.Kh. (1982). *Dynamics of Pulse Systems and Neural Networks*, Leningrad Univ. Publish. House, Leningrad (in Russian).

Gelig, A.Kh. and A.N. Churilov (1993). *Oscillations and Stability of Nonlinear Pulse Systems*, St. Petersburg Univ. Publishing House, St. Petersburg (in Russian).

Gu, D.W., M.C. Tsai, S.D. O'Young and I. Postlethwaite (1989). *State-space Formulae for Discrete-Time H^∞-Optimization*, Int. J. Control, **49**(5), 1683-1724.

Guegan, D. (1992). *Notion de chaos. Approché dynamique et problèmes d'identification*, INRIA Rapport de recherche No. 1623 (Programme 5 - Traitement du signal, Automatique et Productique).

Gumowski, I. and C. Mira (1980). *Reccurrences and discrete dynamical systems*, Springer Verlag, Berlin.

Haddad, W. and D. Bernstein (1994). Explicit construction of quadratic Liapunov functions for the small gain, positivity, circle and Popov theorems and their application to robust stability. Part II: discrete time theory, *Int. J. of Robust and Nonlinear Control*, **4**, 249-265.

Haddad, W. and V. Kapila (1995). Absolute stability criteria for multiple slope-restricted monotonic nonlinearities, *IEEE Trans. on Aut. Control*, **AC-40**(2), 361-365.

Halanay, A. (1960). Generalization of a theorem of Persidskii (in Romanian), *Comunicările Academiei R.P.R.*, **12**, 1065-1068.

Halanay, A. (1963). Quelques questions de la théorie de la stabilité pour les systèmes aux différences finies, *Archive Rat. Mech. Anal.* **12**(2), 150-154.

Halanay, A. (1966). *Differential Equations. Stability. Oscillations. Time Lags*, Academic Press, New York.

Halanay, A. (1967). Invariant Manifolds for Systems with Time Lag, *Differential Equations and Dynamical Systems*, (Hale & La Salle eds.) 199-213, Academic Press, New York.

Halanay, A. (1969). Almost periodic solutions for a class of nonlinear systems with time lags, *Rev. Roumaine Math. Pures Appl.* **XIV**(9), 1269-1276.

Halanay, A. and V. Ionescu (1994). *Time-Varying Discrete Linear Systems*, Birkhäuser Verlag, Boston.

Halanay, A. and Vl. Răsvan (1977). Periodic and almost periodic solutions for a class of systems described by coupled delay-differential and difference equations, *Nonlinear Analysis, Theory, Methods & Applications*, **1**(3), 197-206.

Halanay, A. and Vl. Răsvan (1979). Frequency domain conditions for forced oscillations in difference systems, *Rev. Roumaine Sci. Techn. Série Electrotechn. et Energ.* **24**(1), 141-148.

Halanay, A. and Vl. Răsvan (1980). Stabilization of a class of bilinear control systems with applications to steam turbine regulation, *Tôhoku Math. Journ.* **32**(2), 299-308.

Halanay, A. and Vl. Răsvan (1990). Absolute stability of discrete systems with slope restricted nonlinearity, *Rev. Roumaine Sci. Techn. Série Electrotechn. et Energ.* **35**(1), 101-111.

Halanay, A. and Vl. Răsvan (1991). Absolute stability of feedback systems with several differentiable nonlinearities, *Int. J. Systems Sci.* **22**(10), 1911-1927.

Haykin, S. (1994). *Neural Networks, a Comprehensive Foundation*, Macmillan, New York.

Hill, D. and P. Moylan (1974). Implications of passivity in a class of nonlinear systems, *IEEE Trans. Aut. Control*, **AC-19**, 373-381.

Hinrichsen, D. and A.J. Pritchard (1988). Robustness Measures for Linear State Space Systems under Complex and Real Parameter Perturbations, *Perspectives in Control Theory. Proc. Sielpia Conference 1988*, 56-74.

Hinrichsen. D. and A.J. Pritchard (1989). Real and Complex Stability Radii: a Survey, *Proc. Workshop Control of Uncertain*

Systems, Bremen, Progress in System and Control Theory, **6**, 119-162, Birkhäuser, Boston.

Hinrichsen, D., B. Kelb and A. Linemann (1990). An algorithm for the computation of the structured stability radius with applications, *Automatica*, **25**, 771-775.

Jury, E. and B.W. Lee (1964). On the stability of a certain class of nonlinear sampled-data systems, *IEEE Trans. on Aut. Contr.* **AC-9**(1), 51-62.

Kalman, R.E. (1957). Physical and mathematical mechanisms of instability in nonlinear automatic control systems, *Trans. ASME,* **79**, 553-556.

Kalman, R.E. (1963). Liapunov functions for the problem of Lurie in automatic control, *Proc. Nat. Acad. Sci. USA*, **49**, 201-205.

Kalman, R.E. (1970). A New Algebraic Method in Stability Theory, *Proc. 5th Int'l Conference on Nonlinear Oscillations*, vol. 2, Ukrainian SSR Publ. House, Kiev.

Kamen, E.W. (1990). *Introduction to Signals and Systems.* Macmillan, New York.

Kapila, V. and W. Haddad (1996). A Multivariable Extension of the Tsypkin Criterion Using a Lyapunov Function Approach, *IEEE Trans. on Aut. Contr.* **AC-41**(1), 149-152.

Kocic, V.L. and G.S. Ladas (1993). *Global Behavior of Nonlinear Difference Equations of Higher Order*, Kluwer Academic Publishers, Dordrecht.

Kosko, B. (1988). Bidirectional associative memories, *IEEE Trans. on Syst. Man. & Cybernetics*, **SMC-18**, 49-60.

Kosko, B. (1992). *Neural Networks and Fuzzy Systems*, Prentice Hall International Inc., Englewood Cliffs.

Kurzweil, J. (1967). Invariant Manifolds for Flows, *Differential Equations and Dynamical Systems* (Hale & La Salle eds.) 431-468, Academic Press, New York.

Kwakernaak, H. and R. Sivan (1991). *Modern Signals and Systems*, Prentice Hall, Englewood Cliffs.

La Salle, J.P. (1976). *The Stability of Dynamical Systems*, SIAM Regional Conference Series in Applied Mathematics, Providence.

Lellouche, G. (1970). A frequency criterion for oscillatory solutions. *SIAM J. Control*, **8**, 202-206.

Likhtarnikov, A.L. (1977). A criterion of absolute stability for nonlinear operator equations, *Izvestia Akad. Nauk SSSR Ser. Matem.* **41**(5), 1064-1083 (in Russian).

Likhtarnikov, A.L., V.I. Ponomarenko and V.A. Yakubovich (1976). A discrete frequency theorem for the case of Hilbert spaces of states and controls II, *Vestnik Leningrad Univ. (Mat.)* 19, 297-305 (in Russian).

Likhtarnikov, A.L. and V.A. Yakubovich (1977). The frequency domain theorem for one-parameter semigroups, *Izvestia Akad. Nauk SSSR Ser. Matem.* **41**(5), 895-911 (in Russian).

Lin, W. and C.J. Byrnes (1994). KYP lemma, state feedback and dynamic output feedback in discrete time bilinear systems, *Syst. & Control Letters*, **23**, 127-136.

Lin, W. and C.J. Byrnes (1995). Passivity and absolute stabilization for a class of discrete time nonlinear systems, *Automatica*, **31**. 263-267.

Mahmoud, M.S. and M.G. Singh (1984). *Discrete Systems. Analysis, Control and Optimization*, Springer Verlag, Berlin.

Mal'kin, I.G. (1954). On resonance in quasi-harmonic systems, *Prikl. Mat. Mekh.* (PMM), **XVIII**(4) (in Russian).

May, R.M. (1976). Simple mathematical models with complicated dynamics, *Nature*, **261**, 459-467.

May, R.M. (1995). Necessity and Chance: Deterministic chaos in ecology and evolution, *Bull. Amer. Math. Soc. (New Series)* **32**(3). 291-308.

Maynard-Smith, J. (1974). *Models in Ecology*, Cambridge at the University Press, Cambridge.

Milman, V.D. and A.D. Myshkis (1960). Stability of motion in the presence of shocks, *Sib. Mat. Ž.* **1**(2), 233-237 (in Russian).

Motscha, M. (1988). An algorithm to compute the complex stability radius, *Intern. Journ. Control*, **48**, 2417-2428.

Neimark, Iu. I. (1978). *Dynamical Systems and Controlled Processes*, Nauka Publ. House, Moscow (in Russian).

Nudelman, A.A. and P.A. Svartsman (1975). Existence of solutions to some operator inequalities, *Sib. Mat. Ž.* **XVI**(3) (in Russian).

Olech, Cz. (1957). Sur un problème de M.G. Sansone lié à la théorie du synchrotrone, *Ann. di Mat. Pura ed Appl.* **44**, 317-330.

Pliss, V.A. (1958). *Some Problems of the Theory of Stability of Motion in the Large*, Leningrad Univ. Publish. House. Leningrad (in Russian).

Popov, V.M. (1959). Stability criteria for nonlinear systems of automatic control based on the utilization of the Laplace transform, *St. cerc. energ.* **9**(1), 119-135 (in Romanian).

Popov, V.M. (1961). On the absolute stability of non-linear systems of automatic control, *Avtom. i telemekh.* **22**(8), 961-979 (in Russian).

Popov, V.M. (1964). Hyperstability and optimality of automatic systems with several control functions, *Rev. Roumaine Sci. Techn. Série Electrotechn. et Energ.* **9**(4).

Popov, V.M. (1973) *Hyperstability of Control Systems*, Springer Verlag, Berlin.

Popov, V.M. (1974). Dichotomy and stability by frequency domain methods, *Proc. IEEE*, **62**(5), 548-562.

Popov, V.M. (1977a). Unitary Treatment of Various Types of Systems in Stability Theory, *Dynamical Systems. Proc. Univ. of Florida Int'l Symposium*, 251-263, Academic Press, New York.

Popov, V.M. (1977b). Applications of the saturability technique in the problem of stability of nonlinear systems, *Nonlinear Analysis, Theory, Methods and Applications*, **1**(5), 571-581.

Racoveanu, N. (1964). Traducteur LC à grande déviation de fréquence, *Mes. régul. autom.* **25**, 78-82.

Ragazzini, J.R. and G.F. Franklin (1958). *Sampled Data Control Systems*, Mc Graw Hill, New York.

Răsvan, Vl. (1978). Some system theory ideas connected with the stability problem, *Journ. Cybernetics*, **8**(2), 203-215.

Sansone, G. (1957). Sopra una equatione che si presenta nelle determinazioni della orbite in un sincrotrone, *Rend. Accad. Naz. Lincei*, **8**, 1-74.

Schaft, A.J. van der (1993). Nonlinear State Space H_∞ Control Theory, *Proceedings European Control Conference, Groningen, Plenary Lectures*, 153-190.

Schwabik, St. (1985). Generalized differential equations (fundamental results), *Rozpr. Cesk. Akad. Ved.* **95**(6).

Sharkovsky, A.N., Iu.L. Maistrenko and E.Iu. Romanenko (1986). *Difference Equations and their Applications*, Nauka Publ. House. Moscow (in Russian).

Singh, V. (1984). A stability inequality for nonlinear systems with slope restricted nonlinearity, *IEEE Trans. on Aut. Control*, **AC-29**, 743-744.

Slemrod, M. (1978). Stabilization of Bilinear Control Systems with Application to Nonconservative Problems in Elasticity, *SIAM J. Contr. Optim.* **16**(1), 131-141.

Stallard, F.W. (1955). *Differential Systems with Interface Conditions*. Oak Ridge Nat. Lab. Report ONRL 1876.

Stallard, F.W. (1962). Functions of bounded variations as solutions of differential systems, *Proc. Amer. Math. Soc.* **13**, 366-373.

Stoorvogel, A.A. (1990). *The H^∞-Control Problem: A State-Space Approach*, Ph. D. Thesis, University of Eindhoven, The Netherlands.

Svirezev, Iu.M. and D.O. Logofet (1978). *Stability of Biological Communities*, Nauka Publ. House. Moscow (in Russian).

Szegö, G. (1963a). Sur la stabilité absolue d'un système non-linéaire discret, *C. R. Acad. Sci. Paris*, **257**, 1749-1751.

Szegö, G. (1963b). On the absolute stability of sampled data control systems, *Proc. Nat. Acad. Sci. USA*, **50**(3).

Szegö, G. and R. Kalman (1963). Sur la stabilité absolue d'un système d'équations aux différences finies, *C. R. Acad. Sci. Paris*, **257**, 388-390.

Szegö, G. and J.B. Pearson (1964). On the absolute stability of sampled-data systems, the indirect control, *IEEE Trans. on Aut. Control*, **AC-9**(4).

Thomson, W. and P.G. Tait (1879). *Natural Philosophy*, Cambridge at the Univ. Press, Cambridge.

Tsypkin, Ya.Z. (1958). *Theory of Pulse Systems*, Fizmatgiz Publ. House, Moscow (in Russian).

Tsypkin, Ya.Z. (1962). On the overall stability of nonlinear automatic systems with impulses, *Dokl. A. N. S. S. S. R.* **145**(1) (in Russian).

Venkatesh, Y.V. (1988). Riesz-Thorin Theorem and l^p-Stability of Nonlinear Time-Varying Discrete Systems, *Journ. Math. Anal. Appl.* **135**(2), 627-643.

Walther, A. (1928). Fastperiodische Folgen und Potenzreihen mit fastperiodischen Koeffizienten, *Abh. Math. Sem. Hamburg Universität*, **VI**, 217-234.

Wexler, D. (1966). Sur une équation différentielle linéaire aux impulsions, *Journ. of Diff. Eqs.* **2**(1), 1-11.

Willems, J.C. (1972). Dissipative dynamical systems, *Arch. Rat. Mech. Anal.* **45**(5), 321-393.

Williams, A.B. (1981). *Electronic Filter Design Handbook*, McGraw Hill, New York.

Yakubovich, V.A. (1962a). Solutions of some matrix inequalities occuring in the theory of automatic control, *Dokl. A. N. S. S. S. R.* **143**(6) (in Russian).

Yakubovich, V.A. (1962b). Frequency Domain Conditions for the Absolute Stability of Nonlinear Control Systems, *Proc. Conference on Applied theory of stability of motion and analytical mechanics*, 135-142, Kazan, USSR (in Russian).

Yakubovich, V.A. (1964). The method of matrix inequalities in the theory of the stability of controlled nonlinear systems I -

Absolute stability of forced oscillations, *Avtom. i Telemekh.* **XXV**(7), 1017-1029 (in Russian).

Yakubovich, V.A. (1965a). Method of matrix inequalities in the theory of stability of nonlinear controlled systems II. Absolute stability in a class of nonlinearities with derivative conditions. *Avtom. i. Telemekh.* **XXVI**(4), 577-590 (in Russian).

Yakubovich, V.A. (1965b). Frequency domain conditions of absolute stability and dissipativity of control systems with a differentiable nonlinearity, *Dokl. AN SSSR*, **160**(2), 298-301 (in Russian).

Yakubovich, V.A. (1967). Frequency conditions of absolute stability of controlled systems with several nonlinear non-stationary blocks, *Avtom. i Telemekh.* 6, 5-30 (in Russian).

Yakubovich, V.A. (1968). Absolute stability of pulse systems with several nonlinear or linear non-stationary blocks II, *Avtom. i Telemekh.* 2, 81-101 (in Russian).

Yakubovich, V.A. (1970). Solution of an algebraic problem of control theory, *Dokl. A. N. SSSR*, **194**(3) (in Russian).

Yakubovich, V.A. (1971). S-procedure in the nonlinear control theory, *Vestnik Leningrad Univ. (Mat)*, 1 (in Russian).

Yakubovich, V.A. (1973). Frequency domain theorem in control theory, *Sib. Mat. Ž.* **XIV**(5), (in Russian).

Yakubovich, V.A. (1974). A frequency domain theorem for the case in which the state space and the control space are Hilbert spaces I, *Sib. Mat. Ž.* **15**(3), 639-668 (in Russian).

Yakubovich, V.A. (1975). A frequency domain theorem for the case in which the state space and the control space are Hilbert spaces II, *Sib. Mat. Ž.* **16**(5), 1081-1102 (in Russian).

Yakubovich, V.A. (1977a). Frequency Domain Methods for Qualitative Study of Nonlinear Control Systems, *VII - Internationale Konferenz ü. nichtlin. Schwingungen. Bd. I.1.* Akademie Verlag, Berlin (in Russian).

Yakubovich, V.A. (1977b). Frequency Domain Conditions for Oscillations in Systems with a Differentiable Nonlinearity. in *Problems of asymptotic theory for nonlinear oscillations.* 264-269, Naukova Dumka, Kiev (in Russian).

Yoshizawa, T. (1966). *Stability Theory by Liapunov's Second Method,*
The Mathem. Society of Japan, Tokyo.

Yoshizawa, T. (1975). *Stability Theory and the Existence of Periodic
Solutions and Almost Periodic Solutions,* Applied Math. Sci.
Series No. 14, Springer Verlag, Berlin.

INDEX

279

Printed and bound by CPI Group (UK) Ltd, Croydon, CR0 4YY

23/10/2024

01777667-0005